T0156056

FIELD THEORY
A Path Integral Approach
Second Edition

World Scientific Lecture Notes in Physics

ISSN: 1793-1436

*Published titles**

Vol. 65: Universal Fluctuations: The Phenomenology of Hadronic Matter
R Botet and M Ploszajczak

Vol. 66: Microcanonical Thermodynamics: Phase Transitions in "Small" Systems
D H E Gross

Vol. 67: Quantum Scaling in Many-Body Systems
M A Continentino

Vol. 69: Deparametrization and Path Integral Quantization of Cosmological Models
C Simeone

Vol. 70: Noise Sustained Patterns: Fluctuations and Nonlinearities
Markus Loecher

Vol. 71: The QCD Vacuum, Hadrons and Superdense Matter (2nd ed.)
Edward V Shuryak

Vol. 72: Massive Neutrinos in Physics and Astrophysics (3rd ed.)
R Mohapatra and P B Pal

Vol. 73: The Elementary Process of Bremsstrahlung
W Nakel and E Haug

Vol. 74: Lattice Gauge Theories: An Introduction (3rd ed.)
H J Rothe

Vol. 75: Field Theory: A Path Integral Approach (2nd ed.)
A Das

Vol. 76: Effective Field Approach to Phase Transitions and Some Applications
to Ferroelectrics (2nd ed.)
J A Gonzalo

Vol. 77: Principles of Phase Structures in Particle Physics
H Meyer-Ortmanns and T Reisz

Vol. 78: Foundations of Quantum Chromodynamics: An Introduction to Perturbation
Methods in Gauge Theories (3rd ed.)
T Muta

Vol. 79: Geometry and Phase Transitions in Colloids and Polymers
W Kung

Vol. 80: Introduction to Supersymmetry (2nd ed.)
H J W Müller-Kirsten and A Wiedemann

Vol. 81: Classical and Quantum Dynamics of Constrained Hamiltonian Systems
H J Rothe and K D Rothe

Vol. 82: Lattice Gauge Theories: An Introduction (4th ed.)
H J Rothe

*For the complete list of published titles, please visit
http://www.worldscibooks.com/series/lnp_series.shtml

World Scientific Lecture Notes in Physics – Vol. 75

FIELD THEORY
A Path Integral Approach

Second Edition

ASHOK DAS
University of Rochester, USA

World Scientific

NEW JERSEY · LONDON · SINGAPORE · BEIJING · SHANGHAI · HONG KONG · TAIPEI · CHENNAI

Published by

World Scientific Publishing Co. Pte. Ltd.
5 Toh Tuck Link, Singapore 596224
USA office: 27 Warren Street, Suite 401-402, Hackensack, NJ 07601
UK office: 57 Shelton Street, Covent Garden, London WC2H 9HE

Library of Congress Cataloging-in-Publication-Data
Das, Ashok, 1953–
    Field theory : a path integral approach, 2nd ed. / Ashok Das.
    p. cm. -- (World Scientific lecture notes in physics ; v. 75)
    Includes bibliographical references and index.
    ISBN-13 978-981-256-847-2
    ISBN-10 981-256-847-6
    ISBN-13 978-981-256-848-9 (pbk)
    ISBN-10 981-256-848-4 (pbk)
    1. Path integrals. 2. Quantum field theory.
    QC174.52.P37 D37 2006
    530.14'3
                                                2006283970

British Library Cataloguing-in-Publication Data
A catalogue record for this book is available from the British Library.

First published 2006
Reprinted 2012

Printed in Singapore

To
Lakshmi
and
Gouri

Preface to the First Edition

Traditionally, field theory had its main thrust of development in high energy physics. Consequently, the conventional field theory courses are taught with a heavy emphasis on high energy physics. Over the years, however, it has become quite clear that the methods and techniques of field theory are widely applicable in many areas of physics. The canonical quantization methods, which is how conventional field theory courses are taught, do not bring out this feature of field theory. A path integral description of field theory is the appropriate setting for this. It is with this goal in mind, namely, to make graduate students aware of the applicability of the field theoretic methods to various areas, that the Department of Physics and Astronomy at the University of Rochester introduced a new one semester course on field theory in Fall 1991.

This course was aimed at second year graduate students who had already taken a one year course on nonrelativistic quantum mechanics but had not necessarily specialized into any area of physics and these lecture notes grew out of this course which I taught. In fact, the lecture notes are identical to what was covered in the class. Even in the published form, I have endeavored to keep as much of the detailed derivations of various results as I could — the idea being that a reader can then concentrate on the logical development of concepts without worrying about the technical details. Most of the concepts were developed within the context of quantum mechanics — which the students were expected to be familiar with — and subsequently these concepts were applied to various branches of physics. In writing these lecture notes, I have added some references at the end of every chapter. They are only intended to be suggestive. There is so much literature that is available in this subject that it would have been impossible to include all of them. The references are not meant to be complete and I apologize to many whose works I have not cited in the references. Since this was developed as a course for general students, the many interesting topics of gauge theories are also not covered in these lectures. It simply would have been impossible to do justice to these topics within a one semester course.

There are many who were responsible for these lecture notes. I would like to thank our chairman, Paul Slattery, for asking me to teach and design a syllabus for this course. The students deserve the most credit for keeping all the derivations complete and raising many issues which I, otherwise, would have taken for granted. I am grateful to my students Paulo Bedaque and Wen-Jui Huang as well as to Dr. Zhu Yang for straightening out many little details which were essential in presenting the material in a coherent and consistent way. I would also like to thank Michael Begel for helping out in numerous ways, in particular, in computer-generating all the figures in the book. The support of many colleagues was also vital for the completion of these lecture notes. Judy Mack, as always, has done a superb job as far as the appearance of the book is concerned and I sincerely thank her. Finally, I am grateful to Ammani for being there.

Ashok Das,
Rochester.

Preface to the Second Edition

This second edition of the book is an expanded version which contains a chapter on path integral quantization of gauge theories as well as a chapter on anomalies. In addition, chapter 6 (Supersymmetry) has been expanded to include a section on supersymmetric singular potentials. While these topics were not covered in the original course on path integrals, they are part of my lectures in other courses that I have taught at the University of Rochester and have been incorporated into this new edition at the request of colleagues from all over the world. There are many people who have helped me to complete this edition of the book and I would like to thank, in particular, Judy Mack, Arsen Melikyan, Dave Munson and J. Boersma for all their assistance.

Ashok Das,
Rochester.

Contents

Preface to the First Edition   vii

Preface to the Second Edition   ix

1. Introduction   1

    1.1   Particles and Fields   1
    1.2   Metric and Other Notations   2
    1.3   Functionals   3
    1.4   Review of Quantum Mechanics   7
    1.5   References   10

2. Path Integrals and Quantum Mechanics   11

    2.1   Basis States   11
    2.2   Operator Ordering   13
    2.3   The Classical Limit   20
    2.4   Equivalence with the Schrödinger Equation   22
    2.5   Free Particle   25
    2.6   References   30

3. Harmonic Oscillator   31

    3.1   Path Integral for the Harmonic Oscillator   31
    3.2   Method of Fourier Transform   33
    3.3   Matrix Method   36
    3.4   The Classical Action   45

3.5   References 51

4. Generating Functional   53

4.1   Euclidean Rotation 53
4.2   Time Ordered Correlation Functions 59
4.3   Correlation Functions in Definite States 61
4.4   Vacuum Functional 64
4.5   Anharmonic Oscillator 71
4.6   References 73

5. Path Integrals for Fermions   75

5.1   Fermionic Oscillator 75
5.2   Grassmann Variables 78
5.3   Generating Functional 83
5.4   Feynman Propagator 86
5.5   The Fermion Determinant 91
5.6   References 95

6. Supersymmetry   97

6.1   Supersymmetric Oscillator 97
6.2   Supersymmetric Quantum Mechanics 102
6.3   Shape Invariance 105
6.4   Example 110
6.5   Supersymmetry and Singular Potentials 111
        6.5.1   Regularized Superpotential 115
        6.5.2   Alternate Regularization 117
6.6   References 118

7. Semi-Classical Methods   121

7.1   WKB Approximation 121
7.2   Saddle Point Method 127
7.3   Semi-Classical Methods in Path Integrals 130
7.4   Double Well Potential 134
7.5   References 142

8. Path Integral for the Double Well 143

   8.1   Instantons 143
   8.2   Zero Modes 150
   8.3   The Instanton Integral 154
   8.4   Evaluating the Determinant 158
   8.5   Multi-Instanton Contributions 163
   8.6   References 166

9. Path Integral for Relativistic Theories 167

   9.1   Systems with Many Degrees of Freedom 167
   9.2   Relativistic Scalar Field Theory 170
   9.3   Feynman Rules 181
   9.4   Connected Diagrams 184
   9.5   References 186

10. Effective Action 187

   10.1   The Classical Field 187
   10.2   Effective Action 193
   10.3   Loop Expansion 200
   10.4   Effective Potential at One Loop 203
   10.5   References 208

11. Invariances and Their Consequences 209

   11.1   Symmetries of the Action 209
   11.2   Noether's Theorem 212
         11.2.1 Example 215
   11.3   Complex Scalar Field 218
   11.4   Ward Identities 222
   11.5   Spontaneous Symmetry Breaking 226
   11.6   Goldstone Theorem 235
   11.7   References 236

12. Gauge Theories 239

   12.1   Maxwell Theory 239
   12.2   Non-Abelian Gauge Theory 246
   12.3   Path Integral for Gauge Theories 255

12.4   BRST Invariance 266
12.5   Ward Identities 274
12.6   References 278

13. Anomalies   279

13.1   Anomalous Ward Identity 279
13.2   Schwinger Model 289
13.3   References 307

14. Systems at Finite Temperature   309

14.1   Statistical Mechanics 309
14.2   Critical Exponents 314
14.3   Harmonic Oscillator 318
14.4   Fermionic Oscillator 324
14.5   References 326

15. Ising Model   327

15.1   One Dimensional Ising Model 327
15.2   The Partition Function 332
15.3   Two Dimensional Ising Model 337
15.4   Duality 339
15.5   High and Low Temperature Expansions 343
15.6   Quantum Mechanical Model 349
15.7   Duality in the Quantum System 356
15.8   References 358

Index   359

Chapter 1

Introduction

1.1 Particles and Fields

Classically, there are two kinds of dynamical systems that we encounter. First, there is the motion of a particle or a rigid body (with a finite number of degrees of freedom) which can be described by a finite number of coordinates. And then, there are physical systems where the number of degrees of freedom is nondenumerably (noncountably) infinite. Such systems are described by fields. Familiar examples of classical fields are the electromagnetic fields described by $\mathbf{E}(\mathbf{x}, t)$ and $\mathbf{B}(\mathbf{x}, t)$ or equivalently by the potentials $(\phi(\mathbf{x}, t), \mathbf{A}(\mathbf{x}, t))$. Similarly, the motion of a one-dimensional string is also described by a field $\phi(\mathbf{x}, t)$, namely, the displacement field. Thus, while the coordinates of a particle depend only on time, fields depend continuously on some space variables as well. Therefore, a theory described by fields is usually known as a $D + 1$ dimensional field theory where D represents the number of spatial dimensions on which the field variables depend. For example, a theory describing the displacements of the one-dimensional string would constitute a 1+1 dimensional field theory whereas the more familiar Maxwell's equations (in four dimensions) can be regarded as a 3+1 dimensional field theory. In this language, then, it is clear that a theory describing the motion of a particle can be regarded as a special case, namely, we can think of such a theory as a 0+1 dimensional field theory.

1

1.2   Metric and Other Notations

In these lectures, we will discuss both non-relativistic as well as relativistic theories. For the relativistic case, we will use the Bjorken-Drell convention. Namely, the contravariant coordinates are assumed to be

$$x^{\mu} = (t, \mathbf{x}), \quad \mu = 0, 1, 2, 3, \tag{1.1}$$

while the covariant coordinates have the form

$$x_{\mu} = \eta_{\mu\nu} x^{\nu} = (t, -\mathbf{x}). \tag{1.2}$$

Here we have assumed the speed of light to be unity ($c = 1$). The covariant metric, therefore, follows to have a diagonal form with the signatures

$$\eta_{\mu\nu} = (+, -, -, -). \tag{1.3}$$

The inverse or the contravariant metric clearly also has the same form, namely,

$$\eta^{\mu\nu} = (+, -, -, -). \tag{1.4}$$

The invariant length is given by

$$x^2 = x^{\mu} x_{\mu} = \eta^{\mu\nu} x_{\mu} x_{\nu} = \eta_{\mu\nu} x^{\mu} x^{\nu} = t^2 - \mathbf{x}^2. \tag{1.5}$$

The gradients are similarly obtained from Eqs. (1.1) and (1.2) to be

$$\partial_{\mu} = \frac{\partial}{\partial x^{\mu}} = \left(\frac{\partial}{\partial t}, \boldsymbol{\nabla} \right), \tag{1.6}$$

$$\partial^{\mu} = \frac{\partial}{\partial x_{\mu}} = \left(\frac{\partial}{\partial t}, -\boldsymbol{\nabla} \right), \tag{1.7}$$

so that the D'Alembertian takes the form

$$\Box = \partial^{\mu} \partial_{\mu} = \eta^{\mu\nu} \partial_{\mu} \partial_{\nu} = \frac{\partial^2}{\partial t^2} - \boldsymbol{\nabla}^2. \tag{1.8}$$

1.3 Functionals

It is evident that in dealing with dynamical systems, we are dealing with functions of continuous variables. In fact, most of the times, we are really dealing with functions of functions which are otherwise known as functionals. If we are considering the motion of a particle in a potential in one dimension, then the Lagrangian is given by

$$L(x, \dot{x}) = \frac{1}{2} m \dot{x}^2 - V(x), \tag{1.9}$$

where $x(t)$ and $\dot{x}(t)$ denote the coordinate and the velocity of the particle and the simplest functional we can think of is the action functional defined as

$$S[x] = \int_{t_i}^{t_f} dt \, L(x, \dot{x}). \tag{1.10}$$

Note that unlike a function whose value depends on a particular point in the coordinate space, the value of the action depends on the entire trajectory along which the integration is carried out. For different paths connecting the initial and the final points, the value of the action functional will be different.

Thus, a functional has the generic form

$$F[f] = \int dx \, F(f(x)), \tag{1.11}$$

where, for example, we may have

$$F(f(x)) = (f(x))^n. \tag{1.12}$$

Sometimes, one loosely also says that $F(f(x))$ is a functional. The notion of a derivative can be extended to the case of functionals in a natural way through the notion of generalized functions. Thus, one defines the functional derivative or the Gateaux derivative from the linear functional

$$F'[v] = \frac{d}{d\epsilon} F[f + \epsilon v] \Big|_{\epsilon=0} = \int dx \, \frac{\delta F[f]}{\delta f(x)} \, v(x). \tag{1.13}$$

Equivalently, from the working point of view, this simply corresponds to defining

$$\frac{\delta F(f(x))}{\delta f(y)} = \lim_{\epsilon \to 0} \frac{F(f(x) + \epsilon\delta(x-y)) - F(f(x))}{\epsilon}. \qquad (1.14)$$

It now follows from Eq. (1.14) that

$$\frac{\delta f(x)}{\delta f(y)} = \delta(x-y). \qquad (1.15)$$

The functional derivative satisfies all the properties of a derivative, namely, it is linear and associative,

$$\frac{\delta}{\delta f(x)}(F_1[f] + F_2[f]) = \frac{\delta F_1[f]}{\delta f(x)} + \frac{\delta F_2[f]}{\delta f(x)},$$

$$\frac{\delta}{\delta f(x)}(F_1[f]F_2[f]) = \frac{\delta F_1[f]}{\delta f(x)} F_2[f] + F_1[f] \frac{\delta F_2[f]}{\delta f(x)}. \qquad (1.16)$$

It also satisfies the chain rule of differentiation. Furthermore, we now see that given a functional $F[f]$, we can Taylor expand it in the form

$$F[f] = \int dx\, P_0(x) + \int dx_1 dx_2\, P_1(x_1, x_2)\, f(x_2)$$

$$+ \int dx_1 dx_2 dx_3\, P_2(x_1, x_2, x_3)\, f(x_2)f(x_3) + \cdots, \qquad (1.17)$$

where

$$P_0(x) = F(f(x))|_{f(x)=0},$$

$$P_1(x_1, x_2) = \frac{\delta F(f(x_1))}{\delta f(x_2)}\bigg|_{f(x)=0}, \qquad (1.18)$$

$$P_2(x_1, x_2, x_3) = \frac{1}{2!}\frac{\delta^2 F(f(x_1))}{\delta f(x_2)\delta f(x_3)}\bigg|_{f(x)=0},$$

and so on.

As simple examples, let us calculate a few particular functional derivatives.

(i) Let

$$F[f] = \int \mathrm{d}y \, F(f(y)) = \int \mathrm{d}y \, (f(y))^n, \qquad (1.19)$$

where n denotes a positive integer. Then,

$$\frac{\delta F(f(y))}{\delta f(x)} = \lim_{\epsilon \to 0} \frac{F(f(y) + \epsilon \delta(y - x)) - F(f(y))}{\epsilon}$$

$$= \lim_{\epsilon \to 0} \frac{(f(y) + \epsilon \delta(y - x))^n - (f(y))^n}{\epsilon}$$

$$= \lim_{\epsilon \to 0} \frac{(f(y))^n + n\epsilon(f(y))^{n-1}\delta(y - x) + O(\epsilon^2) - (f(y))^n}{\epsilon}$$

$$= n(f(y))^{n-1}\delta(y - x). \qquad (1.20)$$

Therefore, we obtain

$$\frac{\delta F[f]}{\delta f(x)} = \int \mathrm{d}y \, \frac{\delta F(f(y))}{\delta f(x)}$$

$$= \int \mathrm{d}y \, n(f(y))^{n-1}\delta(y - x)$$

$$= n(f(x))^{n-1}. \qquad (1.21)$$

(ii) Let us next consider the one-dimensional action in Eq. (1.10)

$$S[x] = \int_{t_i}^{t_f} \mathrm{d}t' \, L(x(t'), \dot{x}(t')), \qquad (1.22)$$

with

$$L(x(t), \dot{x}(t)) = \frac{1}{2}m(\dot{x}(t))^2 - V(x(t))$$

$$= T(\dot{x}(t)) - V(x(t)). \qquad (1.23)$$

In a straightforward manner, we obtain

$$\frac{\delta V(x(t'))}{\delta x(t)} = \lim_{\epsilon \to 0} \frac{V(x(t') + \epsilon \delta(t' - t)) - V(x(t'))}{\epsilon}$$

$$= V'(x(t'))\delta(t' - t), \qquad (1.24)$$

where we have defined

$$V'(x(t')) = \frac{\partial V(x(t'))}{\partial x(t')}.$$

Similarly,

$$\frac{\delta T(\dot{x}(t'))}{\delta x(t)} = \lim_{\epsilon \to 0} \frac{T(\dot{x}(t') + \epsilon \frac{d}{dt'}\delta(t'-t)) - T(\dot{x}(t'))}{\epsilon}$$

$$= m\dot{x}(t')\frac{d}{dt'}\delta(t'-t). \tag{1.25}$$

It is clear now that

$$\frac{\delta L(x(t'), \dot{x}(t'))}{\delta x(t)} = \frac{\delta(T(\dot{x}(t')) - V(x(t')))}{\delta x(t)}$$

$$= m\dot{x}(t')\frac{d}{dt'}\delta(t'-t) - V'(x(t'))\delta(t'-t). \tag{1.26}$$

Consequently, in this case, we obtain for $t_i \leq t \leq t_f$

$$\frac{\delta S[x]}{\delta x(t)} = \int_{t_i}^{t_f} dt' \, \frac{\delta L(x(t'), \dot{x}(t'))}{\delta x(t)}$$

$$= \int_{t_i}^{t_f} dt' \, (m\dot{x}(t')\frac{d}{dt'}\delta(t'-t) - V'(x(t'))\delta(t'-t))$$

$$= -m\ddot{x}(t) - V'(x(t))$$

$$= -\frac{d}{dt}\frac{\partial L(x(t), \dot{x}(t))}{\partial \dot{x}(t)} + \frac{\partial L(x(t), \dot{x}(t))}{\partial x(t)}. \tag{1.27}$$

The right hand side is, of course, reminiscent of the Euler-Lagrange equation. In fact, we note that

$$\frac{\delta S[x]}{\delta x(t)} = -\frac{d}{dt}\frac{\partial L}{\partial \dot{x}} + \frac{\partial L}{\partial x} = 0, \tag{1.28}$$

gives the Euler-Lagrange equation as a functional extremum of the action. This is nothing other than the principle of least action expressed in a compact notation in the language of functionals.

1.4 Review of Quantum Mechanics

In this section, we will describe very briefly the essential features of quantum mechanics assuming that the readers are familiar with the subject. The conventional approach to quantum mechanics starts with the Hamiltonian formulation of classical mechanics and promotes observables to non-commuting operators. The dynamics, in this case, is given by the time-dependent Schrödinger equation

$$i\hbar\frac{\partial|\psi(t)\rangle}{\partial t} = H|\psi(t)\rangle, \qquad (1.29)$$

where H denotes the Hamiltonian operator of the system. Equivalently, in the one dimensional case, the wave function of a particle satisfies

$$i\hbar\frac{\partial\psi(x,t)}{\partial t} = H(x)\psi(x,t)$$

$$= \left(-\frac{\hbar^2}{2m}\frac{\partial^2}{\partial x^2} + V(x)\right)\psi(x,t), \qquad (1.30)$$

where we have identified

$$\psi(x,t) = \langle x|\psi(t)\rangle, \qquad (1.31)$$

with $|x\rangle$ denoting the coordinate basis states. This, then, defines the time evolution of the system.

The main purpose behind solving the Schrödinger equation lies in determining the time evolution operator which generates the time translation of the system. Namely, the time evolution operator transforms the quantum mechanical state at an earlier time t_2 to a future time t_1 as

$$|\psi(t_1)\rangle = U(t_1,t_2)|\psi(t_2)\rangle. \qquad (1.32)$$

Clearly, for a time independent Hamiltonian, we see from Eq. (1.29) (the Schrödinger equation) that for $t_1 > t_2$,

$$U(t_1,t_2) = e^{-\frac{i}{\hbar}(t_1-t_2)H}. \qquad (1.33)$$

More explicitly, we can write

$$U(t_1, t_2) = \theta(t_1 - t_2)e^{-\frac{i}{\hbar}(t_1 - t_2)H}\,. \qquad (1.34)$$

It is obvious that the time evolution operator is nothing other than the Green's function for the time dependent Schrödinger equation and satisfies

$$\left(i\hbar\frac{\partial}{\partial t_1} - H\right)U(t_1, t_2) = i\hbar\delta(t_1 - t_2)\,. \qquad (1.35)$$

Determining this operator is equivalent to finding its matrix elements in a given basis. Thus, for example, in the coordinate basis defined by

$$X|x\rangle = x|x\rangle\,, \qquad (1.36)$$

we can write

$$\langle x_1|U(t_1, t_2)|x_2\rangle = U(t_1, x_1; t_2, x_2)\,. \qquad (1.37)$$

If we know the function $U(t_1, x_1; t_2, x_2)$ completely, then the time evolution of the wave function can be written as

$$\psi(x_1, t_1) = \int \mathrm{d}x_2\, U(t_1, x_1; t_2, x_2)\psi(x_2, t_2)\,. \qquad (1.38)$$

It is interesting to note that the dependence on the intermediate times drops out in the above equation as can be easily checked.

Our discussion has been within the framework of the Schr-ödinger picture so far where the quantum states $|\psi(t)\rangle$ carry time dependence while the operators are time independent. On the other hand, in the Heisenberg picture, where the quantum states are time independent, using Eq. (1.32) we can identify

$$|\psi\rangle_H = |\psi(t = 0)\rangle_S = |\psi(t = 0)\rangle$$

$$= e^{\frac{i}{\hbar}tH}|\psi(t)\rangle = e^{\frac{i}{\hbar}tH}|\psi(t)\rangle_S\,. \qquad (1.39)$$

In this picture, the operators carry all the time dependence. For example, the coordinate operator in the Heisenberg picture is related to the coordinate operator in the Schrödinger picture through the relation

$$X_H(t) = e^{\frac{i}{\hbar}tH} X e^{-\frac{i}{\hbar}tH} \,. \tag{1.40}$$

The eigenstates of this operator satisfying

$$X_H(t)|x,t\rangle_H = x|x,t\rangle_H \,, \tag{1.41}$$

are then easily seen to be related to the coordinate basis in the Schrödinger picture through

$$|x,t\rangle_H = e^{\frac{i}{\hbar}tH}|x\rangle \,. \tag{1.42}$$

It is clear now that for $t_1 > t_2$ we can write

$$\begin{aligned}
{}_H\langle x_1,t_1|x_2,t_2\rangle_H &= \langle x_1|e^{-\frac{i}{\hbar}t_1 H} e^{\frac{i}{\hbar}t_2 H}|x_2\rangle \\
&= \langle x_1|e^{-\frac{i}{\hbar}(t_1-t_2)H}|x_2\rangle \\
&= \langle x_1|U(t_1,t_2)|x_2\rangle \\
&= U(t_1,x_1;t_2,x_2) \,.
\end{aligned} \tag{1.43}$$

This shows that the matrix elements of the time evolution operator are nothing other than the time ordered transition amplitudes between the coordinate basis states in the Heisenberg picture.

Finally, there is the interaction picture where both the quantum states as well as the operators carry partial time dependence. Without going into any technical detail, let us simply note here that the interaction picture is quite useful in the study of nontrivially interacting theories. In any case, the goal of the study of quantum mechanics in any of these pictures is to construct the matrix elements of the time evolution operator which as we have seen can be identified with transition amplitudes between the coordinate basis states in the Heisenberg picture.

1.5   References

Das, A., "Lectures on Quantum Mechanics", Hindustan Book Agency.

Dirac, P. A. M., "Principles of Quantum Mechanics", Oxford Univ. Press.

Schiff, L. I., "Quantum Mechanics", McGraw-Hill Publishing.

Chapter 2

Path Integrals and Quantum Mechanics

2.1 Basis States

Before going into the derivation of the path integral representation for $U(t_f, x_f; t_i, x_i)$ or the transition amplitude, let us recapitulate some of the basic formulae of quantum mechanics. Consider, for simplicity, a one dimensional quantum mechanical system. The eigenstates of the coordinate operator, as we have seen in Eq. (1.36), satisfy

$$X|x\rangle = x|x\rangle. \tag{2.1}$$

These eigenstates define an orthonormal basis. Namely, they satisfy

$$\langle x|x'\rangle = \delta(x - x'),$$

$$\int dx \, |x\rangle\langle x| = \mathbf{1}. \tag{2.2}$$

Similarly, the eigenstates of the momentum operator satisfying

$$P|p\rangle = p|p\rangle, \tag{2.3}$$

also define an orthonormal basis. Namely, the momentum eigenstates satisfy

$$\langle p|p'\rangle = \delta(p - p'),$$

$$\int dp \, |p\rangle\langle p| = \mathbf{1}. \tag{2.4}$$

The inner product of the coordinate and the momentum basis states gives the matrix elements of the transformation operator between the two basis. In fact, one can readily determine that

$$\langle p|x \rangle = \frac{1}{\sqrt{2\pi\hbar}} e^{-\frac{i}{\hbar}px} = \langle x|p \rangle^* . \tag{2.5}$$

These are the defining relations for Fourier transforms. Namely, using the completeness relations of the basis states, the Fourier transform of functions can be defined as

$$f(x) = \langle x|f \rangle = \int dp \, \langle x|p \rangle \langle p|f \rangle$$

$$= \frac{1}{\sqrt{2\pi\hbar}} \int dp \, e^{\frac{i}{\hbar}px} \, f(p)$$

$$= \frac{1}{\sqrt{2\pi}} \int dk \, e^{ikx} \tilde{f}(k) , \tag{2.6}$$

$$\tilde{f}(k) = \sqrt{\hbar} \, f(p)$$

$$= \frac{\sqrt{\hbar}}{\sqrt{2\pi\hbar}} \int dx \, e^{-\frac{i}{\hbar}px} f(x)$$

$$= \frac{1}{\sqrt{2\pi}} \int dx \, e^{-ikx} f(x) . \tag{2.7}$$

These simply take a function from a given space to its conjugate space or the dual space. Here $k = \frac{p}{\hbar}$ can be thought of as the wave number in the case of a quantum mechanical particle. (Some other authors may define Fourier transform with alternate normalizations. Here, the definition is symmetrical.)

As we have seen in Eq. (1.42), the Heisenberg states are related to the Schrödinger states in a simple way. For the coordinate basis states, for example, we will have

$$|x,t\rangle_H = e^{\frac{i}{\hbar}Ht}|x\rangle .$$

It follows now that the coordinate basis states in the Heisenberg picture satisfy

$$_H\langle x,t|x',t\rangle_H = \langle x|e^{-\frac{i}{\hbar}tH}e^{\frac{i}{\hbar}tH}|x'\rangle$$
$$= \langle x|x'\rangle = \delta(x-x'),\qquad(2.8)$$

and

$$\int dx\,|x,t\rangle_H\,_H\langle x,t| = \int dx\,e^{\frac{i}{\hbar}tH}|x\rangle\langle x|e^{-\frac{i}{\hbar}tH}$$

$$= e^{\frac{i}{\hbar}tH}\int dx\,|x\rangle\langle x|e^{-\frac{i}{\hbar}tH}$$

$$= e^{\frac{i}{\hbar}tH}1e^{-\frac{i}{\hbar}tH}$$

$$= 1.\qquad(2.9)$$

It is worth noting here that the orthonormality as well as the completeness relations hold for the Heisenberg states only at equal times.

2.2 Operator Ordering

In the Hamiltonian formalism, the transition from classical mechanics to quantum mechanics is achieved by promoting observables to operators which are not necessarily commuting. Consequently, the Hamiltonian of the classical system is supposed to go over to the quantum operator

$$H(x,p)\to H(x_{\mathrm{op}},p_{\mathrm{op}})\,.\qquad(2.10)$$

This, however, does not specify what should be done when products of x and p (which are non-commuting as operators) are involved. For example, classically we know that

$$xp = px\,.$$

Therefore, the order of these terms does not matter in the classical Hamiltonian. Quantum mechanically, however, the order of the operators is quite crucial and *a priori* it is not clear what such a term ought to correspond to in the quantum theory. This is the operator ordering problem and, unfortunately, there is no well defined principle which specifies the order of operators in the passage from classical to quantum mechanics. There are, however, a few prescriptions which one uses conventionally. In normal ordering, one orders the products of x's and p's such that the momenta stand to the left of the coordinates. Thus,

$$xp \xrightarrow{\text{N.O.}} px \,,$$

$$px \xrightarrow{\text{N.O.}} px \,,$$

$$x^2 p \xrightarrow{\text{N.O.}} px^2 \,,$$

$$xpx \xrightarrow{\text{N.O.}} px^2 \,, \tag{2.11}$$

and so on. However, the prescription that is much more widely used and is much more satisfactory from various other points of view is the Weyl ordering. Here one symmetrizes the product of operators in all possible combinations with equal weight. Thus,

$$xp \xrightarrow{\text{W.O.}} \frac{1}{2}(xp + px) \,,$$

$$px \xrightarrow{\text{W.O.}} \frac{1}{2}(xp + px) \,,$$

$$x^2 p \xrightarrow{\text{W.O.}} \frac{1}{3}(x^2 p + xpx + px^2) \,,$$

$$xpx \xrightarrow{\text{W.O.}} \frac{1}{3}(x^2 p + xpx + px^2) \,, \tag{2.12}$$

and so on.

For normal ordering, it is easy to see that for any quantum Hamiltonian obtained from the classical Hamiltonian $H(x, p)$

$$\langle x' | H^{\text{N.O.}} | x \rangle = \int dp \langle x' | p \rangle \langle p | H^{\text{N.O.}} | x \rangle$$

$$= \int \frac{dp}{2\pi\hbar} e^{-\frac{i}{\hbar} p(x - x')} H(x, p) \,. \tag{2.13}$$

Here we have used the completeness relations of the momentum basis states given in Eq. (2.4) as well as the defining relations in Eqs. (2.1), (2.3) and (2.5). (The matrix element of the quantum Hamiltonian is a classical function for which the ordering is irrelevant.) To understand Weyl ordering, on the other hand, let us note that the expansion of

$$(\alpha x_{\mathrm{op}} + \beta p_{\mathrm{op}})^N,$$

generates the Weyl ordering of products of the form $x_{\mathrm{op}}^n p_{\mathrm{op}}^m$ naturally if we treat x_{op} and p_{op} as non-commuting operators. In fact, we can easily show that

$$(\alpha x_{\mathrm{op}} + \beta p_{\mathrm{op}})^N = \sum_{n+m=N} \frac{N!}{n!m!} \, \alpha^n \beta^m (x_{\mathrm{op}}^n p_{\mathrm{op}}^m)^{\mathrm{W.O.}} \qquad (2.14)$$

The expansion of the exponential operator

$$e^{(\alpha x_{\mathrm{op}} + \beta p_{\mathrm{op}})},$$

would, of course, generate all such powers and by analyzing the matrix elements of this exponential operator, we will learn about the matrix elements of Weyl ordered Hamiltonians.

From the fact that the commutator of x_{op} and p_{op} is a constant, we obtain using the Baker-Campbell-Hausdorff formula

$$e^{\left(\frac{\alpha x_{\mathrm{op}}}{2}\right)} e^{\beta p_{\mathrm{op}}} e^{\left(\frac{\alpha x_{\mathrm{op}}}{2}\right)} = e^{\left(\frac{\alpha x_{\mathrm{op}}}{2}\right)} e^{\left(\beta p_{\mathrm{op}} + \frac{\alpha x_{\mathrm{op}}}{2} - \frac{i\hbar\alpha\beta}{4}\right)}$$

$$= e^{\left(\alpha x_{\mathrm{op}} + \beta p_{\mathrm{op}} - \frac{i\hbar\alpha\beta}{4} + \frac{i\hbar\alpha\beta}{4}\right)}$$

$$= e^{(\alpha x_{\mathrm{op}} + \beta p_{\mathrm{op}})}. \qquad (2.15)$$

Using this relation, it can now be easily shown that

$$\langle x' | e^{(\alpha x_{\mathrm{op}} + \beta p_{\mathrm{op}})} | x \rangle = \langle x' | e^{\left(\frac{\alpha x_{\mathrm{op}}}{2}\right)} e^{\beta p_{\mathrm{op}}} e^{\left(\frac{\alpha x_{\mathrm{op}}}{2}\right)} | x \rangle$$

$$= \int dp \, \langle x' | e^{\left(\frac{\alpha x_{\mathrm{op}}}{2}\right)} e^{\beta p_{\mathrm{op}}} | p \rangle \langle p | e^{\left(\frac{\alpha x_{\mathrm{op}}}{2}\right)} | x \rangle$$

$$= \int \frac{dp}{2\pi\hbar} \, e^{-\frac{i}{\hbar} p(x - x')} e^{\left(\frac{\alpha(x + x')}{2} + \beta p\right)}. \qquad (2.16)$$

Once again, we have used here the completeness properties given in Eq. (2.4) as well as the defining relations in Eqs. (2.1), (2.3) and (2.5). It follows from this that for a Weyl ordered quantum Hamiltonian, we will have

$$\langle x'|H^{\text{W.O.}}(x_{\text{op}}, p_{\text{op}})|x\rangle = \int \frac{dp}{2\pi\hbar} \, e^{-\frac{i}{\hbar}p(x-x')} H\left(\frac{x+x'}{2}, p\right). \quad (2.17)$$

As we see, the matrix elements of the Weyl ordered Hamiltonian leads to what is known as the mid-point prescription and this is what we will use in all of our discussions.

We are now ready to calculate the transition amplitude. Let us recall that in the Heisenberg picture, for $t_f > t_i$, we have

$$U(t_f, x_f; t_i, x_i) = {}_H\langle x_f, t_f|x_i, t_i\rangle_H.$$

Let us divide the time interval between the initial and the final time into N equal segments of infinitesimal length ϵ. Namely, let

$$\epsilon = \frac{t_f - t_i}{N}. \quad (2.18)$$

In other words, for simplicity, we discretize the time interval and in the end, we are interested in taking the continuum limit $\epsilon \to 0$ and $N \to \infty$ such that Eq. (2.18) holds true. We can now label the intermediate times as, say,

$$t_n = t_i + n\epsilon, \quad n = 1, 2, \ldots, (N-1). \quad (2.19)$$

Introducing complete sets of coordinate basis states for every intermediate time point (see Eq. (2.9)), we obtain

$$U(t_f, x_f; t_i, x_i) = {}_H\langle x_f, t_f|x_i, t_i\rangle_H$$

$$= \lim_{\substack{\epsilon \to 0 \\ N \to \infty}} \int dx_1 \cdots dx_{N-1} \, {}_H\langle x_f, t_f|x_{N-1}, t_{N-1}\rangle_H$$

$$\times \, {}_H\langle x_{N-1}, t_{N-1}|x_{N-2}, t_{N-2}\rangle_H \cdots {}_H\langle x_1, t_1|x_i, t_i\rangle_H. \quad (2.20)$$

In writing this, we have clearly assumed an inherent time ordering from left to right. Let us also note here that while there are N inner products in the above expression, there are only $(N-1)$ intermediate points of integration. Furthermore, we note that any intermediate inner product in Eq. (2.20) has the form

$$
\begin{aligned}
{}_H\langle x_n, t_n | x_{n-1}, t_{n-1}\rangle_H &= \langle x_n | e^{-\frac{i}{\hbar} t_n H} e^{\frac{i}{\hbar} t_{n-1} H} | x_{n-1}\rangle \\
&= \langle x_n | e^{-\frac{i}{\hbar}(t_n - t_{n-1})H} | x_{n-1}\rangle \\
&= \langle x_n | e^{-\frac{i}{\hbar}\epsilon H} | x_{n-1}\rangle \\
&= \int \frac{dp_n}{2\pi\hbar} \, e^{\frac{i}{\hbar} p_n(x_n - x_{n-1}) - \frac{i}{\hbar}\epsilon H\left(\frac{x_n + x_{n-1}}{2}, p_n\right)} .
\end{aligned}
\tag{2.21}
$$

Here we have used the mid-point prescription of Eq. (2.17) corresponding to Weyl ordering.

Substituting this form of the inner product into the transition amplitude, we obtain

$$
U(t_f, x_f; t_i, x_i) = \lim_{\substack{\epsilon \to 0 \\ N \to \infty}} \int dx_1 \cdots dx_{N-1} \frac{dp_1}{2\pi\hbar} \cdots \frac{dp_N}{2\pi\hbar}
$$

$$
\times \, e^{\frac{i}{\hbar} \sum_{n=1}^{N} \left(p_n(x_n - x_{n-1}) - \epsilon H\left(\frac{x_n + x_{n-1}}{2}, p_n\right)\right)} . \tag{2.22}
$$

In writing this, we have identified

$$
x_0 = x_i, \qquad x_N = x_f . \tag{2.23}
$$

This is the crudest form of Feynman's path integral and is defined in the phase space of the system. It is worth emphasizing here that the number of intermediate coordinate integrations differs from the number of momentum integrations and has profound consequences in the study of the symmetry properties of the transition amplitudes. Note that in the continuum limit, namely, for $\epsilon \to 0$, we can write the phase factor of Eq. (2.22) as

$$\lim_{\substack{\epsilon \to 0 \\ N \to \infty}} \frac{i}{\hbar} \sum_{n=1}^{N} \left(p_n (x_n - x_{n-1}) - \epsilon H \left(\frac{x_n + x_{n-1}}{2}, p_n \right) \right)$$

$$= \lim_{\substack{\epsilon \to 0 \\ N \to \infty}} \frac{i}{\hbar} \epsilon \sum_{n=1}^{N} \left(p_n \left(\frac{x_n - x_{n-1}}{\epsilon} \right) - H \left(\frac{x_n + x_{n-1}}{2}, p_n \right) \right)$$

$$= \frac{i}{\hbar} \int_{t_i}^{t_f} dt \, (p\dot{x} - H(x,p))$$

$$= \frac{i}{\hbar} \int_{t_i}^{t_f} dt \, L \,. \tag{2.24}$$

Namely, it is proportional to the action in the mixed variables.

To obtain the more familiar form of the path integral involving the Lagrangian in the configuration space, let us specialize to the class of Hamiltonians which are quadratic in the momentum variables. Namely, let us choose

$$H(x,p) = \frac{p^2}{2m} + V(x) \,. \tag{2.25}$$

In such a case, we have from Eq. (2.22)

$$U(t_f, x_f; t_i, x_i) = \lim_{\substack{\epsilon \to 0 \\ N \to \infty}} \int dx_1 \cdots dx_{N-1} \frac{dp_1}{2\pi\hbar} \cdots \frac{dp_N}{2\pi\hbar}$$

$$\times e^{\frac{i\epsilon}{\hbar} \sum_{n=1}^{N} \left(p_n \left(\frac{x_n - x_{n-1}}{\epsilon} \right) - \frac{p_n^2}{2m} - V \left(\frac{x_n + x_{n-1}}{2} \right) \right)} \,. \tag{2.26}$$

The momentum integrals are Gaussian and, therefore, can be done readily. We note that

$$\int \frac{dp_n}{2\pi\hbar} e^{-\frac{i\epsilon}{\hbar} \left(\frac{p_n^2}{2m} - \frac{p_n(x_n - x_{n-1})}{\epsilon} \right)}$$

$$= \int \frac{dp_n}{2\pi\hbar} e^{-\frac{i\epsilon}{2m\hbar} \left(p_n^2 - \frac{2mp_n(x_n - x_{n-1})}{\epsilon} \right)}$$

$$= \int \frac{\mathrm{d}p_n}{2\pi\hbar} \, e^{-\frac{i\epsilon}{2m\hbar}\left[\left(p_n - \frac{m(x_n - x_{n-1})}{\epsilon}\right)^2 - \left(\frac{m(x_n - x_{n-1})}{\epsilon}\right)^2\right]}$$

$$= \frac{1}{2\pi\hbar} \left(\frac{2\pi m\hbar}{i\epsilon}\right)^{\frac{1}{2}} e^{\frac{im\epsilon}{2\hbar}\left(\frac{x_n - x_{n-1}}{\epsilon}\right)^2}$$

$$= \left(\frac{m}{2\pi i\hbar\epsilon}\right)^{\frac{1}{2}} e^{\frac{im\epsilon}{2\hbar}\left(\frac{x_n - x_{n-1}}{\epsilon}\right)^2}. \tag{2.27}$$

Substituting this back into the transition amplitude in Eq. (2.26), we obtain

$$U(t_f, x_f; t_i, x_i) = \lim_{\substack{\epsilon \to 0 \\ N \to \infty}} \left(\frac{m}{2\pi i\hbar\epsilon}\right)^{\frac{N}{2}}$$

$$\times \int \mathrm{d}x_1 \cdots \mathrm{d}x_{N-1} e^{\frac{i\epsilon}{\hbar}\sum_{n=1}^{N}\left(\frac{m}{2}\left(\frac{x_n - x_{n-1}}{\epsilon}\right)^2 - V\left(\frac{x_n + x_{n-1}}{2}\right)\right)}$$

$$= A \int \mathcal{D}x \, e^{\frac{i}{\hbar}\int_{t_i}^{t_f} \mathrm{d}t\left(\frac{1}{2}m\dot{x}^2 - V(x)\right)}$$

$$= A \int \mathcal{D}x \, e^{\frac{i}{\hbar}S[x]}, \tag{2.28}$$

where A is a constant independent of the dynamics of the system and $S[x]$ is the action for the system given in Eq. (1.10). This is Feynman's path integral for the transition amplitude in quantum mechanics.

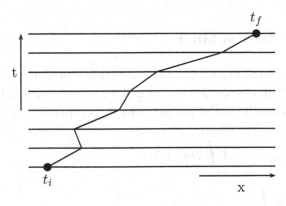

To understand the meaning of this, let us try to understand the meaning of the path integral measure $\mathcal{D}x$. In this integration, the end points are held fixed and only the intermediate coordinates are integrated over the entire space. Any spatial configuration of the intermediate points, of course, gives rise to a trajectory between the initial and the final points. Thus, integrating over all such configurations (that is precisely what the integrations over the intermediate points are supposed to do) is equivalent to summing over all the paths connecting the initial and the final points. Therefore, Feynman's path integral simply says that the transition amplitude between an initial and a final state is the sum over all paths, connecting the two points, of the weight factor $e^{\frac{i}{\hbar}S[x]}$. We know from the study of quantum mechanics that if a process can take place in several distinct ways, then the transition amplitude is the sum of the individual amplitudes corresponding to every possible way the process can take place. The sum over the paths is, therefore, quite expected. However, it is the weight factor $e^{\frac{i}{\hbar}S[x]}$ that is quite crucial and unexpected. Classically, we know that it is the classical action that determines the classical dynamics. Quantum mechanically, however, what we see is that all the paths contribute to the transition amplitude. It is also worth pointing out here that even though we derived the path integral representation for the transition amplitude for a special class of Hamiltonians, the expression holds in general. For Hamiltonians which are not quadratic in the momenta, one should simply be careful in defining the path integral measure $\mathcal{D}x$.

2.3 The Classical Limit

As we have seen in Eq. (2.28), the transition amplitude can be written as a sum over paths and for the case of a one dimensional Hamiltonian which is quadratic in the momentum, it is represented as

$$U(t_f, x_f; t_i, x_i) = A \int \mathcal{D}x \, e^{\frac{i}{\hbar}S[x]} \tag{2.29}$$

$$= \lim_{\substack{\epsilon \to 0 \\ N \to \infty}} A_N \int dx_1 \cdots dx_{N-1} \, e^{\frac{i\epsilon}{\hbar}\sum_{n=1}^{N}\left[\frac{m}{2}\left(\frac{x_n-x_{n-1}}{\epsilon}\right)^2 - V\left(\frac{x_n+x_{n-1}}{2}\right)\right]},$$

where

$$A_N = \left(\frac{m}{2\pi i \hbar \epsilon}\right)^{\frac{N}{2}}.$$

Even though one can be more quantitative in the discussion of the behavior of the transition amplitude, let us try to be qualitative in the following. We note that for paths where

$$x_n \gg x_{n-1},$$

the first term in the exponential would be quite large, particularly since ϵ is infinitesimally small. Therefore, such paths will lead to a very large phase and consequently, the weight factor can easily be positive or negative. In other words, for every such x_n, there would be a nearby x_n differing only slightly which would have a cancelling effect. Thus, in the path integral, all such contributions will average out to zero.

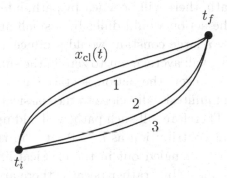

Let us, therefore, concentrate only on paths connecting the initial and the final points that differ from one another only slightly. For simplicity, we only look at continuous paths which are differentiable. (A more careful analysis shows that the paths which contribute non-trivially are the continuous paths which are not necessarily differentiable. But for simplicity of argument, we will ignore this technical point.) The question that we would like to understand is how among all the paths which can contribute to the transition amplitude, it is only the classical path that is singled out in the classical limit, namely, when $\hbar \to 0$. We note here that the weight factor in the path integral, namely, $e^{\frac{i}{\hbar}S[x]}$, is a phase multiplied by a large quantity when $\hbar \to 0$. Mathematically, therefore, it is clear that the dominant contribution to the path integral would arise from paths near the one which extremizes the phase factor. In other words, only the trajectories close to the ones satisfying

$$\frac{\delta S[x]}{\delta x(t)} = 0, \qquad (2.30)$$

would contribute significantly to the transition amplitude in the classical limit. But, from the principle of least action, we know that these are precisely the trajectories which a classical particle would follow, namely, the classical trajectories. Once again, we can see this more intuitively in the following way. Suppose, we are considering a path, say #3, which is quite far away from the classical trajectory. Then, because $\hbar$ is small, the phase along this trajectory will be quite large. For every such path, there will be a nearby path, infinitesimally close, say #2, where the action would differ by a small amount, but since it is multiplied by a large constant would produce a large phase. All such paths, clearly, will average out to zero in the sum. Near the classical trajectory, however, the action is stationary. Consequently, if we choose a path infinitesimally close to the classical path, the action will not change. Therefore, all such paths will add up coherently and give the dominant contribution as $\hbar \to 0$. It is in this way that the classical trajectory is singled out in the classical limit, not because it contributes the most, but rather because there are paths infinitesimally close to it which add coherently. One can, of course, make various estimates as to how far away a path can be from the classical trajectory before its contribution becomes unimportant. But let us not go into these details here.

2.4 Equivalence with the Schrödinger Equation

At this point one may wonder about the Schrödinger equation in the path integral formalism. Namely, it is not clear how we can recover the time dependent Schrödinger equation (see Eq. (1.30)) from the path integral representation of the transition amplitude. Let us recall that the Schrödinger equation is a differential equation. Therefore, it determines infinitesimal changes in the wave function. Consequently, to derive the Schrödinger equation, we merely have to examine the infinitesimal form of the transition amplitude or the path integral. From the explicit form of the transition amplitude in Eq. (2.29), we obtain for infinitesimal ϵ

$$U(t_f = \epsilon, x_f; t_i = 0, x_i)$$

$$= \left(\frac{m}{2\pi i \hbar \epsilon}\right)^{\frac{1}{2}} e^{\frac{i\epsilon}{\hbar}\left(\frac{m}{2}\left(\frac{x_f - x_i}{\epsilon}\right)^2 - V\left(\frac{x_f + x_i}{2}\right)\right)}. \qquad (2.31)$$

We also know from Eq. (1.38) that the transition amplitude is the propagator which gives the propagation of the wave function in the following way,

$$\psi(x, \epsilon) = \int_{-\infty}^{\infty} dx'\, U(\epsilon, x; 0, x')\psi(x', 0). \qquad (2.32)$$

Therefore, substituting the form of the transition amplitude namely, Eq. (2.31) into Eq. (2.32), we obtain

$$\psi(x, \epsilon) = \left(\frac{m}{2\pi i \hbar \epsilon}\right)^{\frac{1}{2}} \int_{-\infty}^{\infty} dx'\, e^{\frac{im}{2\hbar\epsilon}(x-x')^2 - \frac{i\epsilon}{\hbar}V\left(\frac{x+x'}{2}\right)}\psi(x', 0). \qquad (2.33)$$

Let us next change variables to

$$\eta = x' - x, \qquad (2.34)$$

so that we can write

$$\psi(x, \epsilon) = \left(\frac{m}{2\pi i \hbar \epsilon}\right)^{\frac{1}{2}} \int_{-\infty}^{\infty} d\eta\, e^{[\frac{im}{2\hbar\epsilon}\eta^2 - \frac{i\epsilon}{\hbar}V(x+\frac{\eta}{2})]}\psi(x+\eta, 0). \qquad (2.35)$$

It is obvious that because ϵ is infinitesimal, if η is large, then the first term in the exponent would lead to rapid oscillations and all such contributions will average out to zero. The dominant contribution will, therefore, come from the region of integration

$$0 \le |\eta| \le \left(\frac{2\pi \hbar \epsilon}{m}\right)^{\frac{1}{2}}, \qquad (2.36)$$

where the change in the first exponent is of the order of unity. Thus, we can Taylor expand the integrand and since we are interested in the infinitesimal behavior, we can keep terms consistently up to order ϵ. Therefore, we obtain

$$
\psi(x,\epsilon) = \left(\frac{m}{2\pi i\hbar\epsilon}\right)^{\frac{1}{2}} \int_{-\infty}^{\infty} d\eta\, e^{\frac{im}{2\hbar\epsilon}\eta^2} \left(1 - \frac{i\epsilon}{\hbar}V\left(x + \frac{\eta}{2}\right)\right) \psi(x+\eta, 0)
$$

$$
= \left(\frac{m}{2\pi i\hbar\epsilon}\right)^{\frac{1}{2}} \int_{-\infty}^{\infty} d\eta\, e^{\frac{im}{2\hbar\epsilon}\eta^2} \left(1 - \frac{i\epsilon}{\hbar}V(x) + O(\epsilon^2)\right)
$$

$$
\times \left(\psi(x,0) + \eta\psi'(x,0) + \frac{\eta^2}{2}\psi''(x,0) + O(\eta^3)\right)
$$

$$
= \left(\frac{m}{2\pi i\hbar\epsilon}\right)^{\frac{1}{2}} \int_{-\infty}^{\infty} d\eta\, e^{\frac{im}{2\hbar\epsilon}\eta^2} \left[\psi(x,0) - \frac{i\epsilon}{\hbar}V(x)\psi(x,0)\right.
$$

$$
\left. + \eta\psi'(x,0) + \frac{\eta^2}{2}\psi''(x,0) + O(\eta^3,\epsilon^2)\right]. \tag{2.37}
$$

The individual integrations can be easily done and the results are

$$
\int_{-\infty}^{\infty} d\eta\, e^{\frac{im}{2\hbar\epsilon}\eta^2} = \left(\frac{2\pi i\hbar\epsilon}{m}\right)^{\frac{1}{2}},
$$

$$
\int_{-\infty}^{\infty} d\eta\, \eta\, e^{\frac{im}{2\hbar\epsilon}\eta^2} = 0, \tag{2.38}
$$

$$
\int_{-\infty}^{\infty} d\eta\, \eta^2\, e^{\frac{im}{2\hbar\epsilon}\eta^2} = \frac{i\hbar\epsilon}{m}\left(\frac{2\pi i\hbar\epsilon}{m}\right)^{\frac{1}{2}}.
$$

Note that these integrals contain oscillatory integrands and the simplest way of evaluating them is through a regularization. For example,

$$
\int_{-\infty}^{\infty} d\eta\, e^{\frac{im}{2\hbar\epsilon}\eta^2} = \lim_{\delta\to 0^+} \int_{-\infty}^{\infty} d\eta\, e^{\left(\frac{im}{2\hbar\epsilon} - \delta\right)\eta^2}
$$

$$
= \lim_{\delta\to 0^+} \left(\frac{\pi}{\left(\delta - \frac{im}{2\hbar\epsilon}\right)}\right)^{\frac{1}{2}} = \left(\frac{2\pi i\hbar\epsilon}{m}\right)^{\frac{1}{2}}, \tag{2.39}
$$

and so on.

Substituting these back into Eq. (2.37), we obtain

$$\psi\left(x,\epsilon\right) = \left(\frac{m}{2\pi i\hbar\epsilon}\right)^{\frac{1}{2}}\left[\left(\frac{2\pi i\hbar\epsilon}{m}\right)^{\frac{1}{2}}\left(\psi\left(x,0\right) - \frac{i\epsilon}{\hbar}V\left(x\right)\psi\left(x,0\right)\right)\right.$$

$$\left. + \frac{i\hbar\epsilon}{2m}\left(\frac{2\pi i\hbar\epsilon}{m}\right)^{\frac{1}{2}}\psi''\left(x,0\right) + O\left(\epsilon^2\right)\right]$$

$$= \psi\left(x,0\right) + \frac{i\hbar\epsilon}{2m}\psi''\left(x,0\right) - \frac{i\epsilon}{\hbar}V\left(x\right)\psi\left(x,0\right) + O\left(\epsilon^2\right)$$

or, $\psi\left(x,\epsilon\right) - \psi\left(x,0\right) = -\frac{i\epsilon}{\hbar}\left(-\frac{\hbar^2}{2m}\frac{\partial^2}{\partial x^2} + V\left(x\right)\right)\psi\left(x,0\right) + O\left(\epsilon^2\right)$.

$$(2.40)$$

In the limit $\epsilon \to 0$, therefore, we obtain the time dependent Schrödinger equation (Eq. (1.30))

$$i\hbar\frac{\partial\psi\left(x,t\right)}{\partial t} = \left(-\frac{\hbar^2}{2m}\frac{\partial^2}{\partial x^2} + V\left(x\right)\right)\psi\left(x,t\right) .$$

The path integral representation, therefore, contains the Schrödinger equation and is equivalent to it.

2.5   Free Particle

We recognize that the path integral is a functional integral. Namely, the integrand which is the phase factor is a functional of the trajectory between the initial and the final points. Since we do not have a feeling for such quantities, let us evaluate some of these integrals associated with simple systems. The free particle is probably the simplest of quantum mechanical systems. For a free particle in one dimension, the Lagrangian has the form

$$L = \frac{1}{2}m\dot{x}^2 .$$

$$(2.41)$$

Therefore, from our definition of the transition amplitude in Eq. (2.28) or (2.29), we obtain

$$U(t_f, x_f; t_i, x_i)$$

$$= \lim_{\substack{\epsilon \to 0 \\ N \to \infty}} \left(\frac{m}{2\pi i\hbar\epsilon}\right)^{\frac{N}{2}} \int dx_1 \cdots dx_{N-1} e^{\frac{i\epsilon}{\hbar} \sum_{n=1}^{N} \frac{m}{2} \left(\frac{x_n - x_{n-1}}{\epsilon}\right)^2}$$

$$= \lim_{\substack{\epsilon \to 0 \\ N \to \infty}} \left(\frac{m}{2\pi i\hbar\epsilon}\right)^{\frac{N}{2}} \int dx_1 \cdots dx_{N-1} e^{\frac{im}{2\hbar\epsilon} \sum_{n=1}^{N} (x_n - x_{n-1})^2} .$$

$$(2.42)$$

Defining

$$y_n = \left(\frac{m}{2\hbar\epsilon}\right)^{\frac{1}{2}} x_n , \tag{2.43}$$

we have

$$U(t_f, x_f; t_i, x_i) = \lim_{\substack{\epsilon \to 0 \\ N \to \infty}} \left(\frac{m}{2\pi i\hbar\epsilon}\right)^{\frac{N}{2}} \left(\frac{2\hbar\epsilon}{m}\right)^{\frac{N-1}{2}}$$

$$\times \int dy_1 \cdots dy_{N-1} \, e^{i \sum_{n=1}^{N} (y_n - y_{n-1})^2} . \tag{2.44}$$

This is a Gaussian integral which can be evaluated in many different ways. However, the simplest method probably is to work out a few lower order ones and derive a pattern. We note that

$$\int dy_1 \, e^{i[(y_1 - y_0)^2 + (y_2 - y_1)^2]} = \int dy_1 \, e^{i[2(y_1 - \frac{y_0 + y_2}{2})^2 + \frac{1}{2}(y_2 - y_0)^2]}$$

$$= \left(\frac{i\pi}{2}\right)^{\frac{1}{2}} e^{\frac{i}{2}(y_2 - y_0)^2} . \tag{2.45}$$

If we had two intermediate integrations, then we will have

$$\int dy_1 dy_2 \; e^{i\left[(y_1-y_0)^2+(y_2-y_1)^2+(y_3-y_2)^2\right]}$$

$$= \left(\frac{i\pi}{2}\right)^{\frac{1}{2}} \int dy_2 \; e^{i\left[\frac{1}{2}(y_2-y_0)^2+(y_3-y_2)^2\right]}$$

$$= \left(\frac{i\pi}{2}\right)^{\frac{1}{2}} \int dy_2 \; e^{\left[\frac{3i}{2}(y_2-\frac{y_0+2y_3}{3})^2+\frac{i}{3}(y_3-y_0)^2\right]}$$

$$= \left(\frac{i\pi}{2}\right)^{\frac{1}{2}} \left(\frac{2i\pi}{3}\right)^{\frac{1}{2}} e^{\frac{i}{3}(y_3-y_0)^2}$$

$$= \left(\frac{(i\pi)^2}{3}\right)^{\frac{1}{2}} e^{\frac{i}{3}(y_3-y_0)^2} . \tag{2.46}$$

A pattern is now obvious and using this we can write

$$U(t_f, x_f; t_i, x_i)$$

$$= \lim_{\substack{\epsilon \to 0 \\ N \to \infty}} \left(\frac{m}{2\pi i\hbar\epsilon}\right)^{\frac{N}{2}} \left(\frac{2\hbar\epsilon}{m}\right)^{\frac{N-1}{2}} \left(\frac{(i\pi)^{N-1}}{N}\right)^{\frac{1}{2}} e^{\frac{i}{N}(y_N-y_0)^2}$$

$$= \lim_{\substack{\epsilon \to 0 \\ N \to \infty}} \left(\frac{m}{2\pi i\hbar\epsilon}\right)^{\frac{N}{2}} \left(\frac{2\pi i\hbar\epsilon}{m}\right)^{\frac{N-1}{2}} \frac{1}{\sqrt{N}} \; e^{\frac{im}{2\hbar N\epsilon}(x_N-x_0)^2}$$

$$= \lim_{\substack{\epsilon \to 0 \\ N \to \infty}} \left(\frac{m}{2\pi i\hbar N\epsilon}\right)^{\frac{1}{2}} e^{\frac{im}{2\hbar N\epsilon}(x_f-x_i)^2}$$

$$= \left[\frac{m}{2\pi i\hbar(t_f-t_i)}\right]^{\frac{1}{2}} e^{\frac{i}{\hbar}\frac{m(x_f-x_i)^2}{2(t_f-t_i)}} . \tag{2.47}$$

Thus, we see that for a free particle, the transition amplitude can be explicitly evaluated. It has the right behavior in the sense that, we see as $t_f \to t_i$,

$$U(t_f, x_f; t_i, x_i) \to \delta(x_f - x_i), \tag{2.48}$$

which is nothing other than the orthonormality relation for the states in the Heisenberg picture given in Eq. (2.8). Second, all the potentially dangerous singular terms involving ϵ have disappeared. Furthermore this is exactly what one would obtain by solving the Schrödinger equation. It expresses the well known fact that even a well localized wave packet spreads with time. That is, even the simplest of equations has only dispersive solutions.

Let us note here that since

$$S[x] = \int_{t_i}^{t_f} dt \, \frac{1}{2} m\dot{x}^2,$$

the Euler-Lagrange equations give (see Eq. (1.28))

$$\frac{\delta S[x]}{\delta x(t)} = m\ddot{x} = 0. \tag{2.49}$$

This gives as solutions

$$\dot{x}_{\rm cl}(t) = v = \text{constant}. \tag{2.50}$$

Thus, for the classical trajectory, we have

$$S[x_{\rm cl}] = \int_{t_i}^{t_f} dt \, \frac{1}{2} m\dot{x}_{\rm cl}^2 = \frac{1}{2} mv^2(t_f - t_i). \tag{2.51}$$

On the other hand, since v is a constant, we can write

$$x_f - x_i = v(t_f - t_i)$$

$$\text{or,} \quad v = \frac{x_f - x_i}{t_f - t_i}. \tag{2.52}$$

Substituting this back into Eq. (2.51), we obtain

$$S[x_{\rm cl}] = \frac{1}{2} m \left(\frac{x_f - x_i}{t_f - t_i} \right)^2 (t_f - t_i) = \frac{m}{2} \frac{(x_f - x_i)^2}{t_f - t_i}. \tag{2.53}$$

We recognize, therefore, that we can also write the quantum transition amplitude, in this case, simply as

$$U(t_f, x_f; t_i, x_i) = \left(\frac{m}{2\pi i\hbar(t_f - t_i)} \right)^{\frac{1}{2}} e^{\frac{i}{\hbar} S[x_{\rm cl}]}. \tag{2.54}$$

This is a particular characteristic of some quantum systems which can be exactly solved. Namely, for these systems, the transition amplitude can be written in the form

$$U(t_f, x_f; t_i, x_i) = A \, e^{\frac{i}{\hbar} S[x_{cl}]}, \qquad (2.55)$$

where A is a constant.

Finally, let us note from the explicit form of the transition amplitude in Eq. (2.47) that

$$\frac{\partial U}{\partial t_f} = -\frac{U}{2(t_f - t_i)} - \frac{im}{2\hbar} \left(\frac{x_f - x_i}{t_f - t_i} \right)^2 U,$$

$$\frac{\partial U}{\partial x_f} = \frac{im}{\hbar} \left(\frac{x_f - x_i}{t_f - t_i} \right) U,$$

$$\frac{\partial^2 U}{\partial x_f^2} = \frac{im}{\hbar} \frac{U}{t_f - t_i} + \left(\frac{im}{\hbar} \right)^2 \left(\frac{x_f - x_i}{t_f - t_i} \right)^2 U$$

$$= -\frac{2m}{\hbar^2} \left(-i\hbar \frac{U}{2(t_f - t_i)} + \frac{m}{2} \left(\frac{x_f - x_i}{t_f - t_i} \right)^2 U \right)$$

$$= -\frac{2m}{\hbar^2} \left(i\hbar \frac{\partial U}{\partial t_f} \right). \qquad (2.56)$$

Therefore, it follows that

$$i\hbar \frac{\partial U}{\partial t_f} = -\frac{\hbar^2}{2m} \frac{\partial^2 U}{\partial x_f^2}, \qquad (2.57)$$

which is equivalent to saying that the transition amplitude obtained from Feynman's path integral, indeed, solves the Schrödinger equation for a free particle (compare with Eq. (1.35)).

2.6   References

Das, A., "Lectures on Quantum Mechanics", Hindustan Book
    Agency.
Feynman, R. P. and A. R. Hibbs, "Quantum Mechanics and
    Path Integrals", McGraw-Hill Publishing.
Sakita, B., "Quantum Theory of Many Variable Systems and
    Fields", World Scientific Publishing.
Schulman, L. S., "Techniques and Applications of Path Integra-
    tion", John Wiley Publishing.

Chapter 3

Harmonic Oscillator

3.1 Path Integral for the Harmonic Oscillator

As a second example of the path integrals, let us consider the one dimensional harmonic oscillator which we know can be solved exactly. In fact, let us consider the oscillator interacting with an external source described by the Lagrangian

$$L = \frac{1}{2}m\dot{x}^2 - \frac{1}{2}m\omega^2 x^2 + Jx,\tag{3.1}$$

with the action given by

$$S = \int dt\, L.\tag{3.2}$$

Here, for example, we can think of the time dependent external source $J(t)$ as an electric field if the oscillator is supposed to carry an electric charge. The well known results for the free harmonic oscillator can be obtained from this system in the limit $J(t) \to 0$. Furthermore, we know that if the external source were time independent, then the problem can also be solved exactly simply because in this case we can write the Lagrangian of Eq. (3.1) as

$$L = \frac{1}{2}m\dot{x}^2 - \frac{1}{2}m\omega^2 x^2 + Jx$$

$$= \frac{1}{2}m\dot{x}^2 - \frac{1}{2}m\omega^2 \left(x - \frac{J}{m\omega^2}\right)^2 + \frac{J^2}{2m\omega^2}$$

$$= \frac{1}{2}m\dot{\bar{x}}^2 - \frac{1}{2}m\omega^2\bar{x}^2 + \frac{J^2}{2m\omega^2},\tag{3.3}$$

where we have defined

$$\bar{x} = x - \frac{J}{m\omega^2} \, . \tag{3.4}$$

In other words, in such a case, the classical equilibrium position of the oscillator is shifted by a constant amount, namely, the system behaves like a spring suspended freely under the effect of gravity. The system described by Eq. (3.1) is, therefore, of considerable interest because we can obtain various known special cases in different limits.

The Euler-Lagrange equation for the action in Eq. (3.2) gives the classical trajectory and takes the form

$$\frac{\delta S[x]}{\delta x(t)} = 0$$

$$\text{or,} \quad m\ddot{x}_{\rm cl} + m\omega^2 x_{\rm cl} - J = 0 \, , \tag{3.5}$$

and the general form of the transition amplitude, as we have seen in Eq. (2.28), is given by

$$U(t_f, x_f; t_i, x_i) = A \int \mathcal{D}x \; e^{\frac{i}{\hbar}S[x]} \, . \tag{3.6}$$

To evaluate this functional integral, let us note that the action is at most quadratic in the dynamical variables $x(t)$. Therefore, if we define

$$x(t) = x_{\rm cl}(t) + \eta(t) \, , \tag{3.7}$$

then, we can Taylor expand the action about the classical path as

$$S[x] = S[x_{\rm cl} + \eta] = S[x_{\rm cl}] + \int dt \, \eta(t) \left. \frac{\delta S[x]}{\delta x(t)} \right|_{x=x_{\rm cl}}$$

$$+ \frac{1}{2!} \int dt_1 dt_2 \, \eta(t_1)\eta(t_2) \left. \frac{\delta^2 S[x]}{\delta x(t_1)\delta x(t_2)} \right|_{x=x_{\rm cl}} \, . \tag{3.8}$$

We note from Eq. (3.5) that the action is an extremum for the classical trajectory. Therefore, we have

$$\left. \frac{\delta S[x]}{\delta x(t)} \right|_{x=x_{\rm cl}} = 0 \, . \tag{3.9}$$

Consequently, we can also write Eq. (3.8) as

$$S[x] = S[x_{\mathrm{cl}}] + \frac{1}{2!} \int dt_1 dt_2 \, \eta(t_1)\eta(t_2) \left. \frac{\delta^2 S[x]}{\delta x(t_1)\delta x(t_2)} \right|_{x=x_{\mathrm{cl}}} . \tag{3.10}$$

If we evaluate the functional derivatives in Eq. (3.10) for the action in Eq. (3.2), we can also rewrite the action as

$$S[x] = S[x_{\mathrm{cl}}] + \frac{1}{2} \int_{t_i}^{t_f} dt \left(m\dot{\eta}^2 - m\omega^2\eta^2 \right) . \tag{3.11}$$

The variable $\eta(t)$ represents the quantum fluctuations around the classical path, namely, it measures the deviation of a trajectory from the classical trajectory. Since the end points of the trajectories are fixed, the fluctuations satisfy the boundary conditions

$$\eta(t_i) = \eta(t_f) = 0 . \tag{3.12}$$

It is clear that summing over all the paths is equivalent to summing over all possible fluctuations subject to the constraint in Eq. (3.12). Consequently, we can rewrite the transition amplitude in this case as

$$U(t_f, x_f; t_i, x_i) = A \int \mathcal{D}\eta \, e^{\left[\frac{i}{\hbar} S[x_{\mathrm{cl}}] + \frac{i}{2\hbar} \int_{t_i}^{t_f} dt \left(m\dot{\eta}^2 - m\omega^2\eta^2 \right) \right]}$$

$$= A \, e^{\frac{i}{\hbar} S[x_{\mathrm{cl}}]} \int \mathcal{D}\eta \, e^{\frac{i}{2\hbar} \int_{t_i}^{t_f} dt \left(m\dot{\eta}^2 - m\omega^2\eta^2 \right)} . \tag{3.13}$$

This is an integral where the exponent is quadratic in the variables and such an integral can be done in several ways. Since the harmonic oscillator is a fundamental system in any branch of physics, we will evaluate this integral in three different ways so as to develop a feeling for the path integrals.

3.2 Method of Fourier Transform

First of all, we note that the integrand in the exponent of the functional integral does not depend on time explicitly. Therefore, we can redefine the variable of integration as

$$t \rightarrow t - t_i , \tag{3.14}$$

in which case, we can write the transition amplitude as

$$U(t_f, x_f; t_i, x_i) = A \, e^{\frac{i}{\hbar} S[x_{\text{cl}}]} \int \mathcal{D}\eta \, e^{\frac{i}{2\hbar} \int_0^T dt (m\dot{\eta}^2 - m\omega^2 \eta^2)} , \qquad (3.15)$$

where we have identified the time interval with

$$T = t_f - t_i . \qquad (3.16)$$

The variable $\eta(t)$ satisfies the boundary conditions (see Eq. (3.12))

$$\eta(0) = \eta(T) = 0 . \qquad (3.17)$$

Consequently, the value of the fluctuation at any point on the trajectory can be represented as a Fourier series of the form

$$\eta(t) = \sum_n a_n \sin \left(\frac{n\pi t}{T} \right), \quad n = 1, 2, \ldots, N - 1 . \qquad (3.18)$$

We note here that since we have chosen to divide the trajectory into N intervals, namely, since there are $(N-1)$ intermediate time points, there can only be $(N - 1)$ independent coefficients a_n in the Fourier expansion in Eq. (3.18). Substituting this back, we find that

$$\int_0^T dt \, \dot{\eta}^2 = \sum_{n,m} \int_0^T dt \, a_n a_m \left(\frac{n\pi}{T} \right) \left(\frac{m\pi}{T} \right) \cos \left(\frac{n\pi t}{T} \right) \cos \left(\frac{m\pi t}{T} \right)$$

$$= \frac{T}{2} \sum_n \left(\frac{n\pi}{T} \right)^2 a_n^2 , \qquad (3.19)$$

where we have used the orthonormality properties of the cosine functions. Similarly, we also obtain

$$\int_0^T dt \, \eta^2(t) = \sum_{n,m} \int_0^T dt \, a_n a_m \sin \left(\frac{n\pi t}{T} \right) \sin \left(\frac{m\pi t}{T} \right)$$

$$= \frac{T}{2} \sum_n a_n^2 . \qquad (3.20)$$

Furthermore, we note that integrating over all possible configurations of $\eta(t)$ or all possible quantum fluctuations is equivalent to integrating over all possible values of the coefficients of expansion a_n. Thus we can write the transition amplitude also as

$$
U(t_f, x_f; t_i, x_i)
$$

$$
= \lim_{\substack{\epsilon \to 0 \\ N \to \infty}} A' e^{\frac{i}{\hbar} S[x_{\text{cl}}]} \int da_1 \cdots da_{N-1}\, e^{\frac{i}{2\hbar} \sum_{n=1}^{N-1} \left(\frac{T}{2} \left(\frac{n\pi}{T} \right)^2 m a_n^2 - \frac{T}{2} m\omega^2 a_n^2 \right)}
$$

$$
= \lim_{\substack{\epsilon \to 0 \\ N \to \infty}} A' e^{\frac{i}{\hbar} S[x_{\text{cl}}]} \int da_1 \cdots da_{N-1}\, e^{\frac{imT}{4\hbar} \sum_{n=1}^{N-1} \left(\left(\frac{n\pi}{T} \right)^2 - \omega^2 \right) a_n^2}. \quad (3.21)
$$

Here we note that any possible factor arising from the Jacobian in the change of variables from η to the coefficients a_n has been lumped into A' whose form we will determine shortly.

We note here that the transition amplitude, in this case, is a product of a set of decoupled integrals each of which has the form of a Gaussian integral which can be easily evaluated. In fact, the individual integrals have the values (see Eq. (2.39))

$$
\int da_n\, e^{\frac{imT}{4\hbar} \left(\left(\frac{n\pi}{T} \right)^2 - \omega^2 \right) a_n^2}
$$

$$
= \left(\frac{4\pi i\hbar}{mT} \right)^{\frac{1}{2}} \left(\left(\frac{n\pi}{T} \right)^2 - \omega^2 \right)^{-\frac{1}{2}}
$$

$$
= \left(\frac{4\pi i\hbar}{mT} \right)^{\frac{1}{2}} \left(\frac{n\pi}{T} \right)^{-1} \left(1 - \left(\frac{\omega T}{n\pi} \right)^2 \right)^{-\frac{1}{2}}. \quad (3.22)
$$

Substituting this form of the individual integrals into the expression for the transition amplitude in Eq. (3.21), we obtain

$$
U(t_f, x_f; t_i, x_i) = \lim_{\substack{\epsilon \to 0 \\ N \to \infty}} A'' e^{\frac{i}{\hbar} S[x_{\text{cl}}]} \prod_{n=1}^{N-1} \left(1 - \left(\frac{\omega T}{n\pi} \right)^2 \right)^{-\frac{1}{2}}. \quad (3.23)
$$

If we now use the identity,

$$
\lim_{N \to \infty} \prod_{n=1}^{N-1} \left(1 - \left(\frac{\omega T}{n\pi} \right)^2 \right) = \frac{\sin \omega T}{\omega T}, \quad (3.24)
$$

we obtain

$$
U(t_f, x_f; t_i, x_i) = \lim_{\substack{\epsilon \to 0 \\ N \to \infty}} A'' e^{\frac{i}{\hbar} S[x_{\mathrm{cl}}]} \left(\frac{\sin \omega T}{\omega T} \right)^{-\frac{1}{2}} . \qquad (3.25)
$$

We can determine the constant A'' by simply noting that when $\omega = 0$, the harmonic oscillator reduces to a free particle for which we have already evaluated the transition amplitude. In fact, recalling from Eq. (2.54) that

$$
U_{\mathrm{F.P.}}(t_f, x_f; t_i, x_i) = \left(\frac{m}{2\pi i \hbar (t_f - t_i)} \right)^{\frac{1}{2}} e^{\frac{i}{\hbar} S[x_{\mathrm{cl}}]} , \qquad (3.26)
$$

and comparing with Eq. (3.25), we obtain

$$
\lim_{\substack{\epsilon \to 0 \\ N \to \infty}} A'' = \left(\frac{m}{2\pi i \hbar T} \right)^{\frac{1}{2}} . \qquad (3.27)
$$

Therefore, we determine the complete form of the transition amplitude for the harmonic oscillator to be

$$
U(t_f, x_f; t_i, x_i) = \left(\frac{m}{2\pi i \hbar T} \right)^{\frac{1}{2}} \left(\frac{\sin \omega T}{\omega T} \right)^{-\frac{1}{2}} e^{\frac{i}{\hbar} S[x_{\mathrm{cl}}]}
$$

$$
= \left(\frac{m\omega}{2\pi i \hbar \sin \omega T} \right)^{\frac{1}{2}} e^{\frac{i}{\hbar} S[x_{\mathrm{cl}}]} . \qquad (3.28)
$$

It is quite straightforward to see that this expression reduces to the transition amplitude for the free particle in the limit of $\omega \to 0$.

3.3   Matrix Method

If the evaluation of the path integral by the method of Fourier transforms appears less satisfactory, then let us evaluate the integral in the conventional manner by discretizing the time interval. Let us parameterize time on a trajectory as

$$
t_n = t_i + n\epsilon, \quad n = 0, 1, \ldots, N .
$$

Correspondingly, let us define the values of the fluctuations at these points as

$$\eta(t_n) = \eta_n. \tag{3.29}$$

Then, we can write the transition amplitude in Eq. (3.15) in the explicit form

$$U(t_f, x_f; t_i, x_i) = A\, e^{\frac{i}{\hbar}S[x_{\rm cl}]} \int \mathcal{D}\eta \; e^{\frac{i}{2\hbar}\int_{t_i}^{t_f} dt\left(m\dot{\eta}^2 - m\omega^2\eta^2\right)}$$

$$= \lim_{\substack{\epsilon \to 0 \\ N \to \infty}} \left(\frac{m}{2\pi i\hbar\epsilon}\right)^{\frac{N}{2}} e^{\frac{i}{\hbar}S[x_{\rm cl}]} \int d\eta_1 \cdots d\eta_{N-1}$$

$$\times\, e^{\frac{i\epsilon}{2\hbar}\sum_{n=1}^{N}\left(m\left(\frac{\eta_n - \eta_{n-1}}{\epsilon}\right)^2 - m\omega^2\left(\frac{\eta_n + \eta_{n-1}}{2}\right)^2\right)}. \tag{3.30}$$

In this expression, we are supposed to identify

$$\eta_0 = \eta_N = 0, \tag{3.31}$$

corresponding to the boundary conditions in Eq. (3.12), namely,

$$\eta(t_i) = \eta(t_f) = 0.$$

To simplify the integral, let us rescale the variables as

$$\eta_n \to \left(\frac{m}{2\hbar\epsilon}\right)^{\frac{1}{2}} \eta_n. \tag{3.32}$$

The transition amplitude, in this case, will take the form

$$U(t_f, x_f; t_i, x_i)$$

$$= \lim_{\substack{\epsilon \to 0 \\ N \to \infty}} \left(\frac{m}{2\pi i\hbar\epsilon}\right)^{\frac{N}{2}} \left(\frac{2\hbar\epsilon}{m}\right)^{\frac{N-1}{2}} e^{\frac{i}{\hbar}S[x_{\rm cl}]} \int d\eta_1 \cdots d\eta_{N-1}$$

$$\times\, e^{i\sum_{n=1}^{N}\left((\eta_n - \eta_{n-1})^2 - \epsilon^2\omega^2\left(\frac{\eta_n + \eta_{n-1}}{2}\right)^2\right)}. \tag{3.33}$$

If we think of the η_n's (there are $(N-1)$ of them) as forming a column matrix, namely,

$$\eta = \begin{pmatrix} \eta_1 \\ \eta_2 \\ \vdots \\ \eta_{N-1} \end{pmatrix}, \tag{3.34}$$

then, we can also write the transition amplitude in terms of matrices as

$$U(t_f, x_f; t_i, x_i) = \lim_{\substack{\epsilon \to 0 \\ N \to \infty}} \left(\frac{m}{2\pi i \hbar \epsilon} \right)^{\frac{N}{2}} \left(\frac{2\hbar \epsilon}{m} \right)^{\frac{N-1}{2}} e^{\frac{i}{\hbar} S[x_{cl}]}$$

$$\times \int d\eta \, e^{i \eta^T B \eta}. \tag{3.35}$$

Here η^T represents the transpose of the column matrix in Eq. (3.34) and the $(N-1) \times (N-1)$ matrix B has the form

$$B = \begin{pmatrix} 2 & -1 & 0 & 0 & \cdots \\ -1 & 2 & -1 & 0 & \cdots \\ 0 & -1 & 2 & -1 & \cdots \\ \vdots & \vdots & \vdots & \vdots & \end{pmatrix} - \frac{\epsilon^2 \omega^2}{4} \begin{pmatrix} 2 & 1 & 0 & 0 & \cdots \\ 1 & 2 & 1 & 0 & \cdots \\ 0 & 1 & 2 & 1 & \cdots \\ \vdots & \vdots & \vdots & \vdots & \end{pmatrix}. \tag{3.36}$$

This is a symmetric matrix and, therefore, we can write it as

$$B = \begin{pmatrix} x & y & 0 & 0 & \cdots \\ y & x & y & 0 & \cdots \\ 0 & y & x & y & \cdots \\ \vdots & \vdots & \vdots & \vdots & \end{pmatrix}, \tag{3.37}$$

where we have defined

$$x = 2 \left(1 - \frac{\epsilon^2 \omega^2}{4} \right),$$

$$y = - \left(1 + \frac{\epsilon^2 \omega^2}{4} \right). \tag{3.38}$$

The matrix B is clearly Hermitian (both x and y are real) and, therefore, can be diagonalized by a unitary matrix (more precisely by an orthogonal matrix) which we denote by $\mathcal{U}$. In other words,

$$
B_D = \begin{pmatrix} b_1 & 0 & 0 & \cdots \\ 0 & b_2 & 0 & \cdots \\ 0 & 0 & b_3 & \cdots \\ \vdots & \vdots & \vdots & \end{pmatrix} = \mathcal{U} B \mathcal{U}^\dagger .
\tag{3.39}
$$

Therefore, defining

$$
\zeta = \mathcal{U}\eta ,
\tag{3.40}
$$

we obtain

$$
\int d\eta\, e^{i\eta^T B \eta} = \int d\zeta\, e^{i\zeta^T B_D \zeta}
$$

$$
= \int d\zeta_1 \cdots d\zeta_{N-1}\, e^{i\sum_{n=1}^{N-1} b_n \zeta_n^2}
$$

$$
= \prod_{n=1}^{N-1} \left(\frac{i\pi}{b_n}\right)^{\frac{1}{2}}
$$

$$
= (i\pi)^{\frac{N-1}{2}} (\det B)^{-\frac{1}{2}} .
\tag{3.41}
$$

Here we have used the familiar fact that the Jacobian for a change of variable by a unitary matrix is unity. Using this result in Eq. (3.35), therefore, we determine the form of the transition amplitude for the harmonic oscillator to be

$$
U(t_f, x_f; t_i, x_i)
$$

$$
= \lim_{\substack{\epsilon \to 0 \\ N \to \infty}} \left(\frac{m}{2\pi i \hbar \epsilon}\right)^{\frac{N}{2}} \left(\frac{2\hbar\epsilon}{m}\right)^{\frac{N-1}{2}} (i\pi)^{\frac{N-1}{2}} (\det B)^{-\frac{1}{2}} e^{\frac{i}{\hbar} S[x_{\rm cl}]}
$$

$$
= \lim_{\substack{\epsilon \to 0 \\ N \to \infty}} \left(\frac{m}{2\pi i \hbar \epsilon}\right)^{\frac{1}{2}} (\det B)^{-\frac{1}{2}} e^{\frac{i}{\hbar} S[x_{\rm cl}]}
$$

$$
= \lim_{\substack{\epsilon \to 0 \\ N \to \infty}} \left(\frac{m}{2\pi i \hbar \epsilon \det B}\right)^{\frac{1}{2}} e^{\frac{i}{\hbar} S[x_{\rm cl}]} .
\tag{3.42}
$$

It is clear from this analysis that the transition amplitude can be defined only if the matrix B does not have any vanishing eigenvalue.

We note here that the main quantity to calculate in order to evaluate the transition amplitude is

$$\lim_{\substack{\epsilon \to 0 \\ N \to \infty}} \epsilon \det B.$$

Let us note from the special structure of B in Eqs. (3.37) and (3.38), that if we denote the determinant of the $n \times n$ sub-matrices of B as I_n, then it is easy to check that they satisfy the recursion relation

$$I_{n+1} = x I_n - y^2 I_{n-1}, \quad n = 0, 1, 2, \ldots, \tag{3.43}$$

where we restrict

$$I_{-1} = 0, \quad I_0 = 1. \tag{3.44}$$

This recursion relation can be checked trivially for low orders of the matrix determinants. Substituting the form of x and y, we obtain

$$I_{n+1} = 2 \left(1 - \frac{\epsilon^2 \omega^2}{4} \right) I_n - \left(1 + \frac{\epsilon^2 \omega^2}{4} \right)^2 I_{n-1}$$

$$\text{or,} \ \ I_{n+1} - 2 I_n + I_{n-1} = -\frac{\epsilon^2 \omega^2}{2} \left(I_n + I_{n-1} + \frac{\epsilon^2 \omega^2}{8} I_{n-1} \right)$$

$$\text{or,} \ \ \frac{I_{n+1} - 2 I_n + I_{n-1}}{\epsilon^2} = -\frac{\omega^2}{2} \left(I_n + I_{n-1} + \frac{\epsilon^2 \omega^2}{8} I_{n-1} \right).$$

$$\tag{3.45}$$

We are, of course, interested in the continuum limit. In order to do so, let us define a function

$$\phi(t_n - t_i) = \phi(n\epsilon) = \epsilon I_n. \tag{3.46}$$

In the continuum limit, we can think of this as a continuous function of t. In other words, we can identify $t = n\epsilon$ as a continuous variable as $\epsilon \to 0$. We note then, that

$$\lim_{\substack{\epsilon \to 0 \\ N \to \infty}} \epsilon \det B = \lim_{\substack{\epsilon \to 0 \\ N \to \infty}} \epsilon I_{N-1} = \phi(t_f - t_i) = \phi(T). \tag{3.47}$$

We also note from Eq. (3.46) that, in the continuum limit,

$$\phi(0) = \lim_{\epsilon \to 0} \epsilon I_0 = \lim_{\epsilon \to 0} \epsilon = 0\,, \tag{3.48}$$

and similarly,

$$\dot{\phi}(0) = \lim_{\epsilon \to 0} \epsilon \left(\frac{I_1 - I_0}{\epsilon} \right) = \lim_{\epsilon \to 0} \left(2 - \frac{\epsilon^2 \omega^2}{2} - 1 \right) = 1\,. \tag{3.49}$$

Furthermore, from the recursion relation for the I_n's in Eq. (3.45), we conclude that in the limit $\epsilon \to 0$, the function $\phi(t)$ satisfies the differential equation

$$\frac{d^2 \phi(t)}{dt^2} = -\omega^2 \phi(t)\,. \tag{3.50}$$

We recognize this to be the harmonic oscillator equation and the solution subject to the initial conditions (Eqs. (3.48) and (3.49)) is clearly

$$\phi(t) = \frac{\sin \omega t}{\omega}\,. \tag{3.51}$$

It now follows from this that

$$\lim_{\substack{\epsilon \to 0 \\ N \to \infty}} \epsilon \det B = \lim_{\substack{\epsilon \to 0 \\ N \to \infty}} \epsilon I_{N-1} = \phi(T) = \frac{\sin \omega T}{\omega}\,. \tag{3.52}$$

Consequently, for the harmonic oscillator, we obtain the transition amplitude in Eq. (3.42) to be

$$U(t_f, x_f; t_i, x_i) = \lim_{\substack{\epsilon \to 0 \\ N \to \infty}} \left(\frac{m}{2\pi i \hbar \epsilon \det B} \right)^{\frac{1}{2}} e^{\frac{i}{\hbar} S[x_{\mathrm{cl}}]}$$

$$= \left(\frac{m\omega}{2\pi i \hbar \sin \omega T} \right)^{\frac{1}{2}} e^{\frac{i}{\hbar} S[x_{\mathrm{cl}}]}\,. \tag{3.53}$$

This is, of course, what we had already derived in Eq. (3.28) using the method of Fourier transforms.

Let us next describe an alternate way to determine $\det B$ which is quite useful in studying some specific problems. Let us recall from

Eq. (3.43) that the determinant of the $n \times n$ matrices, I_n, satisfy the recursion relations

$$I_{n+1} = xI_n - y^2 I_{n-1} .$$

We note here that we can write these recursion relations also in the simple matrix form as

$$\begin{pmatrix} I_{n+1} \\ I_n \end{pmatrix} = \begin{pmatrix} x & -y^2 \\ 1 & 0 \end{pmatrix} \begin{pmatrix} I_n \\ I_{n-1} \end{pmatrix} . \tag{3.54}$$

Iterating this $(N-2)$ times, for $n = N - 2$, we obtain

$$\begin{pmatrix} I_{N-1} \\ I_{N-2} \end{pmatrix} = \underbrace{\begin{pmatrix} x & -y^2 \\ 1 & 0 \end{pmatrix} \cdots}_{(N-2)\,factors} \begin{pmatrix} I_1 \\ I_0 \end{pmatrix}$$

$$= \begin{pmatrix} x & -y^2 \\ 1 & 0 \end{pmatrix}^{(N-2)} \begin{pmatrix} x \\ 1 \end{pmatrix} . \tag{3.55}$$

We can determine the eigenvalues of the fundamental 2×2 matrix in Eq. (3.55) in a straightforward manner. From

$$\det \begin{vmatrix} x - \lambda & -y^2 \\ 1 & -\lambda \end{vmatrix} = 0 ,$$

we obtain

$$\lambda_\pm = \frac{x \pm \sqrt{x^2 - 4y^2}}{2} . \tag{3.56}$$

Furthermore, the 2×2 matrix can be trivially diagonalized by a similarity transformation. In fact, if we define

$$\mathcal{S} = \begin{pmatrix} c\lambda_+ & d\lambda_- \\ c & d \end{pmatrix} , \tag{3.57}$$

with

$$\mathcal{S}^{-1} = \frac{1}{\lambda_+ - \lambda_-} \begin{pmatrix} \frac{1}{c} & -\frac{\lambda_-}{c} \\ -\frac{1}{d} & \frac{\lambda_+}{d} \end{pmatrix} , \tag{3.58}$$

where c and d are arbitrary parameters, then it is easy to check that

$$\begin{pmatrix} x & -y^2 \\ 1 & 0 \end{pmatrix} = \mathcal{S} \begin{pmatrix} \lambda_+ & 0 \\ 0 & \lambda_- \end{pmatrix} \mathcal{S}^{-1}. \tag{3.59}$$

Using this in Eq. (3.55), we obtain

$$\begin{pmatrix} I_{N-1} \\ I_{N-2} \end{pmatrix} = \mathcal{S} \begin{pmatrix} \lambda_+^{N-2} & 0 \\ 0 & \lambda_-^{N-2} \end{pmatrix} \mathcal{S}^{-1} \begin{pmatrix} x \\ 1 \end{pmatrix}. \tag{3.60}$$

We recall from Eq. (3.56) that

$$x = \lambda_+ + \lambda_- . \tag{3.61}$$

Using this as well as the forms for $\mathcal{S}$ and $\mathcal{S}^{-1}$ in Eqs. (3.57) and (3.58), we obtain from Eq. (3.60)

$$I_{N-1} = \frac{1}{\lambda_+ - \lambda_-} \left(\lambda_+^N - \lambda_-^N \right) . \tag{3.62}$$

We can easily check now that

$$I_{-1} = 0 ,$$
$$I_0 = 1 ,$$
$$I_1 = \frac{1}{\lambda_+ - \lambda_-} \left(\lambda_+^2 - \lambda_-^2 \right)$$
$$= \lambda_+ + \lambda_- = x , \tag{3.63}$$

which is consistent with our earlier observations in Eqs. (3.43) and (3.44).

Let us next note from Eqs. (3.56) and (3.38) that

$$\lambda_+ - \lambda_- = \left(x^2 - 4y^2 \right)^{\frac{1}{2}}$$
$$= \left(4 \left(1 - \frac{\epsilon^2 \omega^2}{4} \right)^2 - 4 \left(1 + \frac{\epsilon^2 \omega^2}{4} \right)^2 \right)^{\frac{1}{2}}$$
$$= \left(-4\epsilon^2 \omega^2 \right)^{\frac{1}{2}} = 2i\epsilon\omega ,$$

$$\lambda_+^N = \left(\frac{x + \sqrt{x^2 - 4y^2}}{2} \right)^N$$

$$= \left(\frac{2\left(1 - \frac{\epsilon^2\omega^2}{4}\right) + 2i\epsilon\omega}{2} \right)^N$$

$$= \left(1 + i\epsilon\omega + O\left(\epsilon^2\right)\right)^N \simeq \left(1 + i\epsilon\omega\right)^N ,$$

$$\lambda_-^N = \left(\frac{x - \sqrt{x^2 - 4y^2}}{2} \right)^N$$

$$= \left(\frac{2(1 - \frac{\epsilon^2\omega^2}{4}) - 2i\epsilon\omega}{2} \right)^N$$

$$= \left(1 - i\epsilon\omega + O\left(\epsilon^2\right)\right)^N \simeq \left(1 - i\epsilon\omega\right)^N . \qquad (3.64)$$

Consequently, substituting these relations into Eq. (3.62), we obtain

$$\lim_{\substack{\epsilon \to 0 \\ N \to \infty}} \epsilon I_{N-1} = \lim_{\substack{\epsilon \to 0 \\ N \to \infty}} \epsilon \frac{1}{\lambda_+ - \lambda_-} \left(\lambda_+^N - \lambda_-^N \right)$$

$$= \lim_{\substack{\epsilon \to 0 \\ N \to \infty}} \epsilon \frac{1}{2i\epsilon\omega} \left((1 + i\epsilon\omega)^N - (1 - i\epsilon\omega)^N \right)$$

$$= \lim_{\substack{\epsilon \to 0 \\ N \to \infty}} \frac{1}{2i\omega} \left(\left(1 + \frac{i\omega T}{N}\right)^N - \left(1 - \frac{i\omega T}{N}\right)^N \right)$$

$$= \frac{1}{2i\omega} \left(e^{i\omega T} - e^{-i\omega T} \right) = \frac{\sin \omega T}{\omega} . \qquad (3.65)$$

Here T is the time interval between the initial and the final times and this is, of course, what we had obtained earlier in Eq. (3.52), namely, that

$$\lim_{\substack{\epsilon \to 0 \\ N \to \infty}} \epsilon \det B = \frac{\sin \omega T}{\omega} .$$

In any case, we obtain the transition amplitude for the harmonic oscillator to be

$$U(t_f, x_f; t_i, x_i) = \left(\frac{m\omega}{2\pi i\hbar \sin \omega T}\right)^{\frac{1}{2}} e^{\frac{i}{\hbar} S[x_{cl}]}. \tag{3.66}$$

3.4   The Classical Action

Once again we see from Eq. (3.28) or (3.66) that the transition amplitude has a generic form similar to the one found for the case of the free particle. Namely, it is proportional to $e^{\frac{i}{\hbar} S[x_{cl}]}$. A complete determination of the transition amplitude in this case, therefore, would require us to evaluate the classical action for the system. This can be done simply in the following way. We recall that the Euler-Lagrange equation for the present system is given by (see Eq. (3.5))

$$m\ddot{x}_{cl} + m\omega^2 x_{cl} - J = 0.$$

In other words, the classical trajectory is a solution of the equation

$$\left(\frac{d^2}{dt^2} + \omega^2\right) x_{cl}(t) = \frac{J(t)}{m}. \tag{3.67}$$

The solution, obviously, consists of a homogeneous and an inhomogeneous part and can be written as

$$x_{cl}(t) = x_H(t) + x_I(t), \tag{3.68}$$

where the homogeneous solution is of the form

$$x_H(t) = Ae^{i\omega t} + Be^{-i\omega t}, \tag{3.69}$$

with A and B arbitrary constants to be determined from the boundary conditions. To determine the inhomogeneous solution, we use the method of Green's function. Here the Green's function for Eq. (3.67) is defined by the equation

$$\left(\frac{d^2}{dt^2} + \omega^2\right) G(t - t') = -\delta(t - t'). \tag{3.70}$$

It is clear that if we know the Green's function $G(t - t')$, then the inhomogeneous solution can be written as

$$x_I(t) = -\int_{t_i}^{t_f} dt' \, G(t - t') \frac{J(t')}{m}. \tag{3.71}$$

The Green's function can be easily determined by transforming to the Fourier space. Defining

$$G(t - t') = \int \frac{dk}{\sqrt{2\pi}} \, e^{-ik(t-t')} G(k),$$

$$\delta(t - t') = \int \frac{dk}{2\pi} \, e^{-ik(t-t')}, \tag{3.72}$$

where $G(k)$ is the Fourier transform of $G(t - t')$, and substituting these into Eq. (3.70) we obtain (Note from Eq. (2.6) that the Fourier transform in time is defined with an opposite phase.)

$$\left(\frac{d^2}{dt^2} + \omega^2 \right) G(t - t') = -\delta(t - t')$$

$$\text{or,} \quad \frac{1}{\sqrt{2\pi}} \left(-k^2 + \omega^2 \right) G(k) = -\frac{1}{2\pi}$$

$$\text{or,} \quad G(k) = \frac{1}{\sqrt{2\pi}} \frac{1}{k^2 - \omega^2}. \tag{3.73}$$

Consequently, the Green's function takes the form

$$G(t - t') = \int \frac{dk}{\sqrt{2\pi}} \, e^{-ik(t-t')} G(k)$$

$$= \frac{1}{2\pi} \int dk \, \frac{e^{-ik(t-t')}}{k^2 - \omega^2}. \tag{3.74}$$

A quick inspection shows that the integrand has poles at $k = \pm\omega$. Therefore, we must specify a contour in the complex k-plane in order to evaluate the integral. Normally, in classical mechanics, the Green's functions that are of fundamental interest are the retarded and the advanced Green's functions. But a Green's function that is of fundamental significance in quantum theories is the Feynman Green's function and corresponds to choosing a contour as shown below.

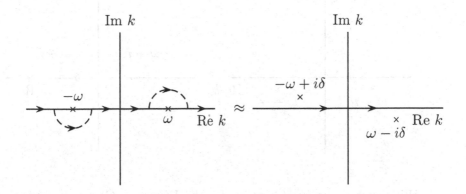

Equivalently, it corresponds to defining (see Eq. (3.73))

$$G_F(k) = \lim_{\epsilon \to 0^+} \frac{1}{\sqrt{2\pi}} \frac{1}{k^2 - \omega^2 + i\epsilon}$$

$$= \lim_{\delta \to 0^+} \frac{1}{\sqrt{2\pi}} \frac{1}{k + \omega - i\delta} \frac{1}{k - \omega + i\delta} , \qquad (3.75)$$

where we have defined

$$\delta = \frac{\epsilon}{2\omega} .$$

In other words, we can think of the Green's function in Eq. (3.75) as the Fourier transform of the function which satisfies the differential equation

$$\lim_{\epsilon \to 0^+} \left(\frac{d^2}{dt^2} + \omega^2 - i\epsilon \right) G_F(t - t') = -\delta(t - t') . \qquad (3.76)$$

We note here, for completeness, that the retarded and the advanced Green's functions, in this language, correspond respectively to choosing the Fourier transforms as

$$G^{R,A}(k) = \lim_{\epsilon \to 0^+} \frac{1}{\sqrt{2\pi}} \frac{1}{(k \pm i\epsilon)^2 - \omega^2} ,$$

with the respective contours

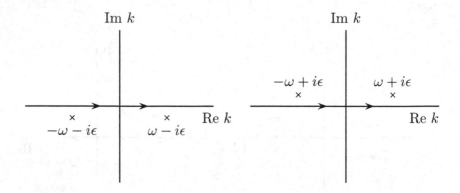

With such a choice of contour for the Feynman Green's function, enclosing the contour in the lower half plane for $t - t' > 0$, we obtain

$$G^{(+)}(t - t') = \lim_{\delta \to 0^+} \frac{1}{2\pi} \int dk \, \frac{e^{-ik(t-t')}}{(k + \omega - i\delta)(k - \omega + i\delta)}$$

$$= \frac{1}{2\pi}(-2\pi i) \frac{e^{-i\omega(t-t')}}{2\omega}$$

$$= \frac{1}{2i\omega} e^{-i\omega(t-t')} . \tag{3.77}$$

On the other hand, for $t - t' < 0$, enclosing the contour in the upper half plane, we find

$$G^{(-)}(t - t') = \lim_{\delta \to 0^+} \frac{1}{2\pi} \int dk \, \frac{e^{-ik(t-t')}}{(k + \omega - i\delta)(k - \omega + i\delta)}$$

$$= \frac{1}{2\pi} 2\pi i \frac{e^{i\omega(t-t')}}{-2\omega}$$

$$= \frac{1}{2i\omega} e^{i\omega(t-t')} . \tag{3.78}$$

Thus, the Feynman Green's function has the form

$$G_F(t - t') = \theta(t - t') \frac{e^{-i\omega(t-t')}}{2i\omega} + \theta(t' - t) \frac{e^{i\omega(t-t')}}{2i\omega}. \tag{3.79}$$

Using this Green's function in Eq. (3.71), we can now obtain the inhomogeneous solution as

$$x_I(t) = -\int_{t_i}^{t_f} dt' \, G_F(t-t') \frac{J(t')}{m}$$

$$= -\frac{1}{m} \left(\int_{t_i}^{t} dt' \frac{e^{-i\omega(t-t')}}{2i\omega} J(t') + \int_{t}^{t_f} dt' \frac{e^{i\omega(t-t')}}{2i\omega} J(t') \right)$$

$$= -\frac{1}{2im\omega} \left(\int_{t_i}^{t} dt' e^{-i\omega(t-t')} J(t') + \int_{t}^{t_f} dt' e^{i\omega(t-t')} J(t') \right).$$

$$(3.80)$$

Thus, substituting Eqs. (3.69) and (3.80) into Eq. (3.68) we can write the classical trajectory as

$$x_{cl}(t) = x_H(t) + x_I(t)$$

$$= Ae^{i\omega t} + Be^{-i\omega t}$$

$$- \frac{1}{2im\omega} \left(\int_{t_i}^{t} dt' e^{-i\omega(t-t')} J(t') + \int_{t}^{t_f} dt' e^{i\omega(t-t')} J(t') \right).$$

$$(3.81)$$

Imposing the boundary conditions

$$x_{cl}(t_i) = x_i$$

$$= Ae^{i\omega t_i} + Be^{-i\omega t_i} - \frac{1}{2im\omega} \int_{t_i}^{t_f} dt' e^{i\omega(t_i-t')} J(t'),$$

$$x_{cl}(t_f) = x_f$$

$$= Ae^{i\omega t_f} + Be^{-i\omega t_f} - \frac{1}{2im\omega} \int_{t_i}^{t_f} dt' e^{-i\omega(t_f-t')} J(t'), \quad (3.82)$$

we see that we can solve for A and B in terms of the initial and the final coordinates of the trajectory as

$$A = \frac{1}{2i \sin \omega T} \left[\left(x_f e^{-i\omega t_i} - x_i e^{-i\omega t_f} \right) \right.$$

$$\left. + \frac{e^{-i\omega t_f}}{m\omega} \int_{t_i}^{t_f} dt' \sin \omega (t' - t_i) J(t') \right],$$

$$B = \frac{1}{2i \sin \omega T} \left[\left(x_i e^{i\omega t_f} - x_f e^{i\omega t_i} \right) \right.$$

$$\left. + \frac{e^{i\omega t_i}}{m\omega} \int_{t_i}^{t_f} dt' \sin \omega (t_f - t') J(t') \right].$$

Substituting these relations into Eq. (3.81), we determine the classical trajectory to be

$$x_{\text{cl}}(t) = \frac{1}{\sin \omega T} \left[x_f \sin \omega (t - t_i) + x_i \sin \omega (t_f - t) \right.$$

$$\left. + \frac{1}{2m\omega} \int_{t_i}^{t_f} dt' J(t') (e^{-i\omega T} \cos \omega (t - t') - \cos \omega (t_f + t_i - t - t')) \right]$$

$$- \frac{1}{2im\omega} \left(\int_{t_i}^{t} dt' J(t') e^{-i\omega (t-t')} + \int_{t}^{t_f} dt' J(t') e^{i\omega (t-t')} \right). \qquad (3.83)$$

We can now derive the classical action from Eqs. (3.2) and (3.1) in a straightforward manner to be

$$S[x_{\text{cl}}] = \frac{m\omega}{2 \sin \omega T} \left[(x_i^2 + x_f^2) \cos \omega T - 2x_i x_f \right]$$

$$+ \frac{x_i}{\sin \omega T} \int_{t_i}^{t_f} dt J(t) \sin \omega (t_f - t) + \frac{x_f}{\sin \omega T} \int_{t_i}^{t_f} dt J(t) \sin \omega (t - t_i)$$

$$- \frac{1}{m\omega \sin \omega T} \int_{t_i}^{t_f} dt \int_{t_i}^{t} dt' J(t) \sin \omega (t_f - t) \sin \omega (t' - t_i) J(t').$$

$$(3.84)$$

This, therefore, completes the derivation of the transition amplitude for the harmonic oscillator interacting with a time dependent external source.

3.5 References

Feynman, R. P. and A. R. Hibbs, "Quantum Mechanics and Path Integrals", McGraw-Hill Publishing.

Kleinert, H., "Path Integrals", World Scientific Publishing.

Schulman, L. S., "Techniques and Applications of Path Integration", John Wiley Publishing.

Chapter 4

Generating Functional

4.1 Euclidean Rotation

We have seen in Eq. (2.39) that the standard Gaussian integral (where the exponent is quadratic), namely,

$$\int_{-\infty}^{\infty} dx \, e^{i\alpha x^2} = \left(\frac{i\pi}{\alpha}\right)^{\frac{1}{2}},$$

generalizes in the case of a $n \times n$ matrix as (see Eq. (3.41))

$$\int_{-\infty}^{\infty} d\eta \, e^{i\eta^T A\eta} = \left(\frac{(i\pi)^n}{\det A}\right)^{\frac{1}{2}}, \qquad (4.1)$$

provided A is a Hermitian matrix. In fact, we will now see explicitly that this result holds true even when we replace the matrix A by a Hermitian operator. In other words, we will see that we can write

$$\int \mathcal{D}\eta \, e^{i \int_{t_i}^{t_f} dt \, \eta(t)O(t)\eta(t)} = N \left[\det O(t)\right]^{-\frac{1}{2}}, \qquad (4.2)$$

where $O(t)$ is a Hermitian operator and N a normalization constant whose explicit form is irrelevant.

To establish the identification, let us go back to the harmonic oscillator and note that the quantity of fundamental importance in this case was the integral (see Eq. (3.13))

$$\int \mathcal{D}\eta \; e^{\frac{i}{\hbar} \int_{t_i}^{t_f} dt \; (\frac{1}{2}m\dot{\eta}^2 - \frac{1}{2}m\omega^2\eta^2)}$$

$$= \int \mathcal{D}\eta \; e^{\frac{im}{2\hbar} \int_{t_i}^{t_f} dt \; (\dot{\eta}^2 - \omega^2\eta^2)}$$

$$= \int \mathcal{D}\eta \; e^{-\frac{im}{2\hbar} \int_{t_i}^{t_f} dt \; \eta(t)(\frac{d^2}{dt^2} + \omega^2)\eta(t)} , \tag{4.3}$$

with the boundary conditions

$$\eta(t_i) = \eta(t_f) = 0. \tag{4.4}$$

The value of this integral was determined earlier (see Eq. (3.25)) to be

$$A'' \left(\frac{\sin \omega T}{\omega T} \right)^{-\frac{1}{2}} ,$$

where $T = t_f - t_i$ represents the total time interval. We evaluated this integral earlier by carefully discretizing the time interval and calculating the determinant of a matrix (see Eqs. (3.30) and (3.35)) whose matrix elements were nothing other than the discrete form of the matrix elements of the operator in the exponent in Eq. (4.3). This would already justify our claim. But let us, in fact, calculate the determinant of the operator in the exponent of Eq. (4.3) explicitly and compare with the result obtained earlier.

The first problem that we face in evaluating the functional integral is that the exponential in the integrand is oscillatory and, therefore, we have to define the integral in some manner. One can, of course, use the same trick as we employed in defining ordinary oscillatory Gaussian integrals (see Eq. (2.39)). Namely, let us define

$$\int \mathcal{D}\eta \; e^{\frac{im}{2\hbar} \int_{t_i}^{t_f} dt \; (\dot{\eta}^2 - \omega^2\eta^2)}$$

$$= \lim_{\epsilon \to 0^+} \int \mathcal{D}\eta \; e^{-\frac{im}{2\hbar} \int_{t_i}^{t_f} dt \; \eta(t)(\frac{d^2}{dt^2} + \omega^2 - i\epsilon)\eta(t)} . \tag{4.5}$$

This provides proper damping to the integrand and, as we will see, leads to the Feynman Green's functions for the theory. In fact, as we have already seen in Eq. (3.76), the inverse of the operator in the exponent in Eq. (4.5) gives the Feynman Green's function which plays the role of the causal propagator in the quantum theory. It is in this sense that one says that the path integral naturally incorporates the causal boundary condition.

There is an alternate but equivalent way of defining the path integral which is quite pleasing and which gives some sense of rigor to all the manipulations involving the path integral. Very simply, it corresponds to analytically continuing all the integrals to imaginary times in the complex t-plane. More explicitly, we let

$$t \to t' = -i\tau, \quad \tau \text{ real.} \tag{4.6}$$

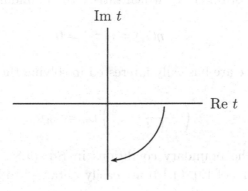

With this analytic continuation, the integral in Eq. (4.3) becomes

$$\int \mathcal{D}\eta \, e^{-\frac{im}{2\hbar} \int_{t_i}^{t_f} \mathrm{d}t \, \eta(t)(\frac{\mathrm{d}^2}{\mathrm{d}t^2}+\omega^2)\eta(t)}$$

$$\to \int \mathcal{D}\eta \, e^{-\frac{m}{2\hbar} \int_{\tau_i}^{\tau_f} \mathrm{d}\tau \, \eta(\tau)(-\frac{\mathrm{d}^2}{\mathrm{d}\tau^2}+\omega^2)\eta(\tau)}$$

$$= N' \int \mathcal{D}\eta \, e^{-\int_{\tau_i}^{\tau_f} \mathrm{d}\tau \, \eta(\tau)(-\frac{\mathrm{d}^2}{\mathrm{d}\tau^2}+\omega^2)\eta(\tau)} . \tag{4.7}$$

Here we have scaled the variables in the last step and N' represents the Jacobian for the change of variables. Furthermore, we have to evaluate the integral in Eq. (4.7) subject to the boundary conditions

$$\eta(\tau_i) = \eta(\tau_f) = 0. \tag{4.8}$$

The right hand side of Eq. (4.7) is now a well defined quantity since the integrand is exponentially damped. (The analytically continued operator has a positive definite spectrum.) We can now evaluate this integral and at the end of our calculations, we are supposed to analytically continue back to real time by letting

$$\tau \to \tau' = it, \quad t \ \text{real}. \tag{4.9}$$

From Eq. (4.2) we see that the quantity which we are interested in is $\det(-\frac{d^2}{d\tau^2}+\omega^2)$. Furthermore, this determinant has to be evaluated in the space of functions which satisfy the boundary conditions

$$\eta(\tau_i) = \eta(\tau_f) = 0.$$

Therefore, we are basically interested in solving the eigenvalue equation

$$\left(-\frac{d^2}{d\tau^2} + \omega^2\right)\psi_n = \lambda_n\psi_n, \tag{4.10}$$

subject to the boundary conditions in Eq. (4.8). The normalized eigenfunctions of Eq. (4.10) are easily obtained to be

$$\psi_n(\tau) = \sqrt{\frac{2}{(\tau_f - \tau_i)}} \, \sin\frac{n\pi(\tau - \tau_i)}{(\tau_f - \tau_i)}, \tag{4.11}$$

with n a positive integer. The corresponding eigenvalues are

$$\lambda_n = \left(\frac{n\pi}{\tau_f - \tau_i}\right)^2 + \omega^2. \tag{4.12}$$

Thus, we see that the determinant of the operator in Eq. (4.7) or

(4.10) has the form

$$\det\left(-\frac{d^2}{d\tau^2}+\omega^2\right) = \prod_{n=1}^{\infty}\lambda_n$$

$$= \prod_{n=1}^{\infty}\left(\left(\frac{n\pi}{\tau_f-\tau_i}\right)^2+\omega^2\right)$$

$$= \prod_{n=1}^{\infty}\left(\frac{n\pi}{\tau_f-\tau_i}\right)^2\prod_{n=1}^{\infty}\left(1+\left(\frac{\omega(\tau_f-\tau_i)}{n\pi}\right)^2\right)$$

$$= B\,\frac{\sinh\omega(\tau_f-\tau_i)}{\omega(\tau_f-\tau_i)}, \qquad (4.13)$$

where we have used a relation similar to the one given in Eq. (3.24) and B is a constant representing the first product whose value can be absorbed into the normalization of the path integral measure. Analytically continuing this back to real time, we obtain

$$\det\left(-\frac{d^2}{d\tau^2}+\omega^2\right) \rightarrow \det\left(\frac{d^2}{dt^2}+\omega^2\right)$$

$$= A\,\frac{\sin\omega(t_f-t_i)}{\omega(t_f-t_i)} = A\,\frac{\sin\omega T}{\omega T}. \qquad (4.14)$$

Here, as before, we have identified $T = t_f - t_i$ with the total time interval. This is, of course, related to the value of the path integral which we had obtained earlier in Eq. (3.28) through a careful evaluation. Therefore, we conclude that

$$\int \mathcal{D}\eta\, e^{\frac{im}{2\hbar}\int_{t_i}^{t_f} dt\,(\dot{\eta}^2-\omega^2)}$$

$$= A''\left(\frac{\sin\omega T}{\omega T}\right)^{-\frac{1}{2}} = N\left(\det\left(\frac{d^2}{dt^2}+\omega^2\right)\right)^{-\frac{1}{2}}. \qquad (4.15)$$

Later, we will generalize this result to field theories or systems with an infinite number of degrees of freedom.

Let us next discuss in some detail the analytic continuation to the imaginary time. Consider a Minkowski space with coordinates

$$x^\mu = (t, \mathbf{x}),$$

where we leave the dimensionality of space-time arbitrary. Then, under the analytic continuation,

$$x^\mu = (t, \mathbf{x}) \rightarrow (-i\tau, \mathbf{x}),$$
$$x^2 = (t^2 - \mathbf{x}^2) \rightarrow -(\tau^2 + \mathbf{x}^2). \tag{4.16}$$

Therefore, we note that τ is nothing other than a Euclidean time. The analytic continuation, consequently, corresponds to a rotation to Euclidean space. (Although, in one dimension, it does not make sense to talk about a Euclidean space, in higher dimensional field theories it is quite meaningful.) The sense of the rotation is completely fixed by the singularity structure of the theory. Let us note that an analytic continuation is meaningful only if no singularity is crossed in the process. We know from our study of the Green's function in the last chapter (see Eq. (3.75)) that in the complex energy plane, the singularities occur at

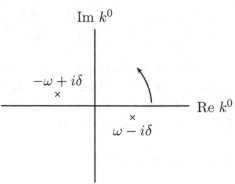

It is clear, therefore, that an analytic continuation from Re k^0 to Im k^0 is meaningful only if the rotation is anticlockwise, namely, only if we let

$$k^0 \rightarrow k'^0 = i\kappa, \quad \kappa \text{ real.} \tag{4.17}$$

Since we can represent

$$k^0 \rightarrow i\hbar \frac{\partial}{\partial t},$$

it follows now that in the complex t-plane, the consistent rotation will be

$$t \rightarrow t' = -i\tau, \quad \tau \text{ real}. \tag{4.18}$$

4.2 Time Ordered Correlation Functions

Let us recapitulate quickly what we have done so far. We have obtained the transition amplitude in the form of a path integral as

$$_H\langle x_f, t_f | x_i, t_i \rangle_H = N \int \mathcal{D}x \, e^{\frac{i}{\hbar} S[x]}. \tag{4.19}$$

Let us next consider a product of operators of the form

$$X_H(t_1) X_H(t_2),$$

and evaluate the matrix element

$$_H\langle x_f, t_f | X_H(t_1) X_H(t_2) | x_i, t_i \rangle_H, \quad t_f \geq t_1 > t_2 \geq t_i.$$

Since $t_1 > t_2$, then we can insert complete sets of coordinate basis states and write

$$_H\langle x_f, t_f | X_H(t_1) X_H(t_2) | x_i, t_i \rangle_H$$

$$= \int dx_1 dx_2 \, _H\langle x_f, t_f | X_H(t_1) | x_1, t_1 \rangle_H$$

$$\times \, _H\langle x_1, t_1 | X_H(t_2) | x_2, t_2 \rangle_H \, _H\langle x_2, t_2 | x_i, t_i \rangle_H \tag{4.20}$$

$$= \int dx_1 dx_2 \, x_1 x_2 \, _H\langle x_f, t_f | x_1, t_1 \rangle_H \, _H\langle x_1, t_1 | x_2, t_2 \rangle_H \, _H\langle x_2, t_2 | x_i, t_i \rangle_H.$$

Here we have used the relation in Eq. (1.41). We note that each inner product in the integrand represents a transition amplitude and, therefore, can be written as a path integral. Combining the products, we can write (for $t_1 > t_2$)

$$_H\langle x_f, t_f | X_H(t_1) X_H(t_2) | x_i, t_i \rangle_H = N \int \mathcal{D}x \, x(t_1) x(t_2) \, e^{\frac{i}{\hbar} S[x]}. \tag{4.21}$$

Here we have used the identification

$$x_1 = x(t_1), \quad x_2 = x(t_2). \tag{4.22}$$

Similarly, we note that for $t_2 > t_1$, we can write

$$_H\langle x_f, t_f | X_H(t_2) X_H(t_1) | x_i, t_i \rangle_H$$

$$= \int dx_1 dx_2 \; _H\langle x_f, t_f | X_H(t_2) | x_2, t_2 \rangle_H$$

$$\times \; _H\langle x_2, t_2 | X_H(t_1) | x_1, t_1 \rangle_H \; _H\langle x_1, t_1 | x_i, t_i \rangle_H$$

$$= N \int \mathcal{D}x \; x(t_1) x(t_2) \; e^{\frac{i}{\hbar} S[x]}. \tag{4.23}$$

In the last step, we have used the fact that factors in the integrand such as $x(t_1)$ and $x(t_2)$ are classical quantities and, therefore, their product is commutative.

Thus, we see from Eqs. (4.21) and (4.23) that the path integral naturally gives the time ordered correlation functions as the moments

$$_H\langle x_f, t_f | T(X_H(t_1) X_H(t_2)) | x_i, t_i \rangle_H = N \int \mathcal{D}x \; x(t_1) x(t_2) \; e^{\frac{i}{\hbar} S[x]}, \tag{4.24}$$

where the time ordering can be explicitly represented as

$$T(X_H(t_1) X_H(t_2))$$

$$= \theta(t_1 - t_2) X_H(t_1) X_H(t_2) + \theta(t_2 - t_1) X_H(t_2) X_H(t_1). \tag{4.25}$$

In fact, it is obvious now that the time ordered product of any set of operators leads to correlation functions in the path integral formalism as

$$_H\langle x_f, t_f | T(O_1(X_H(t_1)) \cdots O_n(X_H(t_n))) | x_i, t_i \rangle_H$$

$$= N \int \mathcal{D}x \; O_1(x(t_1)) \cdots O_n(x(t_n)) e^{\frac{i}{\hbar} S[x]}. \tag{4.26}$$

Furthermore, the beauty of the path integrals lies in the fact that all the factors on the right hand side are c-numbers (classical quantities). There are no operators any more.

4.3   Correlation Functions in Definite States

So far, we have calculated the transition amplitude between two co-ordinate states. In physical applications, however, we are often interested in transitions between physical states. Namely, we would like to know the probability amplitude for a system in a state $|\psi_i\rangle_H$ at time t_i to make a transition to a state $|\psi_f\rangle_H$ at time t_f. This is what the S-matrix elements are supposed to give. Let us note that by definition, this transition amplitude is given by

$$_H\langle\psi_f|\psi_i\rangle_H$$

$$= \int dx_f dx_i \, _H\langle\psi_f|x_f,t_f\rangle_H \, _H\langle x_f,t_f|x_i,t_i\rangle_H \, _H\langle x_i,t_i|\psi_i\rangle_H$$

$$= N \int dx_f dx_i \, \psi_f^*(x_f,t_f)\psi_i(x_i,t_i) \int \mathcal{D}x \, e^{\frac{i}{\hbar}S[x]} \,. \qquad (4.27)$$

Here we have used the usual definition of the wavefunction. Namely,

$$_H\langle x,t|\psi\rangle_H = \psi(x,t) \,.$$

Following our discussion earlier (see Eq. (4.26)), we see that the time ordered correlation functions between such physical states can also be written as

$$_H\langle\psi_f|T(O_1(X_H(t_1))\cdots O_n(X_H(t_n)))|\psi_i\rangle_H$$

$$= N \int dx_f dx_i \, \psi_f^*(x_f,t_f)\psi_i(x_i,t_i)$$

$$\times \int \mathcal{D}x \, O_1(x(t_1))\cdots O_n(x(t_n))e^{\frac{i}{\hbar}S[x]} \,. \qquad (4.28)$$

In dealing with physical systems, we are often interested in calculating expectation values. This is simply obtained by noting that

$$_H\langle\psi_i|T(O_1(X_H(t_1))\cdots O_n(X_H(t_n)))|\psi_i\rangle_H$$

$$= N \int dx_f dx_i \, \psi_i^*(x_f,t_f)\psi_i(x_i,t_i)$$

$$\times \int \mathcal{D}x \, O_1(x(t_1))\cdots O_n(x(t_n))e^{\frac{i}{\hbar}S[x]} \,. \qquad (4.29)$$

Since the states need not necessarily be normalized, we obtain the expectation value to be

$$\langle T(O_1(X_H(t_1)) \cdots O_n(X_H(t_n))) \rangle$$

$$= \frac{{}_H\langle \psi_i | T(O_1(X_H(t_1)) \cdots O_n(X_H(t_n))) | \psi_i \rangle_H}{{}_H\langle \psi_i | \psi_i \rangle_H} \tag{4.30}$$

$$= \frac{\int dx_f dx_i \psi_i^*(x_f, t_f) \psi_i(x_i, t_i) \int \mathcal{D}x \, O_1(x(t_1)) \cdots O_n(x(t_n)) e^{\frac{i}{\hbar}S[x]}}{\int dx_f dx_i \, \psi_i^*(x_f, t_f) \psi_i(x_i, t_i) \int \mathcal{D}x \, e^{\frac{i}{\hbar}S[x]}}.$$

Note that the normalization constant N has cancelled out in the ratio and it is for this reason that we do not often worry about the explicit form of the normalization constant. In most field theoretic questions one is primarily interested in calculating the expectation values of time ordered products in the ground state and, consequently, one tends to be sloppy about this factor in such cases.

From now on let us suppress, for convenience, the subscript H signifying the description in the Heisenberg picture. Let us next note that we can generate the various correlation functions in a simple way in the path integral formalism by adding appropriate external sources. Thus, if we define a modified action of the form

$$S[x, J] = S[x] + \int_{t_i}^{t_f} dt \, x(t) J(t), \tag{4.31}$$

then, clearly,

$$S[x, 0] = S[x],$$

where $S[x]$ defines the dynamics of the system. Let us further define

$$\langle \psi_i | \psi_i \rangle_J = N \int dx_f dx_i \, \psi_i^*(x_f, t_f) \psi_i(x_i, t_i) \int \mathcal{D}x \, e^{\frac{i}{\hbar}S[x,J]}. \tag{4.32}$$

Clearly, then,

$$\langle \psi_i | \psi_i \rangle_{J=0} = N \int dx_f dx_i \, \psi_i^*(x_f, t_f) \psi_i(x_i, t_i) \int \mathcal{D}x \, e^{\frac{i}{\hbar}S[x]}$$

$$= \langle \psi_i | \psi_i \rangle. \tag{4.33}$$

It is clear now from Eqs. (4.31) and (4.32) that $(t_f \geq t_1 \geq t_i)$

$$\frac{\delta \langle \psi_i | \psi_i \rangle_J}{\delta J(t_1)}$$

$$= N \int dx_f dx_i \, \psi_i^*(x_f, t_f) \psi_i(x_i, t_i) \int \mathcal{D}x \, \frac{i}{\hbar} \frac{\delta S[x, J]}{\delta J(t_1)} \, e^{\frac{i}{\hbar} S[x, J]}$$

$$= N \int dx_f dx_i \, \psi_i^*(x_f, t_f) \psi_i(x_i, t_i) \int \mathcal{D}x \, \frac{i}{\hbar} x(t_1) e^{\frac{i}{\hbar} S[x, J]}. \qquad (4.34)$$

It follows, therefore, that

$$\frac{\delta \langle \psi_i | \psi_i \rangle_J}{\delta J(t_1)} \bigg|_{J=0}$$

$$= N \int dx_f dx_i \, \psi_i^*(x_f, t_f) \psi_i(x_i, t_i) \int \mathcal{D}x \, \frac{i}{\hbar} x(t_1) e^{\frac{i}{\hbar} S[x]}$$

$$= \frac{i}{\hbar} \langle \psi_i | X(t_1) | \psi_i \rangle, \qquad (4.35)$$

where we have used the relation in Eq. (4.29). Similarly, we have for $t_f \geq t_1, t_2 \geq t_i$

$$\frac{\delta^2 \langle \psi_i | \psi_i \rangle_J}{\delta J(t_1) \delta J(t_2)} \bigg|_{J=0}$$

$$= N \int dx_f dx_i \, \psi_i^*(x_f, t_f) \psi_i(x_i, t_i)$$

$$\times \int \mathcal{D}x \, \left(\frac{i}{\hbar}\right)^2 \frac{\delta S[x, J]}{\delta J(t_1)} \frac{\delta S[x, J]}{\delta J(t_2)} e^{\frac{i}{\hbar} S[x, J]} \bigg|_{J=0}$$

$$= N \int dx_f dx_i \, \psi_i^*(x_f, t_f) \psi_i(x_i, t_i) \int \mathcal{D}x \, \left(\frac{i}{\hbar}\right)^2 x(t_1) x(t_2) e^{\frac{i}{\hbar} S[x]}$$

$$= \left(\frac{i}{\hbar}\right)^2 \langle \psi_i | T(X(t_1) X(t_2)) | \psi_i \rangle. \qquad (4.36)$$

In general, it is quite straightforward to show that

$$\frac{\delta^n \langle \psi_i | \psi_i \rangle_J}{\delta J(t_1) \cdots \delta J(t_n)} \bigg|_{J=0} = \left(\frac{i}{\hbar}\right)^n \langle \psi_i | T(X(t_1) \cdots X(t_n)) | \psi_i \rangle. \qquad (4.37)$$

Consequently, we can write

$$\langle T(X(t_1) \cdots X(t_n)) \rangle$$

$$= \frac{\langle \psi_i | T(X(t_1) \cdots X(t_n)) | \psi_i \rangle}{\langle \psi_i | \psi_i \rangle}$$

$$= \frac{(-i\hbar)^n}{\langle \psi_i | \psi_i \rangle_J} \left. \frac{\delta^n \langle \psi_i | \psi_i \rangle_J}{\delta J(t_1) \cdots \delta J(t_n)} \right|_{J=0} . \tag{4.38}$$

It is for this reason that $\langle \psi_i | \psi_i \rangle_J$ is also known as the generating functional for the time ordered correlation functions.

4.4   Vacuum Functional

An object of great interest in quantum theories is the vacuum to vacuum transition amplitude in the presence of an external source. The simplest way to obtain this is to go back to the transition amplitude in the coordinate space.

$$\langle x_f, t_f | x_i, t_i \rangle_J = N \int \mathcal{D}x \, e^{\frac{i}{\hbar} S[x,J]}$$

$$= N \int \mathcal{D}x \, e^{\frac{i}{\hbar} S[x] + \frac{i}{\hbar} \int_{t_i}^{t_f} dt \, J(t)x(t)} . \tag{4.39}$$

It is clear from our earlier discussion in Eq. (4.26) that we can also think of this quantity as the matrix element of the operator

$$\langle x_f, t_f | x_i, t_i \rangle_J = N \int \mathcal{D}x \, e^{\frac{i}{\hbar} S[x,J]}$$

$$= \langle x_f, t_f | T \left(e^{\frac{i}{\hbar} \int_{t_i}^{t_f} dt \, X(t)J(t)} \right) | x_i, t_i \rangle . \tag{4.40}$$

Let us next take the limit

$$t_i \to -\infty, \quad t_f \to \infty .$$

That is, let us calculate the amplitude for the system to make a transition from the coordinate state in the infinite past labeled by the coordinate x_i to the coordinate state in the infinite future labeled by x_f in the presence of an external source which switches on adiabatically. We will consider this limit by assuming that the external source is nonzero within a large but finite interval of time. That is, let us assume that

$$J(t) = 0, \quad \text{for } |t| > \tau, \tag{4.41}$$

and we will take the limit $\tau \to \infty$ at the end. In such a case, we can write

$$\lim_{\substack{t_i \to -\infty \\ t_f \to \infty}} \langle x_f, t_f | x_i, t_i \rangle_J = N \int \mathcal{D}x \, e^{\frac{i}{\hbar} \int_{-\infty}^{\infty} dt \, (L(x, \dot{x}) + Jx)}. \tag{4.42}$$

Alternatively, we can write from Eq. (4.40)

$$\lim_{\substack{t_i \to -\infty \\ t_f \to \infty}} \langle x_f, t_f | x_i, t_i \rangle_J = \lim_{\tau \to \infty} \lim_{\substack{t_i \to -\infty \\ t_f \to \infty}} \langle x_f, t_f | T \left(e^{\frac{i}{\hbar} \int_{-\tau}^{\tau} dt \, JX} \right) | x_i, t_i \rangle. \tag{4.43}$$

Let us further assume that the ground state energy of our Hamiltonian is normalized to zero so that

$$H|0\rangle = 0,$$

$$H|n\rangle = E_n|n\rangle, \quad E_n > 0. \tag{4.44}$$

(We wish to point out here that in a relativistic field theory, Lorentz invariance requires

$$P_\mu |0\rangle = 0,$$

which leads to a vanishing ground state energy. In quantum mechanics, however, the ground state energy does not vanish in general and in such a case, the asymptotic limits are not well defined and the derivation becomes involved. We, therefore, choose a derivation parallel to that of a relativistic quantum field theory and assume that the ground state energy is zero.) Although for simplicity of discussion we have assumed the energy eigenstates to be discrete, it is not essential for our arguments. Introducing complete sets of energy eigenstates into the transition amplitude, we obtain

$$\lim_{\substack{t_i \to -\infty \\ t_f \to \infty}} \langle x_f, t_f | x_i, t_i \rangle_J$$

$$= \lim_{\tau \to \infty} \lim_{\substack{t_i \to -\infty \\ t_f \to \infty}} \sum_{n,m} \langle x_f, t_f | n \rangle \langle n | T \left(e^{\frac{i}{\hbar} \int_{-\tau}^{\tau} dt \, JX} \right) | m \rangle \langle m | x_i, t_i \rangle$$

$$= \lim_{\tau \to \infty} \lim_{\substack{t_i \to -\infty \\ t_f \to \infty}} \sum_{n,m} \langle x_f | e^{-\frac{i}{\hbar} H t_f} | n \rangle$$

$$\times \, \langle n | T \left(e^{\frac{i}{\hbar} \int_{-\tau}^{\tau} dt \, JX} \right) | m \rangle \langle m | e^{\frac{i}{\hbar} H t_i} | x_i \rangle$$

$$= \lim_{\tau \to \infty} \lim_{\substack{t_i \to -\infty \\ t_f \to \infty}} \sum_{n,m} e^{-\frac{i}{\hbar} E_n t_f + \frac{i}{\hbar} E_m t_i} \langle x_f | n \rangle$$

$$\times \langle n | T \left(e^{\frac{i}{\hbar} \int_{-\tau}^{\tau} dt \, JX} \right) | m \rangle \langle m | x_i \rangle \,, \tag{4.45}$$

where we have used Eqs. (1.42) and (4.44). In the limit $t_i \to -\infty$ and $t_f \to \infty$, the exponentials oscillate out to zero except for the ground state. One can alternately see this also by analytically continuing to the imaginary time axis (Euclidean space in the case of field theories). Thus, in this asymptotic limit, we obtain

$$\lim_{\substack{t_i \to -\infty \\ t_f \to \infty}} \langle x_f, t_f | x_i, t_i \rangle_J = \lim_{\tau \to \infty} \langle x_f | 0 \rangle \langle 0 | T \left(e^{\frac{i}{\hbar} \int_{-\tau}^{\tau} dt \, JX} \right) | 0 \rangle \langle 0 | x_i \rangle$$

$$= \langle x_f | 0 \rangle \langle 0 | x_i \rangle \langle 0 | T \left(e^{\frac{i}{\hbar} \int_{-\infty}^{\infty} dt \, JX} \right) | 0 \rangle \,. \tag{4.46}$$

Consequently, we can write

$$\langle 0 | T \left(e^{\frac{i}{\hbar} \int_{-\infty}^{\infty} dt \, JX} \right) | 0 \rangle = \lim_{\substack{t_i \to -\infty \\ t_f \to \infty}} \frac{\langle x_f, t_f | x_i, t_i \rangle_J}{\langle x_f | 0 \rangle \langle 0 | x_i \rangle} \,. \tag{4.47}$$

The left hand side is independent of the end points and, therefore, the right hand side must also be independent of the end points. Furthermore, the right hand side has the structure of a path integral and we can write Eq. (4.47) also as

$$\langle 0|T\left(e^{\frac{i}{\hbar}\int_{-\infty}^{\infty} dt\, JX}\right)|0\rangle = \langle 0|0\rangle_J = N\int \mathcal{D}x\, e^{\frac{i}{\hbar}S[x,J]}\,, \qquad (4.48)$$

without the end point constraints and with

$$S[x,J] = \int_{-\infty}^{\infty} dt\,(L(x,\dot{x}) + Jx)\,.$$

Let us note that if we define

$$Z[J] = \langle 0|0\rangle_J = N\int \mathcal{D}x\, e^{\frac{i}{\hbar}S[x,J]}\,, \qquad (4.49)$$

then, it follows from Eq. (4.38) that

$$\frac{(-i\hbar)^n}{Z[J]}\left.\frac{\delta^n Z[J]}{\delta J(t_1)\cdots\delta J(t_n)}\right|_{J=0} = \langle T(X(t_1)\cdots X(t_n))\rangle\,. \qquad (4.50)$$

Namely, $Z[J]$ generates time ordered correlation functions or the Green's functions in the vacuum. If one knows all the vacuum Green's functions, one can construct the S-matrix of the theory and, therefore, solve the theory. In quantum field theory, therefore, these correlation functions or the vacuum Green's functions play a central role. $Z[J]$ is correspondingly known as the vacuum functional or the generating functional for vacuum Green's functions.

In quantum mechanics, we are often interested in various statistical deviations from the mean values. This can be obtained in the path integral formalism in the following way. Let us define

$$Z[J] = e^{\frac{i}{\hbar}W[J]}$$

$$\text{or,}\quad W[J] = -i\hbar\ln Z[J]\,. \qquad (4.51)$$

We have already seen in the case of the free particle as well as the harmonic oscillator that the path integral for the transition amplitude is proportional to the exponential of the classical action (see Eqs. (2.54) and (3.28)). It is for this reason that $W[J]$ is also called an effective action. Let us note that by definition

$$\left.\frac{\delta W[J]}{\delta J(t_1)}\right|_{J=0} = (-i\hbar)\frac{1}{Z[J]}\left.\frac{\delta Z[J]}{\delta J(t_1)}\right|_{J=0} = \langle X(t_1)\rangle\,, \qquad (4.52)$$

where $\langle \cdots \rangle$ stands for the vacuum expectation value from now on. Next, we note that

$$(-i\hbar) \left. \frac{\delta^2 W[J]}{\delta J(t_1)\delta J(t_2)} \right|_{J=0}$$

$$= (-i\hbar)^2 \left. \left(\frac{1}{Z[J]} \frac{\delta^2 Z[J]}{\delta J(t_1)\delta J(t_2)} - \frac{1}{Z^2[J]} \frac{\delta Z[J]}{\delta J(t_1)} \frac{\delta Z[J]}{\delta J(t_2)} \right) \right|_{J=0}$$

$$= \langle T(X(t_1)X(t_2)) \rangle - \langle X(t_1) \rangle \langle X(t_2) \rangle$$

$$= \langle T\left((X(t_1) - \langle X(t_1) \rangle)(X(t_2) - \langle X(t_2) \rangle) \right) \rangle. \tag{4.53}$$

We recognize this to be the second order deviation from the mean and we note that we can similarly, obtain

$$(-i\hbar)^2 \left. \frac{\delta^3 W[J]}{\delta J(t_1)\delta J(t_2)\delta J(t_3)} \right|_{J=0}$$

$$= (-i\hbar)^3 \left(\frac{1}{Z[J]} \frac{\delta^3 Z[J]}{\delta J(t_1)\delta J(t_2)\delta J(t_3)} - \frac{1}{Z^2[J]} \frac{\delta^2 Z[J]}{\delta J(t_1)\delta J(t_2)} \frac{\delta Z[J]}{\delta J(t_3)} \right.$$

$$- \frac{1}{Z^2[J]} \frac{\delta^2 Z[J]}{\delta J(t_3)\delta J(t_1)} \frac{\delta Z[J]}{\delta J(t_2)} - \frac{1}{Z^2[J]} \frac{\delta^2 Z[J]}{\delta J(t_2)\delta J(t_3)} \frac{\delta Z[J]}{\delta J(t_1)}$$

$$\left. + \frac{2}{Z^3} \frac{\delta Z[J]}{\delta J(t_1)} \frac{\delta Z[J]}{\delta J(t_2)} \frac{\delta Z[J]}{\delta J(t_3)} \right) \right|_{J=0}$$

$$= \langle T\left(X(t_1)X(t_2)X(t_3) \right) \rangle - \langle T\left(X(t_1)X(t_2) \right) \rangle \langle X(t_3) \rangle$$

$$- \langle T\left(X(t_3)X(t_1) \right) \rangle \langle X(t_2) \rangle - \langle T\left(X(t_2)X(t_3) \right) \rangle \langle X(t_1) \rangle$$

$$+ 2\langle X(t_1) \rangle \langle X(t_2) \rangle \langle X(t_3) \rangle \tag{4.54}$$

$$= \langle T\left((X(t_1) - \langle X(t_1) \rangle)(X(t_2) - \langle X(t_2) \rangle)(X(t_3) - \langle X(t_3) \rangle) \right) \rangle.$$

We can go on and the expressions start to take a more complicated form starting with the fourth functional derivative of the effective action $W[J]$. However, $W[J]$ can still be shown to generate various statistical deviations and their moments. In quantum field theory, $W[J]$ is known as the generating functional for the connected vacuum Green's functions.

Let us next go back to the example of the harmonic oscillator which we have studied in some detail. In this case, we have

$$Z[J] = N \int \mathcal{D}x \, e^{\frac{i}{\hbar} S[x,J]} \, ,$$

where

$$S[x, J] = \int_{-\infty}^{\infty} dt \left(\frac{1}{2} m \dot{x}^2 - \frac{1}{2} m \omega^2 x^2 + Jx \right) . \tag{4.55}$$

Obviously, in this case, we have

$$\langle X(t_1) \rangle = (-i\hbar) \frac{1}{Z[J]} \frac{\delta Z[J]}{\delta J(t_1)} \bigg|_{J=0}$$

$$= \frac{N \int \mathcal{D}x \, x(t_1) \, e^{\frac{i}{\hbar} S[x]}}{N \int \mathcal{D}x \, e^{\frac{i}{\hbar} S[x]}} = 0 . \tag{4.56}$$

This vanishes because the integrand in the numerator is odd. Therefore, for the harmonic oscillator, we obtain from Eqs. (4.50) and (4.53) that

$$\langle T(X(t_1)X(t_2)) \rangle = (-i\hbar)^2 \frac{1}{Z[J]} \frac{\delta^2 Z[J]}{\delta J(t_1) \delta J(t_2)} \bigg|_{J=0}$$

$$= (-i\hbar) \frac{\delta^2 W[J]}{\delta J(t_1) \delta J(t_2)} \bigg|_{J=0} . \tag{4.57}$$

Let us also note that because the action for the harmonic oscillator is quadratic in the variables, we can write

$$Z[J] = N \int \mathcal{D}x \, e^{\frac{i}{\hbar} \int_{-\infty}^{\infty} dt (\frac{1}{2} m \dot{x}^2 - \frac{1}{2} m \omega^2 x^2 + Jx)} \tag{4.58}$$

$$= \lim_{\epsilon \to 0^+} N \int \mathcal{D}x \, e^{-\frac{im}{2\hbar} \int_{-\infty}^{\infty} dt \, (x(t)(\frac{d^2}{dt^2} + \omega^2 - i\epsilon)x(t) - \frac{2}{m} J(t)x(t))} .$$

Let us recall that (see Eq. (3.76))

$$\lim_{\epsilon \to 0^+} \left(\frac{d^2}{dt^2} + \omega^2 - i\epsilon \right) G_F(t - t') = -\delta(t - t').$$

Using this, we can define

$$\tilde{x}(t) = x(t) + \frac{1}{m} \int_{-\infty}^{\infty} dt'\, G_F(t - t')J(t'), \qquad (4.59)$$

and the generating functional will then take the form

$$Z[J] = \lim_{\epsilon \to 0^+} N \int \mathcal{D}\tilde{x}\, e^{-\frac{im}{2\hbar} \int_{-\infty}^{\infty} dt\, \tilde{x}(t)(\frac{d^2}{dt^2} + \omega^2 - i\epsilon)\tilde{x}(t)}$$

$$\times\, e^{-\frac{i}{2\hbar m} \iint_{-\infty}^{\infty} dt\, dt'\, J(t)G_F(t-t')J(t')}$$

$$= \lim_{\epsilon \to 0^+} N \left[\det\left(\frac{d^2}{dt^2} + \omega^2 - i\epsilon \right) \right]^{-\frac{1}{2}}$$

$$\times\, e^{-\frac{i}{2\hbar m} \iint_{-\infty}^{\infty} dt\, dt'\, J(t)G_F(t-t')J(t')}$$

$$= Z[0]\, e^{-\frac{i}{2\hbar m} \iint_{-\infty}^{\infty} dt\, dt'\, J(t)G_F(t-t')J(t')}. \qquad (4.60)$$

We now obtain in a straightforward manner

$$\frac{\delta^2 Z[J]}{\delta J(t_1)\delta J(t_2)}\bigg|_{J=0} = -\frac{i}{\hbar m} G_F(t_1 - t_2)Z[0]. \qquad (4.61)$$

Consequently, for the harmonic oscillator, we have (see Eq. (4.57))

$$\langle T(X(t_1)X(t_2)) \rangle = (-i\hbar)^2 \frac{1}{Z[J]} \frac{\delta^2 Z[J]}{\delta J(t_1)\delta J(t_2)}\bigg|_{J=0}$$

$$= (-i\hbar)^2 \left(-\frac{i}{\hbar m} \right) G_F(t_1 - t_2) = \frac{i\hbar}{m} G_F(t_1 - t_2). \qquad (4.62)$$

In other words, the two-point time ordered vacuum correlation function, in the present case, gives the Feynman Green's function. This is a general feature of all quantum mechanical theories, namely, that the two-point connected vacuum Green's function is nothing other than the Feynman propagator of the theory.

4.5 Anharmonic Oscillator

Just as there are a handful of quantum mechanical problems which can be solved analytically, similarly, there are only a few path integrals that can be exactly evaluated. (Fortunately, there is a one to one correspondence between the quantum mechanical problems that can be analytically solved and the path integrals that can be exactly evaluated.) The Gaussian (recall the free particle and the harmonic oscillator) is the simplest of the path integrals which can be exactly evaluated. However, we also know that if we perturb the harmonic oscillator even slightly by an additional potential, say a quartic potential, the problem cannot be analytically solved. In other words, the quantum mechanical system corresponding to the Lagrangian

$$L = \frac{1}{2}m\dot{x}^2 - \frac{1}{2}m\omega^2 x^2 - \frac{\lambda}{4}x^4, \quad \lambda > 0, \tag{4.63}$$

is impossible to solve exactly even when $\lambda \ll 1$. In such a case, we, of course, use perturbation theory and calculate corrections to the unperturbed system in a power series in λ. In the Feynman path integral approach, the manifestation of this lies in the fact that the path integral corresponding to the Lagrangian for this anharmonic oscillator cannot be evaluated exactly and has to be calculated perturbatively in the following way. Let us introduce an external source and write the vacuum functional for this theory as

$$Z[J] = N \int \mathcal{D}x \, e^{\frac{i}{\hbar}S[x,J]}$$

$$= N \int \mathcal{D}x \, e^{\frac{i}{\hbar}\int_{-\infty}^{\infty} dt(\frac{1}{2}m\dot{x}^2 - \frac{1}{2}m\omega^2 x^2 - \frac{\lambda}{4}x^4 + Jx)}. \tag{4.64}$$

If we write

$$S[x, J] = S_0[x, J] - \int_{-\infty}^{\infty} dt \, \frac{\lambda}{4} x^4, \tag{4.65}$$

where

$$S_0[x, J] = \int_{-\infty}^{\infty} dt \left(\frac{1}{2}m\dot{x}^2 - \frac{1}{2}m\omega^2 x^2 + Jx \right), \tag{4.66}$$

is the action for the harmonic oscillator in the presence of a source, then we note that

$$\frac{\delta S_0[x, J]}{\delta J(t)} = x(t). \tag{4.67}$$

In other words, operationally we can identify

$$\frac{\delta}{\delta J(t)} \rightarrow x(t), \tag{4.68}$$

when this acts on $S_0[x, J]$. Now, we can use this identification to write

$$Z[J] = N \int \mathcal{D}x \left(e^{-\frac{i\lambda}{4\hbar} \int_{-\infty}^{\infty} dt\, x^4} \right) e^{\frac{i}{\hbar} S_0[x, J]}$$

$$= N \int \mathcal{D}x \left(e^{-\frac{i\lambda}{4\hbar} \int_{-\infty}^{\infty} dt(-i\hbar\frac{\delta}{\delta J(t)})^4} \right) e^{\frac{i}{\hbar} S_0[x, J]}$$

$$= \left(e^{-\frac{i\lambda}{4\hbar} \int_{-\infty}^{\infty} dt(-i\hbar\frac{\delta}{\delta J(t)})^4} \right) N \int \mathcal{D}x\, e^{\frac{i}{\hbar} S_0[x, J]}$$

$$= \left(e^{-\frac{i\lambda}{4\hbar} \int_{-\infty}^{\infty} dt(-i\hbar\frac{\delta}{\delta J(t)})^4} \right) Z_0[J], \tag{4.69}$$

where $Z_0[J]$ is the vacuum functional for the harmonic oscillator interacting with an external source. We have already seen in Eq. (4.60) that it has the form

$$Z_0[J] = Z_0[0]\, e^{-\frac{i}{2\hbar m} \int\int_{-\infty}^{\infty} dt'\, dt''\, J(t') G_F(t'-t'') J(t'')}. \tag{4.70}$$

Substituting this back then, we obtain

$$Z[J] = Z_0[0] \left[e^{-\frac{i\lambda}{4\hbar} \int_{-\infty}^{\infty} dt(-i\hbar\frac{\delta}{\delta J(t)})^4} \right]$$

$$\times e^{-\frac{i}{2\hbar m} \int\int_{-\infty}^{\infty} dt'\, dt''\, J(t') G_F(t'-t'') J(t'')}. \tag{4.71}$$

It is clear that if λ is small, i.e., for weak coupling, we can Taylor expand the first exponential and we will be able to obtain the vacuum functional as a power series in λ. Consequently, all the vacuum Green's functions can also be calculated perturbatively. This is perturbation theory in the framework of Feynman's path integral.

4.6 References

Coleman, S., "Secret Symmetry", Erice Lectures (1973).
Huang, K., "Quarks, Leptons and Gauge Fields", World Scientific Publishing.
Schwinger, J., Phys. Rev. 82, 914 (1951); *ibid* 91, 713 (1953).

Chapter 5

Path Integrals for Fermions

5.1 Fermionic Oscillator

As we know, there are two kinds of particles in nature, namely, bosons and fermions. They are described by quantum mechanical operators with very different properties. The operators describing bosons, for example, obey commutation relations whereas the fermionic operators (i.e., operators describing fermions) satisfy anti-commutation relations. As a preparation for such systems, let us study a prototype example, namely, the fermionic oscillator.

There are many ways to introduce the fermionic oscillator. Let us discuss one that is the most intuitive. Let us recall that the bosonic harmonic oscillator in one dimension with a natural frequency ω has a Hamiltonian which, written in terms of creation and annihilation operators, takes the form

$$H_B = \frac{\omega}{2} \left(a_B^\dagger a_B + a_B a_B^\dagger \right). \tag{5.1}$$

Here, for simplicity, we are assuming that $\hbar = 1$. The creation and the annihilation operators are supposed to satisfy the commutation relations

$$\left[a_B, a_B^\dagger \right] = 1, \tag{5.2}$$

with all others vanishing. The symmetric structure of the Hamiltonian, in this case, is a reflection of the fact that we are dealing with Bose particles and, consequently, the states must have a symmetric form.

Fermionic systems, on the other hand, have an inherent antisymmetry. Therefore, let us try a Hamiltonian for a fermionic oscillator with frequency ω of the form

$$H_F = \frac{\omega}{2} \left(a_F^\dagger a_F - a_F a_F^\dagger \right) . \tag{5.3}$$

If a_F and $a_F^\dagger$ were to satisfy commutation relations like the Bose oscillator, namely, if we had

$$\left[a_F, a_F^\dagger \right] = 1, \tag{5.4}$$

with all others vanishing, then using this, we can rewrite the fermionic Hamiltonian in Eq. (5.3) to be

$$H_F = \frac{\omega}{2} \left(a_F^\dagger a_F - \left(a_F^\dagger a_F + 1 \right) \right) = -\frac{\omega}{2}. \tag{5.5}$$

In other words, in such a case, there would be no dynamics associated with the Hamiltonian. Let us assume, therefore, that the fermionic operators a_F and $a_F^\dagger$ instead satisfy anti-commutation relations. Namely, let

$$[a_F, a_F]_+ \equiv a_F^2 + a_F^2 = 2a_F^2 = 0 ,$$

$$\left[a_F^\dagger, a_F^\dagger \right]_+ \equiv \left(a_F^\dagger \right)^2 + \left(a_F^\dagger \right)^2 = 2 \left(a_F^\dagger \right)^2 = 0 ,$$

$$\left[a_F, a_F^\dagger \right]_+ \equiv a_F a_F^\dagger + a_F^\dagger a_F = 1 = \left[a_F^\dagger, a_F \right]_+ . \tag{5.6}$$

In contrast to the commutators, therefore, the anti-commutators are by definition symmetric.

An immediate consequence of the anti-commutation relations in Eq. (5.6) is that in such a system, the particles must obey Fermi-Dirac statistics. To see this, let us note that if we identify the operators a_F and $a_F^\dagger$ with the annihilation and creation operators for such a system, then we can define a number operator as usual as

$$N_F = a_F^\dagger a_F. \tag{5.7}$$

From the anti-commutation relations in Eq. (5.6) we note that

$$N_F^2 = a_F^\dagger a_F a_F^\dagger a_F$$

$$= a_F^\dagger \left(1 - a_F^\dagger a_F\right) a_F$$

$$= a_F^\dagger a_F = N_F$$

$$\text{or,} \quad N_F \left(N_F - 1\right) = 0. \tag{5.8}$$

Therefore, the eigenvalues of the number operator can only be zero or one. This is the reflection of the Pauli principle or the Fermi statistics, namely, that we can at the most have one fermion in a given quantum state. This shows that the anti-commutation relations are the natural choice for a fermionic system.

Given this, let us rewrite the Hamiltonian for the fermionic oscillator in Eq. (5.3) as

$$H_F = \frac{\omega}{2} \left(a_F^\dagger a_F - \left(1 - a_F^\dagger a_F\right)\right)$$

$$= \omega \left(a_F^\dagger a_F - \frac{1}{2}\right) = \omega \left(N_F - \frac{1}{2}\right). \tag{5.9}$$

Furthermore, the commutation relations between a_F, $a_F^\dagger$ and N_F can now be calculated in a straightforward manner.

$$[a_F, N_F] = \left[a_F, a_F^\dagger a_F\right] = \left[a_F, a_F^\dagger\right]_+ a_F = a_F,$$

$$\left[a_F^\dagger, N_F\right] = \left[a_F^\dagger, a_F^\dagger a_F\right] = -a_F^\dagger \left[a_F^\dagger, a_F\right]_+ = -a_F^\dagger. \tag{5.10}$$

Consequently, if we assume an eigenstate of N_F to be denoted by $|n_F\rangle$, we have

$$N_F|n_F\rangle = n_F|n_F\rangle, \tag{5.11}$$

with $n_F = 0, 1$. The ground state with no quantum is denoted by $|0\rangle$ and satisfies

$$N_F|0\rangle = 0,$$

$$H_F|0\rangle = \omega \left(N_F - \frac{1}{2}\right)|0\rangle = -\frac{\omega}{2}|0\rangle. \tag{5.12}$$

Similarly, the state with one quantum is denoted by $|1\rangle$ and satisfies

$$N_F|1\rangle = |1\rangle \,,$$

$$H_F|1\rangle = \omega \left(N_F - \frac{1}{2} \right) |1\rangle = \frac{\omega}{2}|1\rangle \,. \tag{5.13}$$

The ground state is annihilated by a_F and we have

$$a_F|0\rangle = 0 \,,$$

$$a_F^\dagger|0\rangle = |1\rangle \,. \tag{5.14}$$

It is clear from the anti-commutation relations in Eq. (5.6) that

$$a_F^\dagger|1\rangle = a_F^\dagger a_F^\dagger|0\rangle = 0 \,. \tag{5.15}$$

Therefore, the Hilbert space, in this case, is two dimensional and we note here that the ground state energy has the opposite sign from the ground state energy of a bosonic oscillator.

5.2 Grassmann Variables

Since fermions have no classical analogue, we cannot directly write down a Lagrangian for the fermionic oscillator with the usual notions of coordinates and momenta. Obviously, we need the notion of anti-commuting classical variables. Such variables have been well studied in mathematics and go under the name of Grassmann variables. As one can readily imagine, they have very uncommon properties and let us note only some of these properties which we will need for our discussions. For example, if $\theta_i, i = 1, 2, \ldots, n$, defines a set of Grassmann variables (classical), then they satisfy

$$\theta_i\theta_j + \theta_j\theta_i = 0, \quad i, j = 1, 2, \ldots, n \,. \tag{5.16}$$

This, in particular, implies that for any given i,

$$\theta_i^2 = 0, \quad i \text{ not summed} \,. \tag{5.17}$$

In other words, the Grassmann variables are nilpotent. This has the immediate consequence that if $f(\theta)$ is a function of only one Grassmann variable, then it has the simple Taylor expansion

$$f(\theta) = a + b\,\theta\,. \tag{5.18}$$

Since θ_i's are anti-commuting, the derivatives have to be defined carefully in the sense that the direction in which the derivatives operate must be specified. Thus, for example, a right derivative for Grassmann variables would give

$$\frac{\partial}{\partial\theta_i}(\theta_j\theta_k) = \theta_j\left(\frac{\partial\theta_k}{\partial\theta_i}\right) - \left(\frac{\partial\theta_j}{\partial\theta_i}\right)\theta_k = \delta_{ik}\theta_j - \delta_{ij}\theta_k\,, \tag{5.19}$$

whereas a left derivative would give

$$\frac{\partial}{\partial\theta_i}(\theta_j\theta_k) = \left(\frac{\partial\theta_j}{\partial\theta_i}\right)\theta_k - \theta_j\left(\frac{\partial\theta_k}{\partial\theta_i}\right) = \delta_{ij}\theta_k - \delta_{ik}\theta_j\,. \tag{5.20}$$

Thus, the sense of the derivative is crucial and in all our discussions, we will use left derivatives.

Let us note that like the Grassmann variables, the derivatives with respect to these variables also anti-commute. Namely,

$$\frac{\partial}{\partial\theta_i}\frac{\partial}{\partial\theta_j} + \frac{\partial}{\partial\theta_j}\frac{\partial}{\partial\theta_i} = 0\,. \tag{5.21}$$

These derivatives, in fact, behave quite like the exterior derivatives in differential geometry. We note in particular that for a fixed i

$$\left(\frac{\partial}{\partial\theta_i}\right)^2 = 0\,. \tag{5.22}$$

In other words, the derivatives, in this case, are nilpotent just like the variables themselves. Furthermore, the conventional commutation relation between derivatives and coordinates now takes the form

$$\left[\frac{\partial}{\partial\theta_i}, \theta_j\right]_+ = \frac{\partial}{\partial\theta_i}\theta_j + \theta_j\frac{\partial}{\partial\theta_i} = \delta_{ij}\,. \tag{5.23}$$

The notion of integration can also be generalized to Grassmann variables. Denoting by D the operation of differentiation with respect to one Grassmann variable and by I the operation of integration, we note that these must satisfy the relations

$$ID = 0,$$

$$DI = 0. \tag{5.24}$$

Namely, the integral of a total derivative must vanish if we ignore surface terms and furthermore, an integral, being independent of the variable, must give zero upon differentiation. Note that since differentiation with respect to a Grassmannn variable is nilpotent (see Eq. (5.22)), it satisfies the above properties and hence for Grassmann variables integration can be naturally identified with differentiation. Namely, in this case, we have

$$I = D. \tag{5.25}$$

Thus, for a function of a single Grassmann variable, we have

$$\int d\theta \, f(\theta) = \frac{\partial f(\theta)}{\partial \theta}. \tag{5.26}$$

This immediately leads to the fundamental result that for Grassmann variables

$$\int d\theta = 0,$$

$$\int d\theta \, \theta = 1. \tag{5.27}$$

This is an essential difference between ordinary variables and Grassmann variables and has far reaching consequences.

An immediate consequence of the definition of the integral in Eq. (5.26) is that if we redefine the variable of integration as

$$\theta' = a \, \theta, \quad a \neq 0, \tag{5.28}$$

then, we obtain

$$\int d\theta \, f(\theta) = \frac{\partial f(\theta)}{\partial \theta} = a \frac{\partial f\left(\frac{\theta'}{a}\right)}{\partial \theta'} = a \int d\theta' \, f\left(\frac{\theta'}{a}\right). \tag{5.29}$$

But this is precisely the opposite of what happens for ordinary variables. Namely, we note that the Jacobian in the case of redefinition of Grassmann variables is the inverse of what one would naively expect for ordinary variables. This result can be easily generalized to integrations involving many Grassmann variables and it can be shown that if

$$\theta'_i = a_{ij}\theta_j \,, \tag{5.30}$$

with $\det a_{ij} \neq 0$ and repeated indices being summed, then

$$\int \prod_{i=1}^{n} d\theta_i \, f(\theta_i) = (\det a_{ij}) \int \prod_{i=1}^{n} d\theta'_i f\left(a_{ij}^{-1}\theta'_j\right). \tag{5.31}$$

We can also define a delta function in the space of Grassmann variables as

$$\delta(\theta) = \theta \,. \tag{5.32}$$

That this satisfies all the properties of the delta function can be easily seen by noting that

$$\int d\theta \, \delta(\theta) = \int d\theta \, \theta = 1, \tag{5.33}$$

which follows from Eq. (5.27). In addition, if

$$f(\theta) = a + b\,\theta \,,$$

then,

$$\int d\theta \, \delta(\theta) \, f(\theta) = \int d\theta \, \theta \, f(\theta) = \int d\theta \, \theta(a + b\theta)$$

$$= \int d\theta \, \theta a = \frac{\partial(\theta a)}{\partial \theta} = a = f(0) \,. \tag{5.34}$$

Here we have used the nilpotency of the Grassmann variables. Furthermore, consistent with the rule for change of variables for the Grassmann variables, if

$$g(\theta) = a\theta \,,$$

then, we obtain

$$\delta(g(\theta)) = a\theta = a\delta(\theta) = \frac{\partial g(\theta)}{\partial \theta}\delta(\theta), \qquad (5.35)$$

where $g(\theta)$ is assumed to be Grassmann odd. An integral representation for the delta function can be obtained simply by noting that if ζ is also a Grassmann variable, then

$$\int d\zeta \, e^{i\zeta\theta} = \int d\zeta \, (1 + i\zeta\theta)$$

$$= \frac{\partial}{\partial \zeta}(1 + i\zeta\theta) = i\theta = i\delta(\theta). \qquad (5.36)$$

Let us next evaluate the basic Gaussian integral for Grassmann variables. Let us consider two sets of independent Grassmann variables, namely, $(\theta_1, \theta_2, \ldots, \theta_n)$ and $(\theta_1^*, \theta_2^*, \ldots, \theta_n^*)$ and analyze the integral

$$I = \int \prod_{i,j} d\theta_i^* \, d\theta_j \, e^{-(\theta_i^* M_{ij}\theta_j + c_i^*\theta_i + \theta_i^* c_i)}, \qquad (5.37)$$

where we are assuming that c_i and c_i^* are independent Grassmann variables. Furthermore, the convention for summation over repeated indices is always assumed. Note that if we make the change of variables (we are assuming that M^{-1} exists)

$$\theta_i' = M_{ij}\theta_j + c_i$$

$$\text{or,} \quad \theta_i = M_{ij}^{-1}(\theta_j' - c_j), \qquad (5.38)$$

and

$$\theta_i^{*'} = \theta_i^* + c_j^* M_{ji}^{-1}, \qquad (5.39)$$

then, we obtain using Eqs. (5.31), (5.38) and (5.39) that

$$I = \int \prod_{i,j} d\theta_i^* \, d\theta_j \, e^{-\left(\theta_i^*(M_{ij}\theta_j + c_i) + c_i^*\theta_i\right)}$$

$$= \det M_{ij} \int \prod_{ij} d\theta_i^* \, d\theta_j' \, e^{-(\theta_i^*\theta_i' + c_i^* M_{ij}^{-1}(\theta_j' - c_j))}$$

$$= \det M_{ij} \int \prod_{ij} \mathrm{d}\theta_i^* \, \mathrm{d}\theta_j' \, e^{-((\theta_i^* + c_j^* M_{ji}^{-1})\theta_i' - c_i^* M_{ij}^{-1} c_j)}$$

$$= \det M_{ij} \int \prod_{ij} \mathrm{d}\theta_i^{*'} \, \mathrm{d}\theta_j' \, e^{-\theta_i^{*'} \theta_i' + c_i^* M_{ij}^{-1} c_j}$$

$$= N \, \det M_{ij} \, e^{c_i^* M_{ij}^{-1} c_j}. \tag{5.40}$$

Here N is a constant and we note that the Gaussian integral in the case of Grassmann variables has the same form as the integral for ordinary variables except for the positive power of the determinant. This leads to an essential difference between quantum mechanical bosonic and fermionic theories.

5.3 Generating Functional

With all this background on Grassmann variables, we can now ask whether it is possible to write a Lagrangian for the fermionic oscillator. Indeed, let us consider the Lagrangian

$$L = \frac{i}{2} \left(\bar{\psi}\dot{\psi} - \dot{\bar{\psi}}\psi \right) - \frac{\omega}{2} \left[\bar{\psi}, \psi \right], \tag{5.41}$$

where ψ and $\bar{\psi}$ are two independent Grassmann variables. Quite often one eliminates a total derivative to write an equivalent Lagrangian also as

$$L = i\bar{\psi}\dot{\psi} - \frac{\omega}{2}[\bar{\psi}, \psi]. \tag{5.42}$$

We will, however, continue with the first form of the Lagrangian, namely, Eq. (5.41). This is a first order Lagrangian and one can define canonical conjugate momenta associated with the Grassmann variables ψ and $\bar{\psi}$ as usual

$$\Pi_\psi = \frac{\partial L}{\partial \dot{\psi}} = -\frac{i}{2}\bar{\psi},$$

$$\Pi_{\bar{\psi}} = \frac{\partial L}{\partial \dot{\bar{\psi}}} = -\frac{i}{2}\psi. \tag{5.43}$$

(Such a system is known as a constrained system and one should use the formalism of Dirac quantization to quantize such a system. We will, however, not go into these technical details which are not quite relevant for our present discussion.) With the convention of left derivatives, the proper definition of the Hamiltonian (which is only a function of coordinates and momenta and which also leads to the correct dynamical equations of motion) is

$$
\begin{aligned}
H &= \dot{\psi}\Pi_\psi + \dot{\bar{\psi}}\Pi_{\bar{\psi}} - L \\
&= -\frac{i}{2}\dot{\psi}\bar{\psi} - \frac{i}{2}\dot{\bar{\psi}}\psi - \frac{i}{2}\left(\bar{\psi}\dot{\psi} - \dot{\bar{\psi}}\psi\right) + \frac{\omega}{2}[\bar{\psi},\psi] \\
&= \frac{i}{2}\left(\bar{\psi}\dot{\psi} - \dot{\bar{\psi}}\psi\right) - \frac{i}{2}\left(\bar{\psi}\dot{\psi} - \dot{\bar{\psi}}\psi\right) + \frac{\omega}{2}[\bar{\psi},\psi] \\
&= \frac{\omega}{2}[\bar{\psi},\psi]\,.
\end{aligned}
\tag{5.44}
$$

It is clear, therefore, that this simple Lagrangian will yield the Hamiltonian of Eq. (5.3) for the fermionic oscillator if we identify

$$
\begin{aligned}
\psi &= a_F\,, \\
\bar{\psi} &= a_F^\dagger\,.
\end{aligned}
\tag{5.45}
$$

With this identification of ψ and $\bar{\psi}$ with the annihilation and the creation operators respectively, the hermiticity properties for these variables now follow. Namely, we note that

$$
\begin{aligned}
\psi^\dagger &= \bar{\psi}\,, \\
\bar{\psi}^\dagger &= \psi\,.
\end{aligned}
\tag{5.46}
$$

We also note that with this convention, the number operator defined as

$$
N_F = a_F^\dagger a_F = \bar{\psi}\psi\,,
\tag{5.47}
$$

is Hermitian since

$$
N_F^\dagger = \left(\bar{\psi}\psi\right)^\dagger = \psi^\dagger\bar{\psi}^\dagger = \bar{\psi}\psi\,.
\tag{5.48}
$$

Since Grassmann variables have no classical analogue, even when we are dealing with the reality questions of ordinary Grassmann variables (not operators), we follow the above prescription in defining complex conjugation. Namely, we define for any pair of Grassmann variables η and χ,

$$(\eta\chi)^* = \chi^*\eta^*. \tag{5.49}$$

In other words, even classically, we continue to treat Grassmann variables like operators. This is the only way a consistent transition is possible from a classical to a quantum Lagrangian involving fermions.

With this prescription, let us note that the Lagrangian for the fermionic oscillator given in Eq. (5.41) is Hermitian (real).

$$L^\dagger = \left(\frac{i}{2} \left(\bar{\psi}\dot{\psi} - \dot{\bar{\psi}}\psi \right) - \frac{\omega}{2}[\bar{\psi}, \psi] \right)^\dagger$$

$$= -\frac{i}{2} \left(\dot{\psi}^\dagger \bar{\psi}^\dagger - \psi^\dagger \dot{\bar{\psi}}^\dagger \right) - \frac{\omega}{2}[\psi^\dagger, \bar{\psi}^\dagger]$$

$$= -\frac{i}{2} \left(\dot{\bar{\psi}}\psi - \bar{\psi}\dot{\psi} \right) - \frac{\omega}{2}[\bar{\psi}, \psi]$$

$$= \frac{i}{2} \left(\bar{\psi}\dot{\psi} - \dot{\bar{\psi}}\psi \right) - \frac{\omega}{2}[\bar{\psi}, \psi]$$

$$= L. \tag{5.50}$$

With this, then, we can write the vacuum functional for the fermionic oscillator as

$$Z[\eta, \bar{\eta}] = \langle 0|0 \rangle_{\eta, \bar{\eta}} = N \int \mathcal{D}\bar{\psi}\mathcal{D}\psi \, e^{\frac{i}{\hbar}S[\psi, \bar{\psi}, \eta, \bar{\eta}]}, \tag{5.51}$$

where we have denoted the sources for ψ and $\bar{\psi}$ by $\bar{\eta}$ and η respectively. The complete action for the oscillator, in this case, has the form

$$S[\psi, \bar{\psi}, \eta, \bar{\eta}] = S[\psi, \bar{\psi}] + \int_{-\infty}^{\infty} dt \, (\bar{\eta}\psi + \bar{\psi}\eta), \tag{5.52}$$

with

$$S[\psi, \bar{\psi}] = \int dt \, L = \int dt \left(\frac{i}{2} \left(\bar{\psi}\dot{\psi} - \dot{\bar{\psi}}\psi \right) - \frac{\omega}{2}[\bar{\psi}, \psi] \right). \tag{5.53}$$

Once again, we will assume the hermiticity conditions for the sources similar to the ones given in Eq. (5.46), namely,

$$\eta^\dagger = \bar\eta\,,$$

$$\bar\eta^\dagger = \eta\,, \tag{5.54}$$

in order that the complete action in Eq. (5.52) is Hermitian.

Just as an ordinary derivative with respect to a Grassmann variable is directional, similarly, there are right and left functional derivatives with respect to fermionic variables. The definition of the functional derivative is still the same as given in Eq. (1.14), namely,

$$\frac{\delta F\left(\psi(t)\right)}{\delta\psi(t')} = \lim_{\epsilon\to 0}\frac{F\left(\psi(t) + \epsilon\delta(t - t')\right) - F\left(\psi(t)\right)}{\epsilon}\,. \tag{5.55}$$

However, since ϵ is now a Grassmann variable, the position of ϵ^{-1} in the expression defines the direction of the derivative. (Incidentally, one can think of ϵ^{-1} simply as $\frac{\partial}{\partial\epsilon}$. Secondly, we note here that for polynomial functionals, the limit $\epsilon \to 0$ is redundant since $\epsilon^2 = 0$ and the highest power of ϵ in the expansion of the functional is linear.) Thus, a left functional derivative corresponds to defining

$$\frac{\delta F\left(\psi(t)\right)}{\delta\psi(t')} = \lim_{\epsilon\to 0}\epsilon^{-1}\left[F\left(\psi(t) + \epsilon\delta(t - t')\right) - F\left(\psi(t)\right)\right]\,, \tag{5.56}$$

whereas a right functional derivative would be defined as

$$\frac{\delta F\left(\psi(t)\right)}{\delta\psi(t')} = \lim_{\epsilon\to 0}\left[F\left(\psi(t) + \epsilon\delta(t - t')\right) - F\left(\psi(t)\right)\right]\epsilon^{-1}\,. \tag{5.57}$$

As we have mentioned earlier, we will always work with left derivatives even when dealing with functionals involving fermionic variables.

5.4　Feynman Propagator

Let us next go back to the Lagrangian in Eq. (5.41) for the fermionic oscillator and note that when fermions are involved, we essentially have a matrix structure. This is another reflection of the fact that the Grassmann variables inherently behave like operators. Let us define the following two component matrices.

$$\Psi = \begin{pmatrix} \psi \\ \bar{\psi} \end{pmatrix},$$

$$\bar{\Psi} = \Psi^\dagger \sigma_3 = \underbrace{\bar{\psi} \quad -\psi},$$

$$\Theta = \begin{pmatrix} \eta \\ \bar{\eta} \end{pmatrix},$$

$$\bar{\Theta} = \Theta^\dagger \sigma_3 = \underbrace{\bar{\eta} \quad -\eta}. \tag{5.58}$$

Here σ_3 denotes the Pauli matrix, namely,

$$\sigma_3 = \begin{pmatrix} 1 & 0 \\ 0 & -1 \end{pmatrix}.$$

We note, then, that

$$\bar{\Psi}\sigma_3 \, i\frac{d}{dt}\Psi = \underbrace{\bar{\psi} \quad -\psi} \begin{pmatrix} i\frac{d}{dt} & 0 \\ 0 & -i\frac{d}{dt} \end{pmatrix} \begin{pmatrix} \psi \\ \bar{\psi} \end{pmatrix}$$

$$= \underbrace{\bar{\psi} \quad -\psi} \begin{pmatrix} i\dot{\psi} \\ -i\dot{\bar{\psi}} \end{pmatrix} = i\left(\bar{\psi}\dot{\psi} + \psi\dot{\bar{\psi}} \right)$$

$$= i\left(\bar{\psi}\dot{\psi} - \dot{\bar{\psi}}\psi \right),$$

$$\bar{\Psi}\Psi = \underbrace{\bar{\psi} \quad -\psi} \begin{pmatrix} \psi \\ \bar{\psi} \end{pmatrix}$$

$$= (\bar{\psi}\psi - \psi\bar{\psi}) = [\bar{\psi}, \psi],$$

$$\bar{\Psi}\Theta = \underbrace{\bar{\psi} \quad -\psi} \begin{pmatrix} \eta \\ \bar{\eta} \end{pmatrix} = (\bar{\psi}\eta - \psi\bar{\eta})$$

$$= (\bar{\psi}\eta + \bar{\eta}\psi) = \bar{\Theta}\Psi, \tag{5.59}$$

where we have used the anti-commuting properties of the Grassmann variables.

It is now straightforward to show that the action for the fermionic oscillator in Eq. (5.52) can be written as

$$
S = \int_{-\infty}^{\infty} dt \left(\frac{i}{2} \bar{\Psi} \sigma_3 \frac{d}{dt} \Psi - \frac{\omega}{2} \bar{\Psi} \Psi + \bar{\Theta} \Psi \right)
$$

$$
= \int_{-\infty}^{\infty} dt \left(\frac{1}{2} \bar{\Psi} (i\sigma_3 \frac{d}{dt} - \omega) \Psi + \bar{\Theta} \Psi \right). \tag{5.60}
$$

We note here that, using Eq. (5.59), we could have written the source term also as

$$
\int_{-\infty}^{\infty} dt \ \bar{\Psi} \Theta \,.
$$

In other words, Θ and $\bar{\Theta}$ are not really independent and, therefore, we can write the vacuum functional as ($\hbar{=}1$)

$$
Z[\Theta] = N \int \mathcal{D}\Psi \, e^{iS}
$$

$$
= N \int \mathcal{D}\Psi \, e^{i \int_{-\infty}^{\infty} dt \left(\frac{1}{2} \bar{\Psi} (i\sigma_3 \frac{d}{dt} - \omega)\Psi + \bar{\Theta}\Psi \right)}. \tag{5.61}
$$

We know that we can evaluate the generating functional if we know the Green's function associated with the operator in the quadratic term in the exponent (see Eq. (5.40)). Let us, therefore, study the equation

$$
\left(i\sigma_3 \frac{d}{dt} - \omega \right) G(t - t') = \delta(t - t'). \tag{5.62}
$$

This is clearly a matrix equation and it can be solved easily in the momentum space. Thus, we define

$$
G(t - t') = \int \frac{dk}{\sqrt{2\pi}} G(k) \, e^{-ik(t-t')} \,,
$$

$$
\delta(t - t') = \frac{1}{2\pi} \int dk \, e^{-ik(t-t')} \,, \tag{5.63}
$$

where $G(k)$ is a matrix in the Fourier transformed space. Substituting these expressions back into Eq. (5.62), we obtain

$$\frac{1}{\sqrt{2\pi}}(\sigma_3 k - \omega)G(k) = \frac{1}{2\pi}$$

$$\text{or, } G(k) = \frac{1}{\sqrt{2\pi}} \frac{1}{\sigma_3 k - \omega}$$

$$= \frac{1}{\sqrt{2\pi}} \frac{\sigma_3 k + \omega}{k^2 - \omega^2}. \tag{5.64}$$

Consequently, the Green's function in Eq. (5.62) has the form

$$G(t - t') = \int \frac{dk}{\sqrt{2\pi}} G(k) \, e^{-ik(t-t')}$$

$$= \frac{1}{2\pi} \int dk \, \frac{\sigma_3 k + \omega}{k^2 - \omega^2} \, e^{-ik(t-t')}. \tag{5.65}$$

The singularity structure of the integrand is obvious and the Feynman prescription, in this case, will lead to the propagator (see Eq. (3.75))

$$G_F(t - t') = \lim_{\eta \to 0^+} \frac{1}{2\pi} \int dk \, \frac{\sigma_3 k + \omega}{k^2 - \omega^2 + i\eta} \, e^{-ik(t-t')}$$

$$= \lim_{\eta \to 0^+} \frac{1}{2\pi} \int dk \, \frac{\sigma_3 k + \omega}{k^2 - (\omega - \frac{i\eta}{2\omega})^2} \, e^{-ik(t-t')}. \tag{5.66}$$

This can also be written in the alternate form

$$G_F(t - t') = \lim_{\epsilon \to 0^+} \frac{1}{2\pi} \int dk \, \frac{1}{\sigma_3 k - \omega + i\epsilon} \, e^{-ik(t-t')}, \tag{5.67}$$

and satisfies the equation (see Eq. (3.76))

$$\lim_{\epsilon \to 0^+} \left(i\sigma_3 \frac{d}{dt} - \omega + i\epsilon \right) G_F(t - t') = \delta(t - t'). \tag{5.68}$$

This defines the Feynman propagator in the present case.

Going back to the vacuum functional, we note that we can write

$$Z[\Theta] = \lim_{\epsilon \to 0^+} N \int \mathcal{D}\Psi \, e^{i \int dt(\frac{1}{2}\bar{\Psi}(i\sigma_3 \frac{d}{dt} - \omega + i\epsilon)\Psi + \bar{\Theta}\Psi)} . \qquad (5.69)$$

Therefore, the 1-point function can be obtained to be

$$\frac{\delta Z[\Theta]}{\delta\bar{\Theta}(t_1)}\bigg|_{\Theta=\bar{\Theta}=0} = \lim_{\epsilon \to 0^+} N \int \mathcal{D}\Psi \, (i\Psi(t_1)) \, e^{\frac{i}{2} \int dt \, \bar{\Psi}(i\sigma_3 \frac{d}{dt} - \omega + i\epsilon)\Psi}$$

$$= 0 . \qquad (5.70)$$

This is because the integrand in Eq. (5.70) is odd under

$$\Psi \to -\Psi, \quad \bar{\Psi} \to -\bar{\Psi} . \qquad (5.71)$$

Similarly, we can obtain from Eq. (5.69)

$$\frac{\delta Z[\Theta]}{\delta\Theta(t_1)}\bigg|_{\Theta=\bar{\Theta}=0} = 0 . \qquad (5.72)$$

Thus, we see that if we write

$$Z[\Theta] = e^{iW[\Theta]} , \qquad (5.73)$$

then, in this case,

$$(-i)\frac{\delta^2 W[\Theta]}{\delta\bar{\Theta}(t_1)\delta\Theta(t_2)}\bigg|_{\Theta=\bar{\Theta}=0} = (-i)^2 \frac{1}{Z[\Theta]} \frac{\delta^2 Z[\Theta]}{\delta\bar{\Theta}(t_1)\delta\Theta(t_2)}\bigg|_{\Theta=\bar{\Theta}=0}$$

$$= -\langle T\left(\Psi(t_1)\bar{\Psi}(t_2)\right)\rangle . \qquad (5.74)$$

(Compare this with Eq. (4.57) for $\hbar = 1$.) Incidentally, time ordering, in the case of fermionic variables, is defined as

$$T\left(\Psi(t_1)\bar{\Psi}(t_2)\right) = \theta(t_1 - t_2)\Psi(t_1)\bar{\Psi}(t_2) - \theta(t_2 - t_1)\bar{\Psi}(t_2)\Psi(t_1) . \qquad (5.75)$$

The relative negative sign between the two terms in Eq. (5.75) arises from the change in the order of the fermionic variables, which anti-commute, in the second term. Going back to the vacuum functional, we note that since the exponent is quadratic in the variables (namely, it is a Gaussian integral), it can be explicitly evaluated using Eq. (5.40) to be

$$
Z[\Theta] = \lim_{\epsilon \to 0^+} N \int \mathcal{D}\Psi e^{i \int dt(\frac{1}{2}\bar{\Psi}(i\sigma_3 \frac{d}{dt} - \omega + i\epsilon)\Psi + \bar{\Theta}\Psi)}
$$

$$
= \tilde{N} e^{-\frac{i}{2} \iint dt_1 dt_2 \, \bar{\Theta}(t_1) G_F(t_1-t_2)\Theta(t_2)}
$$

$$
= Z[0] e^{-\frac{i}{2} \iint dt_1 dt_2 \, \bar{\Theta}(t_1) G_F(t_1-t_2)\Theta(t_2)} , \qquad (5.76)
$$

where $Z[0]$ represents the value of the functional in the absence of sources. It is obvious now that

$$
\frac{\delta^2 Z[\Theta]}{\delta\bar{\Theta}(t_1)\delta\Theta(t_2)}\bigg|_{\Theta=\bar{\Theta}=0} = iG_F(t_1 - t_2) Z[0]. \qquad (5.77)
$$

Therefore, we have, using Eq. (5.74)

$$
\langle T\left(\Psi(t_1)\bar{\Psi}(t_2)\right)\rangle = -(-i)^2 \frac{1}{Z[\Theta]} \frac{\delta^2 Z[\Theta]}{\delta\bar{\Theta}(t_1)\delta\Theta(t_2)}\bigg|_{\Theta=\bar{\Theta}=0}
$$

$$
= -(-i)^2 \frac{1}{Z[0]} iG_F(t_1 - t_2)Z[0]
$$

$$
= i\, G_F(t_1 - t_2). \qquad (5.78)
$$

This again shows (see Eq. (4.62)) that the time ordered two-point correlation function in the vacuum gives the Feynman Green's function. As we have argued earlier, this is a general feature of all quantum theories.

5.5 The Fermion Determinant

The fermion action following from Eq. (5.41) or Eq. (5.42) is quadratic in the dynamical variables just like the action of the bosonic oscillator in Eq. (3.2). Therefore, the generating functional can be easily evaluated. In this section, we will evaluate the generating functional for the fermions in the absence of any sources. For simplicity, let us take the dynamical Lagrangian of Eq. (5.42). Then, we can write the generating functional, in the absence of sources, to be

$$Z[0] = \tilde{N} \int \mathcal{D}\bar{\psi}\mathcal{D}\psi \; e^{iS[\psi,\bar{\psi}]}$$

$$= \tilde{N} \int \mathcal{D}\bar{\psi}\mathcal{D}\psi \; e^{i \int_{t_i}^{t_f} dt \, (i\bar{\psi}\dot{\psi} - \omega\bar{\psi}\psi)} . \tag{5.79}$$

Here we have used the anti-commuting properties of the Grassmann variables to rewrite the commutator of the fermionic variables in a simpler form. The constant $\tilde{N}$, representing the normalization of the path integral measure is arbitrary at this point and would be appropriately chosen later. We can once again define

$$t_f - t_i = T,$$

as the time interval and translate the time coordinate to write the generating functional of Eq. (5.79) also as

$$Z[0] = \tilde{N} \int \mathcal{D}\bar{\psi}\mathcal{D}\psi \; e^{i \int_0^T dt \, (i\bar{\psi}\dot{\psi} - \omega\bar{\psi}\psi)} . \tag{5.80}$$

To evaluate the path integral, we should discretize the time interval as in Eq. (2.18). Thus, defining the intermediate time points to be

$$t_n = n\epsilon, \quad n = 1, 2, \ldots, N-1,$$

where the infinitesimal interval is defined to be

$$\epsilon = \frac{T}{N},$$

we can write the path integral to be

$$Z[0] = \lim_{\substack{\epsilon \to 0 \\ N \to \infty}} \tilde{N} \int d\bar{\psi}_1 \cdots d\bar{\psi}_{N-1} d\psi_1 \cdots d\psi_{N-1}$$

$$\times e^{i\epsilon \sum_{n=1}^{N}(i\bar{\psi}_n(\frac{\psi_n - \psi_{n-1}}{\epsilon}) - \omega\bar{\psi}_n(\frac{\psi_n + \psi_{n-1}}{2}))} . \tag{5.81}$$

Here we have used the mid-point prescription of Weyl ordering as discussed in Eqs. (2.17) and (2.21) along with the earlier observation that the variable $\bar{\psi}$ represents the momentum conjugate to ψ.

The exponent in Eq. (5.81) can be written out in detail as

$$-\left[\sum_{n=1}^{N-1} (1 + \frac{i\epsilon\omega}{2})\bar{\psi}_n\psi_n + \left(1 + \frac{i\epsilon\omega}{2}\right)\bar{\psi}_N\psi_N\right.$$

$$-\sum_{n=2}^{N-1}\left(1 - \frac{i\epsilon\omega}{2}\right)\bar{\psi}_n\psi_{n-1} - \left(1 - \frac{i\epsilon\omega}{2}\right)\bar{\psi}_1\psi_0$$

$$\left. -\left(1 - \frac{i\epsilon\omega}{2}\right)\bar{\psi}_N\psi_{N-1}\right]. \qquad (5.82)$$

Thus, defining $(N-1)$ component matrices

$$\psi = \begin{pmatrix} \psi_1 \\ \psi_2 \\ \vdots \\ \psi_{N-1} \end{pmatrix},$$

$$\bar{\psi} = \begin{pmatrix} \bar{\psi}_1 \\ \bar{\psi}_2 \\ \vdots \\ \bar{\psi}_{N-1} \end{pmatrix},$$

$$J = -\left(1 - \frac{i\epsilon\omega}{2}\right)\begin{pmatrix} \psi_0 \\ 0 \\ \vdots \\ 0 \end{pmatrix},$$

$$\bar{J} = -\left(1 - \frac{i\epsilon\omega}{2}\right)\begin{pmatrix} 0 \\ \vdots \\ 0 \\ \bar{\psi}_N \end{pmatrix}, \qquad (5.83)$$

we can write the path integral of Eq. (5.81) also as

$$Z[0] = \lim_{\substack{\epsilon\to 0 \\ N\to\infty}} \tilde{N} \int d\bar{\psi}d\psi \; e^{-(\bar{\psi}^T B\psi + \bar{J}^T \psi + \bar{\psi}^T J + (1 + \frac{i\epsilon\omega}{2})\bar{\psi}_N\psi_N)}, \qquad (5.84)$$

where we have defined a $(N-1) \times (N-1)$ matrix B as

$$B = \begin{pmatrix} x & 0 & 0 & 0 & \cdots \\ y & x & 0 & 0 & \cdots \\ 0 & y & x & 0 & \cdots \\ \vdots & \vdots & \vdots & \vdots & \vdots \end{pmatrix}, \tag{5.85}$$

with

$$x = \left(1 + \frac{i\epsilon\omega}{2}\right),$$

$$y = -\left(1 - \frac{i\epsilon\omega}{2}\right). \tag{5.86}$$

The path integral in Eq. (5.84) can now be easily evaluated using Eq. (5.40) and the result is

$$Z[0] = \lim_{\substack{\epsilon \to 0 \\ N \to \infty}} \tilde{N} \, \det B \, e^{(J^T B^{-1} J - (1 + \frac{i\epsilon\omega}{2})\bar{\psi}_N \psi_N)}$$

$$= \lim_{\substack{\epsilon \to 0 \\ N \to \infty}} \tilde{N} \, \det B \, e^{(\bar{J}_{N-1} B_{N-1,1}^{-1} J_1 - (1 + \frac{i\epsilon\omega}{2})\bar{\psi}_N \psi_N)}. \tag{5.87}$$

The matrix B has a very simple structure and we can easily evaluate the determinant as well as the appropriate element of the inverse matrix which have the following forms.

$$\det B = x^{N-1} = \left(1 + \frac{i\epsilon\omega}{2}\right)^{N-1},$$

$$B_{N-1,1}^{-1} = (-1)^N \frac{y^{N-2}}{x^{N-1}} = \frac{\left(1 - \frac{i\epsilon\omega}{2}\right)^{N-2}}{\left(1 + \frac{i\epsilon\omega}{2}\right)^{N-1}}. \tag{5.88}$$

In the continuum limit of $\epsilon \to 0$ and $N \to \infty$ such that $N\epsilon = T$, the path integral, therefore, has the form

$$Z[0] = \tilde{N} \, e^{\frac{i\omega T}{2}} \, e^{(e^{-i\omega T}\bar{\psi}_N \psi_0 - \bar{\psi}_N \psi_N)}$$

$$= \tilde{N} \, e^{\frac{i\omega T}{2}} \, e^{(e^{-i\omega T}\bar{\psi}_f \psi_i - \bar{\psi}_f \psi_f)}. \tag{5.89}$$

Here we have identified

$$\psi_0 = \psi_i, \quad \psi_N = \psi_f,$$

$$\bar{\psi}_0 = \bar{\psi}_i, \quad \bar{\psi}_N = \bar{\psi}_f. \tag{5.90}$$

We choose, for simplicity and for future use, the normalization of the path integral measure to be $\tilde{N} = 1$ so that the free fermion path integral takes the form

$$Z[0] = e^{\frac{i\omega T}{2}} \, e^{\left(e^{-i\omega T} \bar{\psi}_f \psi_i - \bar{\psi}_f \psi_f\right)}. \tag{5.91}$$

5.6 References

Berezin, F., "The Method of Second Quantization", Academic Press.

DeWitt, B., "Supermanifolds", Cambridge Univ. Press.

Chapter 6

Supersymmetry

6.1 Supersymmetric Oscillator

We have seen in Chapters 3 and 5 that a bosonic oscillator in one dimension with a natural frequency ω is described by the Hamiltonian

$$H_B = \frac{\omega}{2} \left(a_B^\dagger a_B + a_B a_B^\dagger \right) = \omega \left(a_B^\dagger a_B + \frac{1}{2} \right), \qquad (6.1)$$

while a fermionic oscillator with a natural frequency ω is described by the Hamiltonian

$$H_F = \frac{\omega}{2} \left(a_F^\dagger a_F - a_F a_F^\dagger \right) = \omega \left(a_F^\dagger a_F - \frac{1}{2} \right). \qquad (6.2)$$

Here we are assuming that $\hbar = 1$. (See Eqs. (5.2) and (5.3).) The creation and the annihilation operators for the bosonic oscillator satisfy

$$\left[a_B, a_B^\dagger \right] = 1, \qquad (6.3)$$

with all others vanishing. For the fermionic oscillator, on the other hand, the creation and the annihilation operators satisfy the anti-commutation relations (see Eq. (5.6))

$$[a_F, a_F]_+ = 0 = \left[a_F^\dagger, a_F^\dagger \right]_+,$$

$$\left[a_F, a_F^\dagger \right]_+ = 1. \qquad (6.4)$$

Let us note here (as we have pointed out earlier in chapter 5) that the ground state energy for the bosonic oscillator is $\frac{\omega}{2}$ whereas that for the fermionic oscillator is $-\frac{\omega}{2}$.

Let us next consider a system consisting of a bosonic and a fermionic oscillator with the same natural frequency ω. This is known as the supersymmetric oscillator. The Hamiltonian for this system follows from Eqs. (6.1) and (6.2) to be

$$H = H_B + H_F = \frac{\omega}{2}\left(a_B^\dagger a_B + a_B a_B^\dagger + a_F^\dagger a_F - a_F a_F^\dagger\right)$$

$$= \omega\left(a_B^\dagger a_B + \frac{1}{2} + a_F^\dagger a_F - \frac{1}{2}\right)$$

$$= \omega\left(a_B^\dagger a_B + a_F^\dagger a_F\right). \tag{6.5}$$

We note from Eq. (6.5) that the constant term in the Hamiltonian for this system has cancelled out. If we define the number operators for the bosonic and the fermionic oscillators as

$$N_B = a_B^\dagger a_B,$$

$$N_F = a_F^\dagger a_F, \tag{6.6}$$

then, we can write the Hamiltonian for the system also as

$$H = \omega(N_B + N_F). \tag{6.7}$$

It is clear from Eq. (6.7) that the energy eigenstates of the system will be the eigenstates of the number operators N_B and N_F. Consequently, let us define

$$|n_B, n_F\rangle = |n_B\rangle \otimes |n_F\rangle, \tag{6.8}$$

where

$$N_B|n_B\rangle = n_B|n_B\rangle, \qquad n_B = 0, 1, 2, \dots,$$

$$N_F|n_F\rangle = n_F|n_F\rangle, \qquad n_F = 0, 1. \tag{6.9}$$

Here we are using our earlier result in Eq. (5.11) that the eigenvalues for the fermion number operator are 0 or 1 consistent with the Pauli principle while the eigenvalues for the bosonic number operator can take any positive semidefinite integer value. From Eqs. (6.7), (6.8) and (6.9) we note that the energy eigenvalues for the supersymmetric oscillator are given by

$$H|n_B, n_F\rangle = E_{n_B, n_F}|n_B, n_F\rangle = \omega(n_B + n_F)|n_B, n_F\rangle, \qquad (6.10)$$

with $n_B = 0, 1, 2 \ldots$ and $n_F = 0, 1$.

We also note from Eq. (6.10) that the ground state energy of the supersymmetric oscillator vanishes, namely,

$$E_{0,0} = 0. \qquad (6.11)$$

Incidentally, the ground state is assumed to satisfy

$$a_B|0\rangle = 0 = a_F|0\rangle. \qquad (6.12)$$

The vanishing of the ground state energy is a general feature of supersymmetric theories and as we will see shortly it is a consequence of the supersymmetry of the system. We also observe from Eq. (6.10) that except for the ground state, all other energy eigenstates of the system are doubly degenerate. Namely, the states $|n_B, 1\rangle$ and $|n_B+1, 0\rangle$ have the same energy for any value of n_B. The degeneracy in the energy value for a bosonic and a fermionic state, as we will see, is again a consequence of the supersymmetry of the system.

Let us next consider the following two fermionic operators in the theory.

$$Q = a_B^\dagger a_F,$$

$$\bar{Q} = a_F^\dagger a_B. \qquad (6.13)$$

We can show using the commutation relations in Eqs. (6.3) and (6.4) that (The bosonic operators commute with the fermionic ones.)

$$[Q, H] = \left[a_B^\dagger a_F, \omega(a_B^\dagger a_B + a_F^\dagger a_F) \right]$$

$$= \omega \left(a_B^\dagger \left[a_B^\dagger, a_B \right] a_F + a_B^\dagger \left[a_F, a_F^\dagger \right]_+ a_F \right)$$

$$= \omega \left(-a_B^\dagger a_F + a_B^\dagger a_F \right)$$

$$= 0, \tag{6.14}$$

and similarly,

$$[\bar{Q}, H] = \left[a_F^\dagger a_B, \omega (a_B^\dagger a_B + a_F^\dagger a_F) \right] = 0. \tag{6.15}$$

The operators, Q and $\bar{Q}$, therefore, define conserved quantities of this system (charges) and would correspond to the generators of symmetries in the theory. We also note that

$$[Q, \bar{Q}]_+ = \left[a_B^\dagger a_F, a_F^\dagger a_B \right]_+$$

$$= a_B^\dagger \left[a_F, a_F^\dagger \right]_+ a_B - a_F^\dagger \left[a_B^\dagger, a_B \right] a_F$$

$$= a_B^\dagger a_B + a_F^\dagger a_F$$

$$= \frac{1}{\omega} H. \tag{6.16}$$

Thus, we see from Eqs. (6.14), (6.15) and (6.16) that the operators Q, $\bar{Q}$ and H define an algebra which involves both commutators and anti-commutators. Such an algebra is known as a graded Lie algebra and defines the infinitesimal form of the supersymmetric transformations. An immediate consequence of the supersymmetry algebra is that if the ground state is invariant under supersymmetry transformations, namely, if (see Eq. (6.12))

$$Q|0\rangle = 0 = \bar{Q}|0\rangle, \tag{6.17}$$

then it follows from Eq. (6.16) that

$$\langle 0|H|0\rangle = \omega \langle 0|Q\bar{Q} + \bar{Q}Q|0\rangle = 0. \tag{6.18}$$

Namely, the ground state energy in a supersymmetric theory vanishes. Furthermore, we note from Eq. (6.13) that $\bar{Q}$ is really the

Hermitian conjugate of Q. Consequently, it follows from Eq. (6.16) that in a supersymmetric theory H is really a positive semidefinite operator and, therefore, its expectation value in any state must be positive semidefinite.

Let us next analyze the effect of Q and $\bar{Q}$ on the energy eigenstates of the system. We note from the commutation rules of the theory that

$$[Q, N_B] = \left[a_B^\dagger a_F, a_B^\dagger a_B \right]$$

$$= a_B^\dagger \left[a_B^\dagger, a_B \right] a_F$$

$$= -a_B^\dagger a_F = -Q,$$

$$[Q, N_F] = \left[a_B^\dagger a_F, a_F^\dagger a_F \right]$$

$$= a_B^\dagger \left[a_F, a_F^\dagger \right]_+ a_F$$

$$= a_B^\dagger a_F = Q, \tag{6.19}$$

and similarly

$$[\bar{Q}, N_B] = \bar{Q},$$

$$[\bar{Q}, N_F] = -\bar{Q}. \tag{6.20}$$

In other words, we can think of Q as raising the bosonic number n_B while lowering the fermionic number n_F by one unit whereas $\bar{Q}$ does the opposite. It now follows that for

$$|n_B, n_F\rangle = \frac{\left(a_B^\dagger \right)^{n_B}}{\sqrt{n_B!}} \left(a_F^\dagger \right)^{n_F} |0\rangle, \tag{6.21}$$

where we recognize $n_F = 0, 1$ and $n_B = 0, 1, 2, \ldots$, we have

$$Q|n_B, n_F\rangle = \begin{cases} \sqrt{n_B + 1}|n_B + 1, n_F - 1\rangle & \text{if } n_F \neq 0, \\ 0 & \text{if } n_F = 0, \end{cases}$$

$$\tag{6.22}$$

$$\bar{Q}|n_B, n_F\rangle = \begin{cases} \frac{1}{\sqrt{n_B}}|n_B - 1, n_F + 1\rangle & \text{if } n_B \neq 0 \text{ or } n_F \neq 1, \\ 0 & \text{if } n_B = 0 \text{ or } n_F = 1. \end{cases}$$

Namely, we note that acting on any state other than the ground state, the operators Q and $\bar{Q}$ take a bosonic state (with $n_F = 0$) to a fermionic state (with $n_F = 1$) or vice versa. This is the manifestation of supersymmetry on the states in the Hilbert space of the Hamiltonian, i.e., the bosonic and the fermionic states are paired. Furthermore, since Q and $\bar{Q}$ commute with the Hamiltonian of the system (see Eqs (6.14) and (6.15)), it now follows that such paired states will be degenerate in energy. Namely

$$H(Q|n_B, n_F\rangle) = Q(H|n_B, n_F\rangle) = E_{n_B, n_F}(Q|n_B, n_F\rangle),$$

$$H(\bar{Q}|n_B, n_F\rangle) = \bar{Q}(H|n_B, n_F\rangle) = E_{n_B, n_F}(\bar{Q}|n_B, n_F\rangle). \tag{6.23}$$

The supersymmetric oscillator is the simplest example of supersymmetric theories. The concept of supersymmetry and graded Lie algebras generalizes to other cases as well and there exist many useful realizations of these algebras in the context of field theories.

6.2   Supersymmetric Quantum Mechanics

Let us next study a general supersymmetric, quantum mechanical theory. From our discussion in the last section, we note that supersymmetry necessarily involves both bosons and fermions and, therefore, let us consider a Lagrangian of the form

$$L = \frac{1}{2}\dot{x}^2 - \frac{1}{2}(f(x))^2 + i\bar{\psi}\dot{\psi} - f'(x)\bar{\psi}\psi. \tag{6.24}$$

Here, for consistency with earlier discussions we have set $m = 1$ for the bosonic part of the Lagrangian. We also note here that $f(x)$ can be any chosen monomial of x at this point. It is clear from Eq. (6.24) that when

$$f(x) = \omega x, \tag{6.25}$$

the Lagrangian of Eq. (6.24) reduces to that of a supersymmetric oscillator discussed in the last section.

In general, we note that under the infinitesimal transformations

$$\delta_\epsilon x = \frac{1}{\sqrt{2}}\, \bar{\psi}\epsilon,$$

$$\delta_\epsilon \psi = -\frac{i}{\sqrt{2}}\, \dot{x}\epsilon - \frac{1}{\sqrt{2}}\, f(x)\epsilon,$$

$$\delta_\epsilon \bar{\psi} = 0,\tag{6.26}$$

and

$$\delta_{\bar{\epsilon}} x = \frac{1}{\sqrt{2}}\, \bar{\epsilon}\psi,$$

$$\delta_{\bar{\epsilon}} \psi = 0,$$

$$\delta_{\bar{\epsilon}} \bar{\psi} = \frac{i}{\sqrt{2}}\, \dot{x}\bar{\epsilon} - \frac{1}{\sqrt{2}}\, f(x)\bar{\epsilon},\tag{6.27}$$

where ϵ and $\bar{\epsilon}$ are infinitesimal Grassmann parameters, the action for the Lagrangian can be seen to remain unchanged. In other words, the transformations in Eqs. (6.26) and (6.27) define symmetries of the system. (See chapter 11 for a detailed discussion of symmetries.) These symmetry transformations mix up the bosonic and the fermionic variables of the theory and, therefore, are reminiscent of the supersymmetry transformations which we discussed earlier. In fact, one can explicitly show that the two sets of transformations in Eqs. (6.26) and (6.27) are generated respectively by the two supersymmetric charges Q and $\bar{Q}$ in the theory. We also note here, without going into detail, that while the Lagrangian in Eq. (6.24) is supersymmetric for any monomial $f(x)$, in the case of even monomials the presence of instantons breaks supersymmetry. (Instantons are discussed in chapter 8.) Consequently, let us consider monomials only of the form.

$$f(x) \sim x^{2n+1}, \qquad n = 0, 1, 2, \ldots .\tag{6.28}$$

With these preparations, let us next look at the generating functional for the supersymmetric quantum mechanical theory in

Eq. (6.24) ($\hbar = 1$)

$$Z = N \int \mathcal{D}\bar{\psi}\mathcal{D}\psi\mathcal{D}x \; e^{iS[x,\psi,\bar{\psi}]} . \qquad (6.29)$$

As we have seen in the last section, the spectrum of a supersymmetric theory has many interesting features. Correspondingly, the generating functional for such a theory is also quite interesting. In particular, let us note from Eq. (6.24) that since the Lagrangian is quadratic in the fermionic variables, the functional integral for these variables can be done easily using our results in chapter 5. (See Eqs. (5.40) and (5.87)). Thus, we can write

$$\int \mathcal{D}\bar{\psi}\mathcal{D}\psi \; e^{i \int dt(\bar{\psi}(i\frac{d}{dt}-f'(x))\psi)} = N' \det\left(i\frac{d}{dt} - f'(x)\right) . \qquad (6.30)$$

Substituting Eq. (6.30) into Eq. (6.29), we obtain

$$Z = N \int \mathcal{D}\bar{\psi}\mathcal{D}\psi\mathcal{D}x \; e^{iS[x,\psi,\bar{\psi}]}$$

$$= N \int \mathcal{D}x \; e^{i \int dt(\frac{1}{2}\dot{x}^2 - \frac{1}{2}(f(x))^2)} \int \mathcal{D}\bar{\psi}\mathcal{D}\psi \; e^{i \int dt(\bar{\psi}(i\frac{d}{dt}-f'(x))\psi)}$$

$$= \tilde{N} \int \mathcal{D}x \det\left(i\frac{d}{dt} - f'(x)\right) e^{i \int dt(\frac{1}{2}\dot{x}^2 - \frac{1}{2}(f(x))^2)} . \qquad (6.31)$$

Let us next note that if we define a new bosonic variable through the relation

$$\rho = i\dot{x} - f(x) , \qquad (6.32)$$

then the Jacobian for this change of variables in Eq. (6.31) will be given by

$$J = \left[\det\left(i\frac{d}{dt} - f'(x)\right)\right]^{-1} . \qquad (6.33)$$

This is precisely the inverse of the determinant in Eq. (6.31).

Furthermore, we note that

$$\int \mathrm{dt}\, \rho^2 = \int \mathrm{dt}\, (i\dot{x} - f(x))^2$$

$$= \int \mathrm{dt}\, \left(-\dot{x}^2 - 2i\dot{x}f(x) + (f(x))^2\right)$$

$$= -\int \mathrm{dt}\, \left(\dot{x}^2 - (f(x))^2\right) - 2i \int_{-\infty}^{\infty} \mathrm{d}x\, f(x)$$

$$= -\int \mathrm{dt}\, \left(\dot{x}^2 - (f(x))^2\right). \tag{6.34}$$

Here we have used the fact that for $f(x)$ of the form in Eq. (6.28), the last integral vanishes. (For even monomials, on the other hand, this does not vanish giving the contribution due to the instantons which breaks supersymmetry.) Substituting Eqs. (6.32), (6.33) and (6.34) into the generating functional in Eq. (6.31), we find

$$Z = \tilde{N} \int \mathcal{D}\rho\, e^{\frac{i}{2} \int \mathrm{dt}\, \rho^2}. \tag{6.35}$$

In other words, we see that the generating functional for a supersymmetric theory can be redefined to have the form of a free bosonic generating functional. This is known as the Nicolai map (namely, Eq. (6.32)) and generalizes to field theories in higher dimensions as well.

6.3 Shape Invariance

As we have noted earlier, there are only a handful of quantum mechanical systems which can be solved analytically. The solubility of such systems now appears to be related to a special symmetry associated with these systems known as shape invariance. This symmetry is also quite useful in the evaluation of the path integrals for such systems.

Let us consider a one dimensional quantum mechanical system

described by the Hamiltonian

$$H_- = \frac{p^2}{2} + U_-(x). \tag{6.36}$$

If we assume the ground state of the system to have vanishing energy, then we can write the Hamiltonian in Eq. (6.36) also in the factorized form

$$H_- = \bar{Q}Q, \tag{6.37}$$

where

$$Q = \frac{1}{\sqrt{2}}(p - iW(x)),$$

$$\bar{Q} = \frac{1}{\sqrt{2}}(p + iW(x)), \tag{6.38}$$

and we identify

$$U_-(x) = \frac{1}{2}(W^2(x) - W'(x)). \tag{6.39}$$

We note that Q and $\bar{Q}$ are Hermitian conjugates of each other and that given these two operators, we can construct a second Hermitian Hamiltonian of the form

$$H_+ = Q\bar{Q} = \frac{p^2}{2} + U_+(x) = \frac{p^2}{2} + \frac{1}{2}(W^2(x) + W'(x)). \tag{6.40}$$

It now follows that if $|\psi\rangle$ is any eigenstate of the Hamiltonian H_- other than the ground state, namely, if

$$H_-|\psi\rangle = \bar{Q}Q|\psi\rangle = \lambda|\psi\rangle, \qquad \lambda \neq 0, \tag{6.41}$$

then it follows that

$$QH_-|\psi\rangle = \lambda(Q|\psi\rangle)$$

$$\text{or,} \quad Q\bar{Q}(Q|\psi\rangle) = \lambda(Q|\psi\rangle)$$

$$\text{or,} \quad H_+(Q|\psi\rangle) = \lambda(Q|\psi\rangle). \tag{6.42}$$

Namely, we note that the two Hamiltonians, H_- and H_+, are almost isospectral in the sense that they share the same energy spectrum except for the ground state energy of H_-.

The potential, of course, depends on some parameters such as the coupling constants. If the potential of the theory is such that we can write

$$U_+(x, a_0) = U_-(x, a_1) + R(a_1), \qquad (6.43)$$

with $R(a_1)$ a constant and the parameters a_0 and a_1 satisfying a known functional relationship

$$a_1 = f(a_0), \qquad (6.44)$$

then we say that the potential is shape invariant. In such a case, we can write using Eq. (6.43)

$$H_-(a_0) = \bar{Q}(a_0)Q(a_0) = \frac{p^2}{2} + U_-(x, a_0),$$

$$H_+(a_0) = Q(a_0)\bar{Q}(a_0) = \frac{p^2}{2} + U_+(x, a_0)$$

$$= \frac{p^2}{2} + U_-(x, a_1) + R(a_1)$$

$$= H_-(a_1) + R(a_1)$$

$$= \bar{Q}(a_1)Q(a_1) + R(a_1). \qquad (6.45)$$

Since we have assumed that the ground state energy of H_- vanishes and since we know that $H_+(a_0)$ and $H_-(a_0)$ are almost isospectral, it follows now from Eq. (6.45) that the energy value for the first excited state of $H_-(a_0)$ must be

$$E_1 = R(a_1). \qquad (6.46)$$

It is also easy to see now that for a shape invariant potential, we can construct a sequence of Hamiltonians such as

$$H^{(0)} = H_-(a_0),$$

$$H^{(1)} = H_+(a_0) = H_-(a_1) + R(a_1),$$

$$\vdots = \vdots$$

$$H^{(s)} = H_+(a_{s-1}) + \sum_{k=1}^{s-1} R(a_k)$$

$$= H_-(a_s) + \sum_{k=1}^{s} R(a_k). \tag{6.47}$$

Here we have identified

$$a_s = f^s(a_0) = f(f \ldots (f(a_0)) \ldots). \tag{6.48}$$

All the Hamiltonians in Eq. (6.47) will be almost isospectral and from this, with a little bit of analysis, we can determine the energy levels of $H_-(a_0)$ to be

$$E_n = \sum_{k=1}^{n} R(a_k). \tag{6.49}$$

Given the sequence of Hamiltonians in Eq. (6.47) we can write down the relation

$$Q(a_s)\bar{Q}(a_s) = \bar{Q}(a_{s+1})Q(a_{s+1}) + R(a_{s+1})$$

or, $Q(a_s)H^{(s)} = H^{(s+1)}Q(a_s). \tag{6.50}$

This defines a recursion relation between the sequence of Hamiltonians. Furthermore, for $t > 0$, defining the time evolution operator for a particular Hamiltonian $H^{(s)}$ in the sequence to be

$$U^{(s)} = e^{-itH^{(s)}}, \tag{6.51}$$

we note that Eq. (6.50) gives

$$Q(a_s)U^{(s)} = Q(a_s)e^{-itH^{(s)}}$$

$$= e^{-itH^{(s+1)}}Q(a_s) = U^{(s+1)}Q(a_s). \tag{6.52}$$

Similarly, by taking the time derivative of Eq. (6.51), we obtain

$$\frac{\partial U^{(s)}}{\partial t} = -iH^{(s)}e^{-itH^{(s)}}$$

$$= -i\left(\bar{Q}(a_s)Q(a_s) + \sum_{k=1}^{s} R(a_k)\right)e^{-itH^{(s)}}$$

$$= -i\sum_{k=1}^{s} R(a_k)U^{(s)} - i\bar{Q}(a_s)U^{(s+1)}Q(a_s)$$

or, $$\left(\frac{\partial}{\partial t} + i\sum_{k=1}^{s} R(a_k)\right)U^{(s)} = -i\bar{Q}(a_s)U^{(s+1)}Q(a_s). \quad (6.53)$$

The relations in Eqs. (6.52) and (6.53) define recursion relations for the time evolution operator and have the coordinate representation of the form

$$\left(\frac{\partial}{\partial x} + W(x, a_s)\right)U^{(s)}(x, y; t)$$

$$= -\left(\frac{\partial}{\partial y} + W(y, a_s)\right)U^{(s+1)}(x, y; t),$$

$$\left(\frac{\partial}{\partial t} + i\sum_{k=1}^{s} R(a_k)\right)U^{(s)}(x, y; t)$$

$$= -\frac{i}{2}\left(\frac{\partial}{\partial x} - W(x, a_s)\right)\left(\frac{\partial}{\partial y} - W(y, a_s)\right)U^{(s+1)}(x, y; t).$$

$$(6.54)$$

It is clear from this discussion that for a shape invariant potential if one of the Hamiltonians in the sequence coincides with a system which we can solve exactly, then using the recursion relations in Eq. (6.54), we can solve for the time evolution operator of the original system. This will determine the path integral for the system.

6.4   Example

Let us consider a quantum mechanical system with

$$W(x, a_0) = a_0 \tanh x, \qquad a_0 = 1. \tag{6.55}$$

From Eqs. (6.39) and (6.40), we find

$$U_-(x, a_0) = \frac{1}{2} \left(W^2(x, a_0) - W'(x, a_0) \right)$$

$$= \frac{1}{2} - \mathrm{sech}^2 x,$$

$$U_+(x, a_0) = \frac{1}{2} \left(W^2(x, a_0) + W'(x, a_0) \right)$$

$$= \frac{1}{2}. \tag{6.56}$$

In this case, therefore, we can identify

$$a_0 = 1, \quad a_1 = a_0 - 1 = 0, \quad R(a_1) = 0. \tag{6.57}$$

The Hamiltonian for the system, in this case, is

$$H^{(0)} = H_-(a_0) = \frac{p^2}{2} - \mathrm{sech}^2 x + \frac{1}{2}, \tag{6.58}$$

and the next Hamiltonian in the sequence is given by (see Eq. (6.47))

$$H^{(1)} = H_+(a_0) = H_-(a_1) + R(a_1) = \frac{p^2}{2} + \frac{1}{2}. \tag{6.59}$$

This is, of course, the free particle Hamiltonian for which we know the transition amplitude to be (see Eq. (2.47))

$$U^{(1)}(x, y; t) = \frac{1}{\sqrt{2\pi i t}} \, e^{\left(i \frac{(x-y)^2}{2t} - \frac{it}{2} \right)}. \tag{6.60}$$

Substituting Eq. (6.60) into Eq. (6.54), it is easy to see that

$$U^{(0)}(x, y; t) = \frac{1}{2} \mathrm{sech} x \, \mathrm{sech} y$$

$$- \frac{1}{2\pi} \int_{-\infty}^{\infty} dk \, \frac{(ik - \tanh x)(ik + \tanh y)}{1 + k^2} \, e^{\left(ik(x-y) - \frac{it}{2}(k^2 + 1) \right)}.$$

This determines the transition amplitude for the original system.

6.5 Supersymmetry and Singular Potentials

Singular potentials within the context of quantum mechanics are interesting because they remind us of the necessity of regularization even in such simple systems. We know from studies in relativistic quantum field theories that a regularization must always be chosen consistent with the symmetries of the theory under study in order to be able to extract meaningful results. An improper choice of regularization can lead to incorrect conclusions about the theory. In simple quantum mechanical systems, a careful analysis of singular systems using a regularization have led to the working rule that the quantum mechanical wave function must vanish at points where the potential becomes singular. It must be emphasized that such a condition should not be thought of as a boundary condition, rather it is a consistency condition which arises from treating the singular potential in a carefully regularized manner.

In going beyond simple quantum mechanical systems, however, such a working rule should not be blindly imposed which can lead to erroneous conclusions. Singular potentials within the context of supersymmetric quantum mechanical systems provide an excellent example of this and we will discuss this in this section. This will also bring out clearly how choosing a regularization consistent with the symmetries of the system under study is important. We will study a simple model, the super-symmetric "half" oscillator, to bring out these features. Other quantum mechanical systems with a complex singularity structure can be found in the literature.

To understand the supersymmetric "half" oscillator or the supersymmetric oscillator on the half line, it is useful to recapitulate briefly the results of the "half" oscillator. Let us consider a particle moving in the potential

$$V(x) = \begin{cases} \frac{1}{2}(\omega^2 x^2 - \omega), & \text{for } x > 0, \\ \infty, & \text{for } x < 0. \end{cases} \tag{6.61}$$

The spectrum of this potential is quite clear intuitively. Namely, because of the infinite barrier, we expect the wave function to vanish at the origin leading to the conclusion that, of all the solutions of the oscillator on the full line, only the odd solutions (of course, on the "half" line there is no notion of even and odd) would survive in this case. While this is quite obvious, let us analyze the problem systematically for later purpose.

First, let us note that singular potentials are best studied in a regularized manner because this is the only way that appropriate boundary conditions can be determined correctly. Therefore, let us consider the particle moving in the regularized potential

$$
V(x) = \begin{cases} \frac{1}{2}(\omega^2 x^2 - \omega), & \text{for } x > 0 \\ \\ \frac{c^2}{2}, & \text{for } x < 0, \end{cases} \tag{6.62}
$$

with the understanding that the limit $|c| \to \infty$ is to be taken at the end. The Schrödinger equation

$$
\left(-\frac{1}{2} \frac{d^2}{dx^2} + V(x) \right) \psi(x) = \epsilon \psi(x),
$$

can now be solved in the two regions. Since $|c| \to \infty$ at the end, for any finite energy solution, we have the asymptotically damped solution, for $x < 0$,

$$
\psi^{(\text{II})}(x) = A \, e^{(c^2 - 2\epsilon)^{\frac{1}{2}} x}. \tag{6.63}
$$

Since the system no longer has reflection symmetry, the solutions, in the region $x > 0$, cannot be classified into even and odd solutions. Rather, the normalizable (physical) solution would correspond to one which vanishes asymptotically. The solutions of the Schrödinger equation, in the region $x > 0$, are known as the parabolic cylinder functions and the asymptotically damped physical solution is given by

$$
\psi^{(\text{I})}(x) = B \, U(-(\frac{\epsilon}{\omega} + \frac{1}{2}), \sqrt{2\omega}\, x). \tag{6.64}
$$

The parabolic cylinder function, $U(a, x)$, of course, vanishes for large values of x. For small values of x, it satisfies

$$U(a, x) \overset{x \to 0}{\to} \frac{\sqrt{\pi}}{2^{\frac{1}{4}(2a+1)} \Gamma(\frac{3}{4} + \frac{a}{2})},$$

$$U'(a, x) \overset{x \to 0}{\to} -\frac{\sqrt{\pi}}{2^{\frac{1}{4}(2a-1)} \Gamma(\frac{1}{4} + \frac{a}{2})}. \tag{6.65}$$

It is now straightforward to match the solutions in Eqs. (6.63, 6.64) and their first derivatives across the boundary at $x = 0$ and their ratio gives

$$\frac{1}{\sqrt{c^2 - 2\epsilon}} = -\frac{1}{2\sqrt{\omega}} \frac{\Gamma(-\frac{\epsilon}{2\omega})}{\Gamma(-\frac{\epsilon}{2\omega} + \frac{1}{2})}. \tag{6.66}$$

It is clear, then, that as $|c| \to \infty$, this can be satisfied only if

$$-\frac{\epsilon}{2\omega} + \frac{1}{2} \overset{|c| \to \infty}{\longrightarrow} -n, \qquad n = 0, 1, 2, \ldots. \tag{6.67}$$

In other words, when the regularization is removed, the energy levels that survive are the odd ones, namely, (remember that the zero point energy is already subtracted out in (6.61) or (6.62))

$$\epsilon_n = \omega(2n + 1). \tag{6.68}$$

The corresponding physical wave functions are nontrivial only on the half line $x > 0$ and have the form

$$\psi_n(x) = B_n U(-(2n + \frac{3}{2}), \sqrt{2\omega}\, x) = \tilde{B}_n e^{-\frac{1}{2}\omega x^2} H_{2n+1}(\sqrt{\omega}\, x). \tag{6.69}$$

Namely, only the odd Hermite polynomials survive leading to the fact that the wave function vanishes at $x = 0$. Thus, we see that the correct boundary condition naturally arises from regularizing the singular potential and studying the problem systematically.

We now turn to the analysis of the supersymmetric oscillator on the half line. One can define a superpotential

$$W(x) = \begin{cases} -\omega x, & \text{for } x > 0, \\ \infty, & \text{for } x < 0, \end{cases} \tag{6.70}$$

which would, naively, lead to the pair of potentials

$$V_\pm(x) = \begin{cases} \frac{1}{2}(\omega^2 x^2 \mp \omega), & \text{for } x > 0, \\ \infty, & \text{for } x < 0. \end{cases} \tag{6.71}$$

Since, this involves singular potentials, we can study it, as before, by regularizing the singular potentials as

$$V_+(x) = \begin{cases} \frac{1}{2}(\omega^2 x^2 - \omega), & \text{for } x > 0, \\ \frac{c_+^2}{2}, & \text{for } x < 0, \end{cases}$$

$$V_-(x) = \begin{cases} \frac{1}{2}(\omega^2 x^2 + \omega), & \text{for } x > 0, \\ \frac{c_-^2}{2}, & \text{for } x < 0, \end{cases} \tag{6.72}$$

with the understanding that $|c_\pm| \to \infty$ at the end.

The earlier analysis can now be repeated for the pair of potentials in Eq. (6.72). It is straightforward and without going into details, let us simply note the results, namely, that, in this case, we obtain

$$\epsilon_{+,n} = \omega(2n+1), \quad \psi_{+,n}(x) = B_{+,n}\, e^{-\frac{1}{2}\omega x^2}\, H_{2n+1}(\sqrt{\omega}\, x),$$

$$\epsilon_{-,n} = 2\omega(n+1), \quad \psi_{-,n}(x) = B_{-,n}\, e^{-\frac{1}{2}\omega x^2}\, H_{2n+1}(\sqrt{\omega}\, x).$$

$$\tag{6.73}$$

Here $n = 0, 1, 2, \ldots$. There are several things to note from this analysis. First, only the odd Hermite polynomials survive as physical solutions since the wave function has to vanish at the origin. This boundary condition arises from a systematic study involving a regularized potential. Second, the energy levels for the supersymmetric pair of Hamiltonians are no longer degenerate. Furthermore, the state with $\epsilon = 0$ no longer belongs to the Hilbert space (since it corresponds to an even Hermite polynomial solution). This leads to the conventional conclusion that supersymmetry is broken in such a case and let us note, in particular, that in such a case, it would appear that the superpartner states do not belong to the physical Hilbert space (Namely, in this case, the supercharge is an odd operator and hence connects even and odd Hermite polynomials. However, the boundary condition selects out only odd Hermite polynomials as belonging to the physical Hilbert space.).

There is absolutely no doubt that supersymmetry is broken in this case. The question that needs to be addressed is whether it is a dynamical property of the system or an artifact of the regularization (and, hence the boundary condition) used. The answer is quite obvious, namely, that supersymmetry is broken mainly because the regularization (and, therefore, the boundary condition) breaks supersymmetry. In other words, for any value of the regularizing parameters, $c_\pm$ (even if $|c_+| = |c_-|$), the pair of potentials in Eq. (6.72) do not define a supersymmetric system and hence the regularization itself breaks supersymmetry. Consequently, the breaking of supersymmetry that results when the regularization is removed cannot be trusted as a dynamical effect.

6.5.1 *Regularized Superpotential*

Another way to understand this is to note that for a supersymmetric system, it is not the potential that is fundamental. Rather, it is the superpotential which gives the pair of supersymmetric potentials through Riccati type relations. It is natural, therefore, to regularize the superpotential which would automatically lead to a pair of regularized potentials which would be supersymmetric for any value of the regularization parameter. Namely, such a regularization will respect supersymmetry and, with such a regularization, it is, then, meaningful to ask if supersymmetry is broken when the regularization parameter is removed at the end. With this in mind, let us look at the regularized superpotential

$$W(x) = -\omega x \theta(x) + c \theta(-x) . \tag{6.74}$$

Here c is the regularization parameter and we are supposed to take $|c| \to \infty$ at the end. Note that the existence of a normalizable ground state, namely, the form of the superpotential in Eq. (6.70) selects out $c > 0$ (otherwise, the regularization would have broken supersymmetry through instanton effects as we have mentioned earlier).

The regularized superpotential now leads to the pair of regularized supersymmetric potentials

$$V_+(x) = \frac{1}{2}\left[(\omega^2 x^2 - \omega)\theta(x) + c^2\theta(-x) - c\delta(x)\right],$$

$$V_-(x) = \frac{1}{2}\left[(\omega^2 x^2 + \omega)\theta(x) + c^2\theta(-x) + c\delta(x)\right], \qquad (6.75)$$

which are supersymmetric for any $c > 0$. Let us note that the difference here from the earlier case where the potentials were directly regularized (see Eq. (6.72)) lies only in the presence of the $\delta(x)$ terms in the potentials. Consequently, the earlier solutions in the regions $x > 0$ and $x < 0$ continue to hold. However, the matching conditions are now different because of the delta function terms. Carefully matching the wave function and the discontinuity of the first derivative across $x = 0$ for each of the wavefunctions and taking their ratio, we obtain the two conditions

$$\frac{1}{(c^2 - 2\epsilon_+)^{1/2} - c} = -\frac{1}{2\sqrt{\omega}}\frac{\Gamma(-\frac{\epsilon_+}{2\omega})}{\Gamma(-\frac{\epsilon_+}{2\omega} + \frac{1}{2})}, \qquad (6.76)$$

$$\frac{1}{(c^2 - 2\epsilon_-)^{1/2} + c} = -\frac{1}{2\sqrt{\omega}}\frac{\Gamma(-\frac{\epsilon_-}{2\omega} + \frac{1}{2})}{\Gamma(-\frac{\epsilon_-}{2\omega} + 1)}. \qquad (6.77)$$

It is now clear that, as $c \to \infty$, (6.76) and (6.77) give respectively

$$\epsilon_{+,n} = 2\omega n, \quad n = 0, 1, 2, \ldots,$$

$$\epsilon_{-,n} = 2\omega(n+1). \qquad (6.78)$$

The corresponding wave functions, in this case, have the forms

$$\psi_{+,n}(x) = B_{+,n}\, e^{-\frac{1}{2}\omega x^2}\, H_{2n}(\sqrt{\omega}\, x),$$

$$\psi_{-,n}(x) = B_{-,n}\, e^{-\frac{1}{2}\omega x^2}\, H_{2n+1}(\sqrt{\omega}\, x). \qquad (6.79)$$

This is indeed quite interesting for it shows that the spectrum of H_+ contains the ground state with vanishing energy. Furthermore, all the other states of H_+ and H_- are degenerate in energy corresponding to even and odd Hermite polynomials as one would expect from superpartner states. Consequently, it is quite clear that if the supersymmetric "half" oscillator is defined carefully by regularizing the superpotential, then, supersymmetry is manifest in the limit of removing the regularization. This should be contrasted with the general belief that supersymmetry is broken in this system (which is a consequence of using boundary conditions or, equivalently, of regularizing the potentials in a manner which violates supersymmetry).

6.5.2 *Alternate Regularization*

Of course, we should worry at this point as to how regularization independent our conclusion really is. Namely, our results appear to follow from the matching conditions in the presence of singular delta potential terms and, consequently, it is worth investigating whether our conclusions would continue to hold with an alternate regularization of the superpotential which would not introduce such singular terms to the potentials. With this in mind, let us choose a regularized superpotential of the form

$$W(x) = -\omega x \theta(x) - \lambda x \theta(-x). \tag{6.80}$$

Here λ is the regularization parameter and we are to take the limit $|\lambda| \to \infty$ at the end. Once again, we note that, although both signs of λ appear to be allowed, existence of a normalizable ground state would select $\lambda > 0$.

This regularized superpotential would now lead to the pair of supersymmetric potentials of the form

$$V_+(x) = \frac{1}{2} \left[(\omega^2 x^2 - \omega)\theta(x) + (\lambda^2 x^2 - \lambda)\theta(-x) \right],$$

$$V_-(x) = \frac{1}{2} \left[(\omega^2 x^2 + \omega)\theta(x) + (\lambda^2 x^2 + \lambda)\theta(-x) \right]. \tag{6.81}$$

There are no singular delta potential terms with this regularization. In fact, the regularization merely introduces a supersymmetric pair of oscillators for $x < 0$ whose frequency is to be taken to infinity at the end.

Since there is a harmonic oscillator potential for both $x > 0$ and $x < 0$, the solutions are straightforward. They are the parabolic cylinder functions which we have mentioned earlier. Now matching the wave function and its first derivative at $x = 0$ for each of the Hamiltonians and taking the ratio, we obtain

$$\frac{1}{\sqrt{\lambda}} \frac{\Gamma(-\frac{\epsilon_+}{2\lambda})}{\Gamma(-\frac{\epsilon_+}{2\lambda} + \frac{1}{2})} = \frac{1}{\sqrt{\omega}} \frac{\Gamma(-\frac{\epsilon_+}{2\omega})}{\Gamma(-\frac{\epsilon_+}{2\omega} + \frac{1}{2})}, \tag{6.82}$$

$$\frac{1}{\sqrt{\lambda}} \frac{\Gamma(-\frac{\epsilon_-}{2\lambda} + \frac{1}{2})}{\Gamma(-\frac{\epsilon_-}{2\lambda} + 1)} = \frac{1}{\sqrt{\omega}} \frac{\Gamma(-\frac{\epsilon_-}{2\omega} + \frac{1}{2})}{\Gamma(-\frac{\epsilon_+}{2\omega} + 1)}. \tag{6.83}$$

It is clear now that, as $\lambda \to \infty$, Eqs. (6.82) and (6.83) give respectively

$$\epsilon_{+,n} = 2\omega n, \quad n = 0, 1, 2, \ldots,$$

$$\epsilon_{-,n} = 2\omega(n + 1). \tag{6.84}$$

The corresponding wave functions are given by

$$\psi_{+,n}(x) = B_{+,n}\, e^{-\frac{1}{2}\omega x^2}\, H_{2n}(\sqrt{\omega}\, x),$$

$$\psi_{-,n}(x) = B_{-,n}\, e^{-\frac{1}{2}\omega x^2}\, H_{2n+1}(\sqrt{\omega}\, x). \tag{6.85}$$

These are, of course, the same energy levels and wave functions as obtained in Eqs. (6.78) and (6.79) respectively showing again that supersymmetry is manifest. Furthermore, this shows that this conclusion is independent of the regularization used as long as the regularization preserves supersymmetry which can be achieved by properly regularizing the superpotential. This analysis can be carried out in a straight forward manner to more complicated superpotentials and the conclusions hold without any change.

6.6 References

Das, A. and W. J. Huang, Phys. Rev. **D41**, 3241 (1990).
Das, A. and S. Pernice, Nucl. Phys. **B505**, 123 (1997).

Das, A. and S. Pernice, Nucl. Phys. **561**, 357 (1999).
Gendenshtein, L., JETP Lett. **38**, 356 (1983).
Nicolai, H., Nucl. Phys. **B170**, 419 (1980).
Witten, E., Nucl. Phys. **B202**, 253 (1982).

Chapter 7

Semi-Classical Methods

7.1 WKB Approximation

As we know, most quantum systems cannot be solved analytically. In such a case, of course, we use perturbation theory and perturbation theory brings out many interesting properties of the system. However, by definition perturbation theory cannot provide information about nonperturbative aspects of the theory. For example, the Born approximations used in scattering theory give more accurate estimates of the scattering amplitudes as we go to higher orders of perturbation, but we cannot obtain information on the bound states of the system from this analysis. Similarly, even though we may be able to obtain the energy levels and the eigenstates for the motion of a particle in a potential well by using perturbation theory, we will never learn about barrier penetration from such an analysis. These are inherently nonperturbative phenomena.

It is, therefore, useful to develop an approximation scheme which brings out some of these nonperturbative characteristics. WKB is such an approximation scheme. The basic idea behind this is quite simple. Let us assume that we have a particle moving in a complicated potential $V(x)$. Then, the stationary states of the system will satisfy the time-independent Schrödinger equation given by

$$
\left(-\frac{\hbar^2}{2m} \frac{\partial^2}{\partial x^2} + V(x) \right) \psi(x) = E\psi(x) .
\tag{7.1}
$$

Here E is a constant representing the energy of the state. We know that if the potential were a constant, namely, if

$$V(x) = V = \text{constant},$$

then, the solutions of Eq. (7.1) will be plane waves (for $E > V$). Namely,

$$\psi(x) = A\, e^{\pm \frac{i}{\hbar} px}, \tag{7.2}$$

where

$$p = \sqrt{2m(E - V)}.$$

When the potential changes with the coordinate, but changes slowly, then it is easy to convince ourselves that within a region where the potential does not change appreciably, the solutions of Eq. (7.1) can still be written as plane waves of the form of Eq. (7.2) with

$$p(x) = \sqrt{2m(E - V(x))}. \tag{7.3}$$

It is clear, therefore, that we can try a general solution to the time-independent Schrödinger equation of the form

$$\psi(x) = N A(x)\, e^{iB(x)}, \tag{7.4}$$

where N is a normalization constant and furthermore, noting that we can write (for non-negative $A(x)$)

$$A(x) = e^{\ln A(x)},$$

we conclude that the general solution of Eq. (7.1) can be represented as a phase where the phase, in general, is complex. Since the Schrödinger operator depends on $\hbar$, the phase clearly will be a function of $\hbar$. With all these information, let us write the general solution of the time-independent Schrödinger equation to have the form

$$\psi(x) = N\, e^{\frac{i}{\hbar} \phi(x)}. \tag{7.5}$$

If we substitute this wave function back into the Schrödinger equation

(Eq. (7.1)), we obtain

$$\frac{\partial^2 \psi(x)}{\partial x^2} + \frac{2m}{\hbar^2}(E - V(x))\psi(x) = 0$$

or, $\left(-\frac{1}{\hbar^2}(\phi'(x))^2 + \frac{i}{\hbar}\phi''(x) + \frac{2m}{\hbar^2}(E - V(x)) \right) e^{\frac{i}{\hbar}\phi(x)} = 0$

or, $\left(-\frac{1}{\hbar^2}(\phi'(x))^2 + \frac{i}{\hbar}\phi''(x) + \frac{2m}{\hbar^2}(E - V(x)) \right) = 0 .$ (7.6)

So far, everything has been exact. Let us next assume a power series expansion for $\phi(x)$ of the form

$$\phi(x) = \phi_0(x) + \hbar\phi_1(x) + \hbar^2\phi_2(x) + \cdots . \tag{7.7}$$

It is clear, then, that $\phi_0(x)$ will correspond to the classical phase since that is what will survive in the limit $\hbar \to 0$. Other terms in the series, therefore, represent quantum corrections to the classical phase.

Substituting the power series back into Eq. (7.6), we obtain

$$\frac{1}{\hbar^2} \left[-(\phi_0'(x) + \hbar\phi_1'(x) + \cdots)^2 + 2m(E - V(x)) \right]$$

$$+ \frac{i}{\hbar} \left(\phi_0''(x) + \hbar\phi_1''(x) + \cdots \right) = 0$$

or, $\frac{1}{\hbar^2} \left(-(\phi_0'(x))^2 + 2m(E - V(x)) \right)$

$$+ \frac{1}{\hbar} \left(i\phi_0''(x) - 2\phi_0'(x)\phi_1'(x) \right) + O(\hbar^0) = 0 . \tag{7.8}$$

For this to be true, the coefficients of the individual terms in Eq. (7.8) must be zero. Equating the coefficient of the $\frac{1}{\hbar^2}$ term to zero, we obtain

$$-(\phi_0'(x))^2 + 2m(E - V(x)) = 0$$

or, $(\phi_0'(x))^2 = 2m(E - V(x)) = p^2(x)$

or, $\phi_0'(x) = \pm p(x)$

$$\text{or,} \quad \phi_0(x) = \pm \int_{x_0}^{x} dx' \, p(x') \,. \tag{7.9}$$

Here $p(x)$ is the momentum of the particle at the point x defined by Eq. (7.3), corresponding to motion with energy E. Furthermore, let us note that even though both the signs of the solution are allowed in Eq. (7.9), consistency of the subsequent relations will pick out only the positive sign. This, therefore, determines the classical phase to be

$$\phi_0(x) = \int_{x_0}^{x} dx' \, p(x') \,. \tag{7.10}$$

If we keep only the leading order term in the expansion of $\phi(x)$, then the wave function would have the form

$$\psi(x) = N \, e^{\frac{i}{\hbar} \phi_0(x)} = N \, e^{\frac{i}{\hbar} \int_{x_0}^{x} dx' \, p(x')} \,. \tag{7.11}$$

The time-dependent stationary wave function, in this case, would be given by

$$\psi(x,t) = e^{-\frac{i}{\hbar} Et} \psi(x)$$

$$= N \, e^{-\frac{i}{\hbar} Et + \frac{i}{\hbar} \int_{x_0}^{x} dx' \, p(x')}$$

$$= N \, e^{\frac{i}{\hbar} \int_0^t dt' \, (p\dot{x} - E)}$$

$$= N \, e^{\frac{i}{\hbar} S[x_{\text{cl}}]} \,. \tag{7.12}$$

This is exactly what we would have expected in the classical limit.

Let us next include the first order correction to the phase. Setting the coefficient of the $\frac{1}{\hbar}$ term in Eq. (7.8) to zero, we obtain

$$i\phi_0''(x) - 2\phi_0'(x)\phi_1'(x) = 0$$

$$\text{or,} \quad \phi_1'(x) = \frac{i}{2} \frac{\phi_0''(x)}{\phi_0'(x)} = \frac{i}{2} \left(\ln \phi_0'(x) \right)'$$

$$\text{or,} \quad \phi_1(x) = \frac{i}{2} \ln \phi_0'(x) = \frac{i}{2} \ln p(x) \,. \tag{7.13}$$

The constant of integration in Eq. (7.13) can be absorbed into the normalization constant N of the wave function. We also note here that this selects out the positive root for $\phi_0(x)$ in Eq. (7.9).

Thus, keeping up to the first order correction to the classical phase, we can write the wave function to be

$$\psi(x) \simeq N\, e^{\frac{i}{\hbar}(\phi_0(x)+\hbar\phi_1(x))}$$

$$= N\, e^{\frac{i}{\hbar}(\int^x \mathrm{d}x'\, p(x')+\frac{i\hbar}{2}\ln p(x))}$$

$$= \frac{N}{\sqrt{p(x)}}\, e^{\frac{i}{\hbar}\int^x \mathrm{d}x'\, p(x')}\,. \tag{7.14}$$

This is known as the WKB approximation for the quantum mechanical wave function of the system. This approximation, clearly, breaks down for small $p(x)$ and in particular when

$$p(x) = 0\,. \tag{7.15}$$

Namely, at the classical turning points, the classical momentum vanishes and, consequently, in these regions the WKB approximation breaks down and we must examine the Schrödinger equation and its solutions more carefully.

From the form of the WKB wave function in Eq. (7.14), we note that

$$\psi^*(x)\psi(x) \propto \frac{1}{p(x)}\,. \tag{7.16}$$

In other words, the probability density, in this case, is inversely proportional to the momentum or the velocity. This is, of course, what we would expect from classical considerations alone. Namely, classically, we would expect qualitatively that the system is more likely to be found at points where its velocity is smaller. Thus, the WKB approximation gives us a quantum wave function which retains some of the classical properties. It is for this reason that the WKB approximation is often also called the semi-classical approximation.

Let us consider a particle moving in a potential of the form shown above. Then, the normalization constant for the WKB wave function can be determined approximately in the following way. Since the wave function damps outside the well, we can write the normalization condition to be approximately

$$\int_{-\infty}^{\infty} dx\, \psi^*(x)\psi(x) \simeq \int_{x_a}^{x_b} dx\, \psi^*(x)\psi(x)$$

$$= |N|^2 \int_{x_a}^{x_b} \frac{dx}{p(x)} = 1. \qquad (7.17)$$

Recalling that the classical period of oscillation is given by

$$T = 2 \int_{x_a}^{x_b} \frac{dx}{v(x)} = 2m \int_{x_a}^{x_b} \frac{dx}{p(x)}, \qquad (7.18)$$

we obtain

$$|N|^2 \frac{T}{2m} = 1$$

$$\text{or,} \quad N = N^* = \sqrt{\frac{2m}{T}}. \qquad (7.19)$$

Therefore, we can write the normalized WKB wave function to be

$$\psi(x) \simeq \sqrt{\frac{2m}{Tp(x)}}\, e^{\frac{i}{\hbar}\int^x dx'\, p(x')} = \sqrt{\frac{\omega}{\pi v(x)}}\, e^{\frac{i}{\hbar}\int^x dx'\, p(x')}, \qquad (7.20)$$

where $\omega = \frac{2\pi}{T}$ denotes the classical angular frequency of motion.

7.2 Saddle Point Method

Let us consider an integral of the form

$$I = \int_{-\infty}^{\infty} dx \, e^{\frac{1}{a} f(x)}, \qquad (7.21)$$

where a is a very small constant. Furthermore, let us assume that the function $f(x)$ has an extremum at $x = x_0$ which is a maximum. In other words, we are assuming that

$$f'(x)\big|_{x=x_0} = 0,$$

$$f''(x)\big|_{x=x_0} < 0. \qquad (7.22)$$

We can now expand the function around this extremum as

$$f(x) = f(x_0) + \frac{1}{2}(x - x_0)^2 f''(x_0) + O((x - x_0)^3). \qquad (7.23)$$

Substituting this back into the integral in Eq. (7.21), we obtain

$$I = \int_{-\infty}^{\infty} dx \, e^{[\frac{1}{a}(f(x_0) + \frac{1}{2}(x-x_0)^2 f''(x_0) + O((x-x_0)^3))]}$$

$$= e^{\frac{1}{a} f(x_0)} \int_{-\infty}^{\infty} dx \, e^{(-\frac{1}{2a}(x-x_0)^2 |f''(x_0)| + O((x-x_0)^3))}$$

$$= e^{\frac{1}{a} f(x_0)} \int_{-\infty}^{\infty} dy \, \sqrt{a} \, e^{(-\frac{1}{2} y^2 |f''(x_0)| + O(\sqrt{a} y^3))}$$

$$\simeq e^{\frac{1}{a} f(x_0)} \sqrt{a} \, \sqrt{\frac{2\pi}{|f''(x_0)|}}$$

$$= \sqrt{\frac{2\pi a}{|f''(x_0)|}} \, e^{\frac{1}{a} f(x_0)}. \qquad (7.24)$$

It is easy to see that the terms we neglected are higher order in a and, therefore, this is the most dominant contribution to the integral.

Note from Eq. (7.23) that if we consider the function in the complex x-plane, then along the imaginary axis it has a minimum at the extremal point. Therefore, the extremal point is really a saddle point in the complex x-plane. Hence the name, the saddle point method. Let us also note that the direction of our integration has been along the direction of steepest descent (along the real axis) and, therefore, this method of evaluating the integral is also referred to as the method of steepest descent. It is worth emphasizing here that if the function in the exponent has several extrema, then the value of the integral will approximately equal the sum of the contributions around each of the extrema.

This method is quite useful in obtaining approximate values of very complicated integrals. As an example, let us analyze the Gamma function for large values of the argument. The Gamma function has the integral representation given by

$$\Gamma(n+1) = n! = \int_0^\infty dx\, x^n\, e^{-x}$$

$$= \int_0^\infty dx\, e^{(-x+n \ln x)}$$

$$= \int_0^\infty dx\, e^{n(\ln x - \frac{x}{n})}. \tag{7.25}$$

Let us next assume that n is very large. This, then, would correspond to $\frac{1}{a}$ in our previous discussion in Eq. (7.21). In the present case,

$$f(x) = \ln x - \frac{x}{n},$$

$$f'(x) = \frac{1}{x} - \frac{1}{n}. \tag{7.26}$$

Requiring the first derivative to vanish gives

$$x_0 = n,$$

$$f(x_0) = \ln x_0 - \frac{x_0}{n} = (\ln n - 1). \qquad (7.27)$$

Furthermore,

$$f''(x) = -\frac{1}{x^2}, \qquad (7.28)$$

and, therefore, we have

$$f''(x_0) = -\frac{1}{n^2} < 0. \qquad (7.29)$$

Therefore, $x_0 = n$ is a maximum of the function and, in this case, is the only extremum. Thus, for large n, we can write

$$\Gamma(n+1) \simeq e^{nf(x_0)} \int_0^\infty dx \, e^{n(-\frac{1}{2n^2}(x-x_0)^2)}$$

$$= e^{nf(x_0)} \int_0^\infty dx \, e^{-\frac{1}{2n}(x-x_0)^2}$$

$$\simeq e^{nf(x_0)} \int_{-\infty}^\infty dx \, e^{-\frac{1}{2n}(x-x_0)^2}$$

$$= \sqrt{2\pi n} \, e^{nf(x_0)}$$

$$= \sqrt{2\pi n} \, e^{n(\ln n - 1)}$$

$$\text{or,} \quad n! \simeq \sqrt{2\pi n} \left(\frac{n}{e}\right)^n, \quad \text{for large } n. \qquad (7.30)$$

(Since $x_0 = n$ is large and positive, we have extended the integration to the negative axis in the intermediate step because the contribution from this region is negligible.) This is known as Stirling's approximation and holds when n is large. Sometimes, this is also written in the form

$$\ln n! \simeq n(\ln n - 1), \quad \text{for large } n. \qquad (7.31)$$

7.3 Semi-Classical Methods in Path Integrals

The saddle point method or the method of steepest descent can be applied to path integrals as well. Note that so far we have only evaluated path integrals which involve quadratic actions. But any realistic theory will involve interactions which are inherently nonlinear. The path integral cannot always be evaluated exactly for such systems and the method of steepest descent gives rise to a very useful approximation in such a case.

Let us consider a general action $S[x]$ which is not necessarily quadratic. The transition amplitude associated with this action, as we have seen, is given by (see Eq. (2.28))

$$\langle x_f, t_f | x_i, t_i \rangle = N \int \mathcal{D}x \, e^{\frac{i}{\hbar} S[x]} . \tag{7.32}$$

Since $\hbar$ is a small parameter, this integral is similar to the one in Eq. (7.21), but is not exactly in the same form. Namely, this integral is oscillatory. However, we know that, in such a case, we can rotate to the Euclidean space as we have discussed earlier and the integrand becomes well behaved. We will continue to use the real time description keeping in mind the fact that in all our discussions, we are assuming that the actual calculations are always done in the Euclidean space and then the results are rotated back to Minkowski space.

Let us recall that the classical trajectory satisfies the equation (see Eq. (1.28) or (3.9))

$$\left. \frac{\delta S[x]}{\delta x(t)} \right|_{x = x_{\text{cl}}} = 0 .$$

Therefore, it provides an extremum of the exponent in the path integral. Furthermore, the action is a minimum for the classical trajectory. Thus, we can expand the action around the classical trajectory. Namely, let us define

$$x(t) = x_{\text{cl}}(t) + \eta(t) . \tag{7.33}$$

Then, we have

$$S[x] = S[x_{\mathrm{cl}} + \eta]$$

$$= S[x_{\mathrm{cl}}] + \frac{1}{2} \iint dt_1 dt_2 \eta(t_1) \frac{\delta^2 S[x_{\mathrm{cl}}]}{\delta x_{\mathrm{cl}}(t_1)\delta x_{\mathrm{cl}}(t_2)} \eta(t_2) + O(\eta^3).$$

(7.34)

Substituting this back into the transition amplitude in Eq. (7.32), we obtain

$$\langle x_f, t_f | x_i, t_i \rangle$$

$$= N \int \mathcal{D}\eta \, e^{\frac{i}{\hbar}\left(S[x_{\mathrm{cl}}] + \frac{1}{2} \iint dt_1 dt_2 \eta(t_1) \frac{\delta^2 S[x_{\mathrm{cl}}]}{\delta x_{\mathrm{cl}}(t_1)\delta x_{\mathrm{cl}}(t_2)} \eta(t_2) + O(\eta^3)\right)}$$

$$\simeq N \, e^{\frac{i}{\hbar}S[x_{\mathrm{cl}}]} \int \mathcal{D}\eta \, e^{\frac{i}{2\hbar} \iint dt_1 dt_2 \eta(t_1) \frac{\delta^2 S[x_{\mathrm{cl}}]}{\delta x_{\mathrm{cl}}(t_1)\delta x_{\mathrm{cl}}(t_2)} \eta(t_2)}$$

$$= \frac{N}{\sqrt{\det\left(\frac{1}{\hbar}\frac{\delta^2 S[x_{\mathrm{cl}}]}{\delta x_{\mathrm{cl}}(t_1)\delta x_{\mathrm{cl}}(t_2)}\right)}} e^{\frac{i}{\hbar}S[x_{\mathrm{cl}}]},$$

(7.35)

where we have used the result of Eq. (4.2). (We are also keeping the $\hbar$ term explicitly to bring out the quantum nature of the calculations.) Clearly, the saddle point method breaks down if

$$\det \frac{\delta^2 S[x_{\mathrm{cl}}]}{\delta x_{\mathrm{cl}}(t_1)\delta x_{\mathrm{cl}}(t_2)} = 0.$$

Normally, this happens when there is some symmetry or its spontaneous breakdown occurring in the theory. When this happens, the path integral has to be evaluated more carefully using the method of collective coordinates as we will discuss in the next chapter.

We note here that the form of the transition amplitude in the saddle point approximation in Eq. (7.35) is surprisingly similar in form to the WKB wave function in Eq. (7.20). (Recall from Eq. (1.35) that the transition amplitude is a Schrödinger wave function for a delta function source.) In fact, the phases are identical (see Eq. (7.12)).

It is the multiplying factors that we readily do not see to be comparable. We will describe below, without going into too much detail, how we can, in fact, relate the two multiplying factors as well. Let us note that a general action has the form

$$S[x] = \int_{t_i}^{t_f} dt \left(\frac{1}{2} m \dot{x}^2 - V(x) \right). \tag{7.36}$$

For simplicity, we will choose $t_i = -\frac{T}{2}$ and $t_f = \frac{T}{2}$ which will also be useful later. In this case, we note that

$$\frac{\delta^2 S[x_{cl}]}{\delta x_{cl}(t_1) \delta x_{cl}(t_2)} = - \left(m \frac{d^2}{dt_1^2} + V''(x_{cl}) \right) \delta(t_1 - t_2). \tag{7.37}$$

Therefore, we see that

$$\det \left(\frac{1}{\hbar} \frac{\delta^2 S[x_{cl}]}{\delta x_{cl}(t_1) \delta x_{cl}(t_2)} \right) \propto \det \left(\frac{1}{\hbar} \left(m \frac{d^2}{dt^2} + V''(x_{cl}) \right) \right). \tag{7.38}$$

Thus, we are interested in evaluating determinants of the form

$$\det \left(\frac{1}{\hbar} \left(m \frac{d^2}{dt^2} + W(x) \right) \right),$$

in the space of functions where $\eta(\frac{T}{2}) = 0 = \eta(-\frac{T}{2})$. We have already evaluated such determinants earlier in connection with the harmonic oscillator (see Eq. (4.10)). Let us recall here some general results which hold for determinants of operators containing bounded potentials. With a little bit of analysis, it is possible to show that, if

$$\frac{1}{\hbar} \left(m \frac{d^2}{dt^2} + W(x) \right) \psi_W^{(\lambda)} = \lambda \psi_W^{(\lambda)}, \tag{7.39}$$

with the initial value conditions $\psi_W^\lambda(-\frac{T}{2}) = 0$ and $\dot{\psi}_W^\lambda(-\frac{T}{2}) = 1$, then

$$\frac{\det(\frac{1}{\hbar}(m \frac{d^2}{dt^2} + W_1(x)) - \lambda)}{\det(\frac{1}{\hbar}(m \frac{d^2}{dt^2} + W_2(x)) - \lambda)} = \frac{\psi_{W_1}^{(\lambda)}(\frac{T}{2})}{\psi_{W_2}^{(\lambda)}(\frac{T}{2})}, \tag{7.40}$$

where $W_1(x)$ and $W_2(x)$ are two bounded potentials. We note that for λ coinciding with an eigenvalue of one of the operators, the left hand side will have a zero or a pole. But so will the right hand side for the same value of λ because in such a case $\psi^{(\lambda)}$ would correspond to an energy eigenfunction satisfying the boundary conditions at the end points. Since both the left and the right hand side of the above equation are entire functions of λ with identical zeroes and poles, they must be equal. It follows from this result that

$$\frac{\det(\frac{1}{\hbar}(m\frac{d^2}{dt^2} + W_1(x)))}{\psi_{W_1}^{(0)}(\frac{T}{2})} = \frac{\det(\frac{1}{\hbar}(m\frac{d^2}{dt^2} + W_2(x)))}{\psi_{W_2}^{(0)}(\frac{T}{2})} = \text{constant}.$$

(7.41)

That is, this ratio is independent of the particular form of the potential $W(x)$ and can be used to define the normalization constant in the path integral. We will define a particular normalization later. But, for the moment, let us use this result to write the transition amplitude in Eq. (7.35) in the form

$$\langle x_f, t_f | x_i, t_i \rangle = \frac{N}{\sqrt{\det\left(\frac{1}{\hbar}\left(m\frac{d^2}{dt^2} + V''(x_{\text{cl}})\right)\right)}} e^{\frac{i}{\hbar}S[x_{\text{cl}}]}$$

$$= \frac{N}{\sqrt{\psi_{V''}^{(0)}(\frac{T}{2})}} e^{\frac{i}{\hbar}S[x_{\text{cl}}]}.$$

(7.42)

Let us note here that by definition (see Eq. (7.39)), $\psi_{V''}^{(0)}(t)$ satisfies the equation (it is an eigenstate with zero eigenvalue)

$$\left(m\frac{d^2}{dt^2} + V''(x_{\text{cl}})\right)\psi_{V''}^{(0)}(t) = 0.$$

(7.43)

We note here that the classical equations (Euler-Lagrange equations) following from our action have the form (see Eq. (1.28))

$$m\frac{d^2 x_{\text{cl}}}{dt^2} + V'(x_{\text{cl}}) = 0.$$

(7.44)

It follows from this that

$$\frac{d}{dt}\left(m\frac{d^2 x_{cl}}{dt^2} + V'(x_{cl})\right) = 0$$

$$\text{or, } m\frac{d^2}{dt^2}\left(\frac{dx_{cl}}{dt}\right) + V''(x_{cl})\frac{dx_{cl}}{dt} = 0$$

$$\text{or, } \left(m\frac{d^2}{dt^2} + V''(x_{cl})\right)\frac{dx_{cl}}{dt} = 0. \tag{7.45}$$

Comparing with Eq. (7.43), we readily identify that

$$\psi_{V''}^{(0)}(t) \propto \frac{dx_{cl}}{dt} \propto p(x_{cl}). \tag{7.46}$$

Consequently, we recognize that we can write the transition amplitude in Eq. (7.42) also in the form

$$\langle x_f, \frac{T}{2} | x_i, -\frac{T}{2} \rangle = \frac{N}{\sqrt{p(x_f)}} e^{\frac{i}{\hbar}S[x_{cl}]}. \tag{7.47}$$

The correspondence with the WKB wave function in Eq. (7.20) is now complete and we recognize that the method of steepest descent merely gives the WKB approximation.

7.4 Double Well Potential

As an application of the WKB method, let us consider a particle moving in one-dimension in an anharmonic potential of the form

$$V(x) = \frac{g^2}{8}(x^2 - a^2)^2, \tag{7.48}$$

where g and a are constants. Consequently, the action has the form

$$S[x] = \int dt \left(\frac{1}{2}m\dot{x}^2 - \frac{g^2}{8}(x^2 - a^2)^2\right). \tag{7.49}$$

This potential, when plotted, has the shape of a double well.

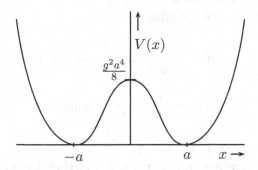

This is a very interesting potential and shows up in all branches of physics in different forms. Note that it is an even potential with a local maximum at the origin. The two minima of the potential are symmetrically located at

$$x = \pm a.\qquad(7.50)$$

The height of the potential at the origin is given by

$$V(x = 0) = \frac{g^2 a^4}{8}.\qquad(7.51)$$

Let us also define here for later use

$$V''(x = \pm a) = g^2 a^2 = m\omega^2,\qquad(7.52)$$

where we can identify ω with the natural angular frequency of harmonic oscillations near the minima. We note from Eq. (7.51) that for infinitely large coupling, the potential separates into two symmetrical wells with an infinite barrier. The motion of the particle is easy to analyze in this case. Each well has quantized levels of energy and if the particle is in one well, then it stays there forever. Namely, there will be no tunneling from one well to the other. Furthermore, from the symmetry of the problem at hand (namely, $x \leftrightarrow -x$), we conclude that both the wells in the present case will have degenerate energy levels. Thus, if $\psi_0(x)$ denotes the ground state wave function of the well in the positive x-axis with energy E_0, then $\psi_0(-x)$ will describe the ground state wave function of the left well with the same energy. In fact, any linear combination of the two wave functions and, in particular, the combinations

$$\psi_1(x) = \frac{1}{\sqrt{2}}(\psi_0(x) + \psi_0(-x)),$$

$$\psi_2(x) = \frac{1}{\sqrt{2}}(\psi_0(x) - \psi_0(-x)), \tag{7.53}$$

will also be degenerate in energy.

When the coupling constant g is finite, then the potential barrier will be finite. Consequently, the particle initially confined to one well can tunnel into the other well and the states of the two wells will mix. The symmetry of the system (Hamiltonian) still dictates that the eigenstates of the Hamiltonian can only be even and odd linear combinations as described above. They will, however, not be degenerate in energy any longer because of tunneling and we wish to calculate the splitting in the energy levels due to tunneling using the WKB approximation.

Let us note that $\psi_1(x)$ is a symmetric wave function whereas $\psi_2(x)$ is anti-symmetric. Consequently, it is obvious that $\psi_1(x)$ would represent the ground state of the system (after taking tunneling into account). Let us write down the time-independent Schrödinger equations that various wave functions satisfy.

$$\frac{\mathrm{d}^2\psi_0(x)}{\mathrm{d}x^2} + \frac{2m}{\hbar^2}(E_0 - V(x))\psi_0(x) = 0, \tag{7.54}$$

$$\frac{\mathrm{d}^2\psi_1(x)}{\mathrm{d}x^2} + \frac{2m}{\hbar^2}(E_1 - V(x))\psi_1(x) = 0, \tag{7.55}$$

$$\frac{\mathrm{d}^2\psi_2(x)}{\mathrm{d}x^2} + \frac{2m}{\hbar^2}(E_2 - V(x))\psi_2(x) = 0. \tag{7.56}$$

Furthermore, let us assume, for simplicity, that the wave functions are all real. If $\psi_0(x)$ denotes the wave function in the well I (that is, the well on the right), then, we can easily see that its value will be vanishingly small in the well II, or the well on the left. Thus, we can normalize the wave function as

$$\int_0^\infty dx\, \psi_0^2(x) \simeq 1.$$ (7.57)

Let us note similarly that if $\psi_0(-x)$ denotes the wave function in the well II, then it would have vanishingly small value in well I. Consequently, a product such as $\psi_0(x)\psi_0(-x)$ will be negligible everywhere. We see from Eq. (7.53), therefore, that

$$\psi_1(x)\psi_0(x) \simeq \frac{1}{\sqrt{2}}\, \psi_0^2(x),$$ (7.58)

so that

$$\int_0^\infty dx\, \psi_1(x)\psi_0(x) \simeq \frac{1}{\sqrt{2}} \int_0^\infty dx\, \psi_0^2(x) \simeq \frac{1}{\sqrt{2}}.$$ (7.59)

Let us next multiply Eq. (7.54) by $\psi_1(x)$ and Eq. (7.55) by $\psi_0(x)$ and subtract one from the other. This gives

$$\psi_1(x)\frac{d^2\psi_0(x)}{dx^2} - \psi_0(x)\frac{d^2\psi_1(x)}{dx^2} + \frac{2m}{\hbar^2}(E_0 - E_1)\psi_1(x)\psi_0(x) = 0.$$ (7.60)

Integrating this equation and using Eq. (7.59), we obtain

$$\int_0^\infty dx\left(\psi_1(x)\frac{d^2\psi_0(x)}{dx^2} - \psi_0(x)\frac{d^2\psi_1(x)}{dx^2}\right)$$

$$+ \frac{2m}{\hbar^2}(E_0 - E_1)\int_0^\infty dx\, \psi_1(x)\psi_0(x) = 0$$

$$\text{or, } \frac{2m}{\hbar^2}(E_0 - E_1)\frac{1}{\sqrt{2}} = -\left.(\psi_1(x)\psi_0'(x) - \psi_0(x)\psi_1'(x))\right|_0^\infty$$

$$= \psi_1(0)\psi_0'(0) - \psi_0(0)\psi_1'(0)$$

$$\text{or, } E_0 - E_1 = \frac{\hbar^2}{m\sqrt{2}}\left(\psi_1(0)\psi_0'(0) - \psi_0(0)\psi_1'(0)\right).$$ (7.61)

Let us note that since

$$\psi_1(x) = \frac{1}{\sqrt{2}}(\psi_0(x) + \psi_0(-x)),$$

$$\psi_1(0) = \sqrt{2}\,\psi_0(0),$$

$$\psi_1'(0) = 0. \tag{7.62}$$

Substituting this back into Eq. (7.61), we obtain

$$E_0 - E_1 = \frac{\hbar^2}{m\sqrt{2}}\,\sqrt{2}\psi_0(0)\psi_0'(0) = \frac{\hbar^2}{m}\psi_0(0)\psi_0'(0). \tag{7.63}$$

It is here that we would like to use the WKB approximation. We recall from Eq. (7.20) that we can write the WKB wave function as

$$\psi_0(0) = \sqrt{\frac{\omega}{\pi v(0)}}\,e^{-\frac{1}{\hbar}\int_0^a dx\,|p(x)|},$$

$$\psi_0'(0) \simeq \frac{mv(0)}{\hbar}\,\psi_0(0). \tag{7.64}$$

Here, we note that

$$v(0) = \sqrt{\frac{2}{m}(V(0) - E_0)}, \quad V(0) \gg E_0. \tag{7.65}$$

Putting these back, we obtain

$$E_0 - E_1 = \frac{\hbar^2}{m}\frac{mv(0)}{\hbar}(\psi_0(0))^2$$

$$= \hbar v(0)\frac{\omega}{\pi v(0)}\,e^{-\frac{2}{\hbar}\int_0^a dx\,|p(x)|}$$

$$= \frac{\hbar\omega}{\pi}\,e^{-\frac{2}{\hbar}\int_0^a dx\,|p(x)|}. \tag{7.66}$$

This is the splitting in the energy level of the true ground state from the case of the infinite well. We can similarly show that

$$E_2 - E_0 = \frac{\hbar\omega}{\pi} e^{-\frac{2}{\hbar} \int_0^a dx \, |p(x)|} , \qquad (7.67)$$

and, consequently,

$$E_2 - E_1 = \frac{2\hbar\omega}{\pi} e^{-\frac{2}{\hbar} \int_0^a dx \, |p(x)|} , \qquad (7.68)$$

which gives the splitting between the two degenerate levels in this approximation.

The damping exponential in Eq. (7.68) reflects the effects of tunneling. In fact, note that because of the reflection symmetry in the problem, we can write

$$e^{-\frac{2}{\hbar} \int_0^a dx \, |p(x)|} = e^{-\frac{1}{\hbar} \int_{-a}^a dx \, |p(x)|} , \qquad (7.69)$$

and this gives the coefficient for tunneling from the minimum at $x = -a$ to the one at $x = a$. This, of course, assumes that the particle under consideration has vanishing energy. In reality, however, we know that the quantum mechanical ground state energy is nonzero in general. In fact, if we approximate the potential near each of the minima by a harmonic oscillator potential, then we can identify the ground state energy of the system with

$$E_0 = \frac{\hbar\omega}{2} . \qquad (7.70)$$

This would, then, imply that the turning points for motion in both the wells, in this case, are given by

$$\frac{1}{2} m\omega^2 (x - a)^2 = E_0 = \frac{\hbar\omega}{2}$$

$$\text{or, } x - a = \pm\sqrt{\frac{\hbar}{m\omega}}$$

$$\text{or, } x_I = a \pm \sqrt{\frac{\hbar}{m\omega}} , \qquad (7.71)$$

and, similarly,

$$\frac{1}{2}\,m\omega^2(x+a)^2 = E_0 = \frac{\hbar\omega}{2}$$

$$\text{or, } x+a = \pm\sqrt{\frac{\hbar}{m\omega}}$$

$$\text{or, } x_{\text{II}} = -a \pm \sqrt{\frac{\hbar}{m\omega}}\,. \tag{7.72}$$

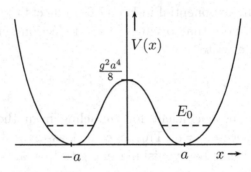

The tunneling from one well to the other in this case, therefore, would correspond to tunneling from $-a + \sqrt{\frac{\hbar}{m\omega}}$ to $a - \sqrt{\frac{\hbar}{m\omega}}$. Correspondingly, for a more accurate estimation of the splitting in the energy levels, we should replace the exponential in Eq. (7.68) by

$$e^{-\frac{2}{\hbar}\int_0^a \mathrm{d}x\,|p(x)|} \longrightarrow e^{-\frac{2}{\hbar}\int_0^{a-\sqrt{\frac{\hbar}{m\omega}}} \mathrm{d}x\,|p(x)|}\,. \tag{7.73}$$

Furthermore, recalling that (see Eq. (7.3))

$$|p(x)| = \sqrt{2m(V(x) - E_0)}\,,$$

we can evaluate the exponent in a straightforward manner as

$$\int_0^{a-\sqrt{\frac{\hbar}{m\omega}}} dx\, |p(x)|$$

$$= \int_0^{a-\sqrt{\frac{\hbar}{m\omega}}} dx\, \sqrt{2m\left(\frac{g^2}{8}(x^2-a^2)^2 - \frac{\hbar\omega}{2}\right)}$$

$$= \int_0^{a-\sqrt{\frac{\hbar}{m\omega}}} dx\, \frac{m\omega(a^2-x^2)}{2a}\left(1 - \frac{4a^2\hbar}{m\omega(a^2-x^2)^2}\right)^{\frac{1}{2}}$$

$$\simeq \int_0^{a-\sqrt{\frac{\hbar}{m\omega}}} dx\, \left(\frac{m\omega(a^2-x^2)}{2a} - \frac{a\hbar}{(a^2-x^2)}\right). \tag{7.74}$$

This integral can be trivially done and has the value

$$\int_0^{a-\sqrt{\frac{\hbar}{m\omega}}} dx\, |p(x)|$$

$$\simeq \frac{m\omega a^2}{3} - \frac{\hbar}{2} + \frac{\hbar}{2}\ln\sqrt{\frac{\hbar}{m\omega a^2}} - \frac{\hbar}{2}\ln\left(2 - \sqrt{\frac{\hbar}{m\omega a^2}}\right) + O(\hbar^{\frac{3}{2}})$$

$$= \frac{1}{2}S_0 + \frac{\hbar}{2}\ln\sqrt{\frac{\hbar}{4m\omega a^2 e^2}} + O(\hbar^{\frac{3}{2}}). \tag{7.75}$$

Here we have defined

$$S_0 = \frac{2m\omega a^2}{3}. \tag{7.76}$$

Substituting Eqs. (7.73) and (7.76) into Eq. (7.68), we obtain the splitting in the energy levels to be

$$E_2 - E_1 \simeq \frac{2\hbar\omega}{\pi}\sqrt{\frac{4m\omega a^2 e^2}{\hbar}}\, e^{-\frac{1}{\hbar}S_0}$$

$$= \frac{4e}{\pi}\sqrt{m\hbar}\omega^{\frac{3}{2}}a\, e^{-\frac{1}{\hbar}S_0}. \tag{7.77}$$

In the next chapter, we will calculate this energy splitting using the path integrals and compare the two results.

7.5   References

Coleman, S., "The Uses of Instantons", Erice Lectures, 1977.

Landau, L. D. and L. M. Lifshitz, "Nonrelativistic Quantum Mechanics", Pergamon Press.

Migdal, A. B. and V. Krainov, "Approximation Methods in Quantum Mechanics", Benjamin Publishing.

Sakita, B., "Quantum Theory of Many Variable Systems and Fields", World Scientific Publishing.

Chapter 8

Path Integral for the Double Well

8.1 Instantons

Let us next try to do the path integral for the double well potential. We recall from Eq. (2.28) that the transition amplitude is defined as

$$\langle x_f, \frac{T}{2} | x_i, -\frac{T}{2} \rangle = \langle x_f | e^{-\frac{i}{\hbar} HT} | x_i \rangle = N \int \mathcal{D}x \, e^{\frac{i}{\hbar} S[x]}, \qquad (8.1)$$

where for a double well potential (see Eq. (7.49)) the action is given by

$$S[x] = \int_{-\frac{T}{2}}^{\frac{T}{2}} dt \, \left(\frac{1}{2} m \dot{x}^2 - V(x) \right)$$

$$= \int_{-\frac{T}{2}}^{\frac{T}{2}} dt \, \left(\frac{1}{2} m \dot{x}^2 - \frac{g^2}{8} (x^2 - a^2)^2 \right). \qquad (8.2)$$

As we have seen earlier in chapter 4, the best way to evaluate the path integral is to go to the Euclidean space. Thus, by rotating to imaginary time

$$t \rightarrow -it,$$

we obtain using Eq. (1.42)

$$\langle x_f | e^{-\frac{1}{\hbar} HT} | x_i \rangle = N \int \mathcal{D}x \, e^{-\frac{1}{\hbar} S_E[x]}, \qquad (8.3)$$

where the Euclidean action has the form

$$S_E[x] = \int_{-\frac{T}{2}}^{\frac{T}{2}} dt \, \left(\frac{1}{2} m \dot{x}^2 + V(x) \right). \qquad (8.4)$$

143

In the semi-classical method, we can evaluate the path integral by the saddle point method. The classical equation which is obtained from the extremum of the action in Eq. (8.4) has the form

$$\frac{\delta S_E[x]}{\delta x(t)} = 0$$

$$\text{or, } m\ddot{x} - V'(x) = 0\,, \tag{8.5}$$

with

$$V(x) = \frac{g^2}{8}(x^2 - a^2)^2\,. \tag{8.6}$$

The Euclidean equations, therefore, correspond to a particle moving in an inverted potential otherwise also known as a double humped potential. The Euclidean energy associated with such a motion is given by

$$E = \frac{1}{2}m\dot{x}^2 - V(x)\,. \tag{8.7}$$

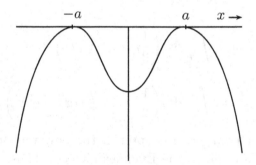

Two solutions to the Euclidean classical equation of motion in Eq. (8.5) with minimum energy are obvious. Namely,

$$x(t) = \pm a\,, \tag{8.8}$$

satisfy the classical equation with $E = 0$. In other words, the particle stays at rest on top of one of the hills in such a case. Quantum mechanically, this would correspond to the case where the particle executes small oscillations at the bottom of either of the wells in the Minkowski space and these small oscillations can be approximated by a harmonic oscillator motion. However, in the large T limit, (namely, when $T \rightarrow \infty$) in which we are ultimately interested in, there are nontrivial solutions to the Euclidean equation of motion in Eq. (8.5) which play a dominant role in evaluating the path integral. Let

$$x_{cl}(t) = \pm a \tanh \frac{\omega(t - t_c)}{2}, \qquad (8.9)$$

where t_c is a constant and we have identified as in Eq. (7.52)

$$m\omega^2 = V''(\pm a) = g^2 a^2. \qquad (8.10)$$

Then, we see that

$$\dot{x}_{cl} = \pm \frac{a\omega}{2} \operatorname{sech}^2 \frac{\omega(t - t_c)}{2}$$

$$= \pm \frac{a\omega}{2}\left(1 - \frac{x_{cl}^2}{a^2}\right) = \mp \frac{\omega}{2a}(x_{cl}^2 - a^2)$$

$$= \mp \frac{g}{2\sqrt{m}}(x_{cl}^2 - a^2) = \mp \sqrt{\frac{2V(x_{cl})}{m}}, \qquad (8.11)$$

$$\ddot{x}_{cl} = \mp \frac{\omega}{2a} 2x_{cl}\dot{x}_{cl}$$

$$= \mp \frac{\omega}{a} x_{cl}\left(\mp \frac{\omega}{2a}(x_{cl}^2 - a^2)\right)$$

$$= \frac{\omega^2}{2a^2} x_{cl}(x_{cl}^2 - a^2) = \frac{g^2}{2m} x_{cl}(x_{cl}^2 - a^2)$$

$$= \frac{1}{m} V'(x_{cl}). \qquad (8.12)$$

Consequently, it follows that

$$m\ddot{x}_{\text{cl}} - V'(x_{\text{cl}}) = 0, \qquad (8.13)$$

and we conclude that the solutions in Eq. (8.9) represent nontrivial solutions of the Euclidean equation of motion in Eq. (8.5).

We also note that, for these solutions,

$$x_{\text{cl}}(t \to -\infty) = \mp a,$$

$$x_{\text{cl}}(t \to \infty) = \pm a. \qquad (8.14)$$

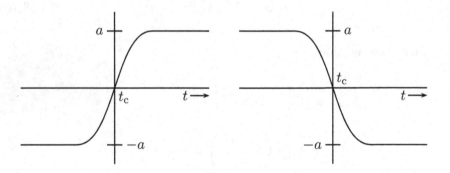

These solutions, therefore, correspond to the particle starting out on one of the hill tops at $t \to -\infty$ and then moving over to the other hill top at $t \to \infty$. Let us also note (see Eq. (8.11)) that for such solutions

$$\frac{1}{2} m\dot{x}_{\text{cl}}^2 = \frac{1}{2} m \times \frac{2}{m} V(x_{\text{cl}}) = V(x_{\text{cl}}). \qquad (8.15)$$

Therefore, for such motions, the Euclidean energy defined in Eq. (8.7) has the value

$$E = \frac{1}{2} m\dot{x}_{\text{cl}}^2 - V(x_{\text{cl}}) = 0. \qquad (8.16)$$

In other words, these also correspond to minimum energy solutions like the trivial motion. The value of the action corresponding to such a classical motion can be easily calculated.

$$S_E[x_{cl}] = S_0 = \int_{-\infty}^{\infty} dt \left(\frac{1}{2} m\dot{x}_{cl}^2 + V(x_{cl}) \right)$$

$$= \int_{-\infty}^{\infty} dt \, m\dot{x}_{cl}^2 = m \int_{\mp a}^{\pm a} dx_{cl} \, \dot{x}_{cl}$$

$$= m \int_{\mp a}^{\pm a} dx_{cl} \left(\mp \frac{\omega}{2a} (x_{cl}^2 - a^2) \right)$$

$$= \mp \frac{m\omega}{2a} \left(\frac{1}{3} x_{cl}^3 - x_{cl} a^2 \right) \Big|_{\mp a}^{\pm a} = \frac{m\omega}{2a} \frac{4a^3}{3}$$

$$= \frac{2}{3} \, m\omega a^2 = \frac{2m^2\omega^3}{3g^2} \,, \tag{8.17}$$

where we have used Eq. (8.11). We note that this action has the same value as the S_0 defined in Eq. (7.76) in connection with the WKB calculation of the splitting of the energy levels for the double well. The solutions in Eq. (8.9) are, therefore, finite action solutions in the Euclidean space and have the graphical form as shown above. They are known respectively as the instanton and the anti-instanton solutions (classical solutions in Euclidean time). If we look at the Lagrangian for such a solution, then using Eq. (8.11) we find that

$$L_E = \frac{1}{2} m \, \dot{x}_{cl}^2 + V(x_{cl}) = m\dot{x}_{cl}^2$$

$$= m \left(\frac{a^2\omega^2}{4} \text{sech}^4 \frac{\omega(t - t_c)}{2} \right)$$

$$= \frac{ma^2\omega^2}{4} \, \text{sech}^4 \frac{\omega(t - t_c)}{2} \,. \tag{8.18}$$

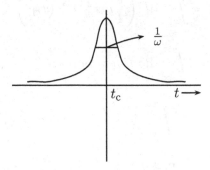

In other words, the Lagrangian is fairly localized around $t = t_c$ with a size of about

$$\Delta t \sim \frac{1}{\omega} = \frac{\sqrt{m}}{ga} \, . \tag{8.19}$$

It is in this sense that one says that instantons are localized solutions in time with a size of about $\frac{1}{\omega}$. The constant t_c which signifies the time when the solution reaches the valley of the Euclidean potential is really arbitrary. This is a direct reflection of the time translation invariance in the theory.

Just as we can have a one instanton or one anti-instanton solution, we can also have multi-instanton solutions in such a theory. However, before going into this, let us calculate the contribution to the transition amplitude coming from the one instanton or anti-instanton trajectory. From our earlier discussion in Eq. (7.35) of the saddle point method in connection with path integrals, we conclude that (O.I. stands for One Instanton)

$$\langle a | e^{-\frac{1}{\hbar} H T} | - a \rangle_{\text{O.I.}}$$

$$= N \int \mathcal{D}x \, e^{-\frac{1}{\hbar} S_E[x]}$$

$$\simeq N \, e^{-\frac{1}{\hbar} S_E[x_{\text{cl}}]} \int \mathcal{D}\eta \, e^{-\frac{1}{2\hbar} \iint dt_1 dt_2 \, \eta(t_1) \frac{\delta^2 S_E[x_{\text{cl}}]}{\delta x_{\text{cl}}(t_1) \delta x_{\text{cl}}(t_2)} \eta(t_2)}$$

$$= \frac{N}{\sqrt{\det\left(\frac{1}{\hbar}\left(-m\frac{d^2}{dt^2} + V''(x_{cl})\right)\right)_{\text{O.I.}}}} e^{-\frac{1}{\hbar}S_0}. \qquad (8.20)$$

Here we have defined

$$x(t) = x_{cl}(t) + \eta(t), \qquad (8.21)$$

and, for the one instanton case, as we have already seen in Eq. (8.9),

$$x_{cl}(t - t_c) = a\,\tanh\frac{\omega(t - t_c)}{2}. \qquad (8.22)$$

Let us next analyze the determinant in Eq. (8.20), for the one instanton case, in a bit more detail. We know from Eq. (8.11) that

$$\frac{dx_{cl}}{dt} = \frac{a\omega}{2}\,\text{sech}^2\frac{\omega(t - t_c)}{2}, \qquad (8.23)$$

satisfies the zero eigenvalue equation (see Eq. (7.45) rotated to imaginary time)

$$\left(-m\frac{d^2}{dt^2} + V''(x_{cl})\right)\frac{dx_{cl}}{dt} = 0. \qquad (8.24)$$

In fact, using Eq. (8.17), we can define the normalized zero eigenvalue solution of Eq. (8.24) as

$$\psi_0(t) = \left(\frac{S_0}{m}\right)^{-\frac{1}{2}}\frac{dx_{cl}}{dt} = \left(\frac{S_0}{m}\right)^{-\frac{1}{2}}\frac{a\omega}{2}\,\text{sech}^2\frac{\omega(t - t_c)}{2}. \qquad (8.25)$$

As we had seen earlier in Eq. (7.41), the determinant in Eq. (8.20) can be obtained from this solution simply as

$$\det\left(\frac{1}{\hbar}\left(-m\frac{d^2}{dt^2} + V''(x_{cl})\right)\right) \propto \psi_0\left(\frac{T}{2}\right), \quad T \to \infty. \qquad (8.26)$$

But from the form of the solution in Eq. (8.9) or (8.14), it is clear that

$$\lim_{T\to\infty}\psi_0\left(\frac{T}{2}\right) \to 0. \qquad (8.27)$$

In other words, in this case, the determinant identically vanishes. The reason for this is obvious, namely, in the present case $\psi_0(t)$ happens to be an exact eigenstate of the operator $(-m\frac{d^2}{dt^2} + V''(x_{cl}))$ with zero eigenvalue. (This means that $\psi_0(\pm\frac{T}{2}) = 0$ for $T \to \infty$.) This is what one means in saying that there is a zero mode in this theory.

8.2   Zero Modes

As we have argued before, a zero mode is present in the theory whenever there is a symmetry operative in the system. To see this, let us recall that the determinant in Eq. (8.20) arose from integrating out the Gaussian fluctuations. Therefore, the term that we need to re-examine is

$$\int \mathcal{D}\eta \, e^{-\frac{1}{2\hbar} \iint dt_1 dt_2 \, \eta(t_1) \frac{\delta^2 S[x_{cl}]}{\delta x_{cl}(t_1)\delta x_{cl}(t_2)} \eta(t_2)} \,. \tag{8.28}$$

Note that, in this case, since $\psi_0(t)$ represents a zero mode of the operator $\frac{\delta^2 S[x_{cl}]}{\delta x_{cl}(t_1)\delta x_{cl}(t_2)}$, if we make a change of the integration variable as

$$\delta\eta(t) = \epsilon\psi_0(t)\,, \tag{8.29}$$

where ϵ is a constant parameter, then the Gaussian does not change. In other words, the transformation in Eq. (8.29) defines a symmetry of the quadratic action. Another way to visualize the trouble is to note that if we were to expand the fluctuations around the classical trajectory in a complete basis of the eigenstates of the operator $\frac{\delta^2 S}{\delta x_{cl}(t_1)\delta x_{cl}(t_2)}$, then we can write

$$\eta(t) = \sum_{n\geq 0} c_n\psi_n(t),$$

$$\mathcal{D}\eta(t) = \prod_{n\geq 0} dc_n\,. \tag{8.30}$$

Substituting this expansion into Eq. (8.28), we obtain

$$\int \mathcal{D}\eta \, e^{-\frac{1}{2\hbar} \iint dt_1 dt_2 \, \eta(t_1) \frac{\delta^2 S[x_{cl}]}{\delta x_{cl}(t_1)\delta x_{cl}(t_2)} \eta(t_2)}$$

$$= \int \prod_{n\geq 0} dc_n \, e^{-\frac{1}{2\hbar} \sum_{n>0} \lambda_n c_n^2}$$

$$= \int dc_0 \int \prod_{n>0} dc_n \, e^{-\frac{1}{2\hbar} \sum_{n>0} \lambda_n c_n^2} , \qquad (8.31)$$

where λ_n denotes the eigenvalues corresponding to the eigenstates ψ_n. Here we note that the zero mode drops out of the exponent and, consequently, there is no Gaussian damping for the dc_0 integration.

In such a case, we have to evaluate the integral more carefully. To understand further the origin of the problem, let us examine a simple two dimensional integral. Let

$$I = \iint_{-\infty}^{\infty} dx_1 \, dx_2 \, e^{\frac{1}{a} f(\mathbf{x})}$$

$$= \iint_{-\infty}^{\infty} dx_1 \, dx_2 \, e^{\frac{1}{a}(\frac{1}{2}\mathbf{x}^2 - g^2(\mathbf{x}^2)^2)} . \qquad (8.32)$$

Here a is a small parameter and we have defined

$$f(\mathbf{x}) = \frac{1}{2}\mathbf{x}^2 - g^2(\mathbf{x}^2)^2 ,$$

$$\mathbf{x}^2 = x_1^2 + x_2^2 . \qquad (8.33)$$

This example is, in fact, quite analogous to the instanton calculation. The classical equations, in this case, lead to the maximum

$$\frac{\partial f}{\partial \mathbf{x}} = \mathbf{x} \left(1 - 4g^2 \mathbf{x}^2 \right) = 0 ,$$

$$\text{or,} \quad |\mathbf{x}_{cl}| = \frac{1}{2g} . \qquad (8.34)$$

It is easy to see that the other solution, namely, the origin, in this case, corresponds to a local minimum. The most general classical solution following from Eq. (8.34) can, therefore, be written as

$$x_{1,\text{cl}} = \frac{1}{2g} \cos\theta,$$

$$x_{2,\text{cl}} = \frac{1}{2g} \sin\theta, \tag{8.35}$$

where θ is an arbitrary constant angular parameter. This is very much like the arbitrary parameter t_c which arises in the case of the instantons. In the present case, the presence of the angular parameter, θ, is a consequence of the rotational invariance of the function $f(\mathbf{x})$ in Eq. (8.33).

Expanding around the classical solution, namely, choosing

$$x_\alpha = x_{\alpha,\text{cl}} + \eta_\alpha, \quad \alpha = 1, 2, \tag{8.36}$$

we have

$$f(\mathbf{x}) = f(\mathbf{x}_{\text{cl}} + \eta) \simeq f(\mathbf{x}_{\text{cl}}) + \frac{1}{2}\eta_\alpha \frac{\partial^2 f(\mathbf{x}_{\text{cl}})}{\partial x_{\text{cl}}^\alpha \partial x_{\text{cl}}^\beta} \eta_\beta + O(\eta^3). \tag{8.37}$$

From the form of $f(\mathbf{x})$ in Eq. (8.33), we note that

$$\frac{\partial^2 f(\mathbf{x}_{\text{cl}})}{\partial x_{\text{cl}}^\alpha \partial x_{\text{cl}}^\beta} = \delta_{\alpha\beta}(1 - 4g^2\mathbf{x}_{\text{cl}}^2) - 8g^2 x_{\text{cl}}^\alpha x_{\text{cl}}^\beta$$

$$= -8g^2 x_{\text{cl}}^\alpha x_{\text{cl}}^\beta. \tag{8.38}$$

In the matrix form, therefore, we can write

$$\frac{\partial^2 f(\mathbf{x}_{\text{cl}})}{\partial x_{\text{cl}}^\alpha \partial x_{\text{cl}}^\beta} = -2 \begin{pmatrix} \cos^2\theta & \cos\theta\sin\theta \\ \cos\theta\sin\theta & \sin^2\theta \end{pmatrix}. \tag{8.39}$$

Thus, if we use the saddle point method naively, we would obtain the value of the integral in Eq. (8.32) to be

$$I \simeq e^{\frac{1}{a}f(\mathbf{x}_{\text{cl}})} \int d\eta_\alpha \, e^{\frac{1}{2a}\eta_\alpha \frac{\partial^2 f(\mathbf{x}_{\text{cl}})}{\partial x_{\text{cl}}^\alpha \partial x_{\text{cl}}^\beta} \eta_\beta}. \tag{8.40}$$

Let us note that the matrix $\frac{\partial^2 f(\mathbf{x}_{\text{cl}})}{\partial x_{\text{cl}}^\alpha \partial x_{\text{cl}}^\beta}$ in Eq. (8.39) has two eigenvalues, $\lambda = 0, -2$. As a result, the Gaussian integral in Eq. (8.40) does not exist. This is very much like the instanton calculation that we did. In fact, it is easy to check from Eq. (8.39) that the eigenstate with zero eigenvalue has the form

$$x_0 = \begin{pmatrix} \sin\theta \\ -\cos\theta \end{pmatrix}. \tag{8.41}$$

Consequently, under a transformation of the variables of integration of the form

$$\delta \begin{pmatrix} \eta_1 \\ \eta_2 \end{pmatrix} = \epsilon x_0 = \epsilon \begin{pmatrix} \sin\theta \\ -\cos\theta \end{pmatrix} = -2g\epsilon \frac{d}{d\theta} \begin{pmatrix} x_{1,\text{cl}} \\ x_{2,\text{cl}} \end{pmatrix}, \tag{8.42}$$

which we recognize as an infinitesimal rotation, the quadratic exponent in Eq. (8.40) does not change. In fact, writing out the exponent completely, we have

$$I \simeq e^{\frac{1}{a} f(\mathbf{x}_{\text{cl}})} \iint_{-\infty}^{\infty} d\eta_1 d\eta_2 \, e^{-\frac{1}{a}(\eta_1 \cos\theta + \eta_2 \sin\theta)^2}. \tag{8.43}$$

Furthermore, redefining the variables as

$$\tilde{\eta}_1 = \eta_1 \cos\theta + \eta_2 \sin\theta,$$

$$\tilde{\eta}_2 = -\eta_1 \sin\theta + \eta_2 \cos\theta, \tag{8.44}$$

we note that we can write the integral in Eq. (8.43) also as

$$I \simeq e^{\frac{1}{a} f(\mathbf{x}_{\text{cl}})} \iint_{-\infty}^{\infty} d\tilde{\eta}_1 d\tilde{\eta}_2 \, e^{-\frac{1}{a}\tilde{\eta}_1^2}$$

$$= e^{\frac{1}{a} f(\mathbf{x}_{\text{cl}})} \int_{-\infty}^{\infty} d\tilde{\eta}_2 \int_{-\infty}^{\infty} d\tilde{\eta}_1 \, e^{-\frac{1}{a}\tilde{\eta}_1^2}. \tag{8.45}$$

The analogy with the instanton case is now complete. There is no damping for the $d\tilde{\eta}_2$ integration. That is the origin of the divergence and it is a consequence of rotational invariance in this case. In this simple example, the solution to the problem is obvious. Namely, since the function $f(\mathbf{x})$ in Eq. (8.33) is rotationally invariant, it is appropriate to use circular (polar) coordinates. The angular integral, in this case, can be trivially performed giving a finite result after which the saddle point approximation can be applied to the radial integral which will have no zero mode. This method generalizes readily to other systems with more degrees of freedom and is known as the method of collective coordinates. This is what we will try to use in order to evaluate the instanton integral.

8.3 The Instanton Integral

In the case of the instanton, we have already seen in Eq. (8.31) that

$$\int \mathcal{D}\eta \, e^{-\frac{1}{2\hbar} \iint dt_1 dt_2 \, \eta(t_1) \frac{\delta^2 S[x_{cl}]}{\delta x_{cl}(t_1)\delta x_{cl}(t_2)} \eta(t_2)}$$

$$= \int dc_0 \int \prod_{n>0} dc_n \, e^{-\frac{1}{2\hbar} \sum_{n>0} \lambda_n \, c_n^2}, \qquad (8.46)$$

which is divergent. The divergence, in the present case, is a consequence of time translation invariance, namely, the position of the center of the instanton can be arbitrary. So, following our earlier discussion, we will like to replace the dc_0 integration by an integration over the position of the center of the instanton. Let us discuss very briefly how this is done. Let us recall that expanding around the instanton trajectory yields

$$x(t) = x_{cl}(t - t_c) + \eta(t - t_c)$$

$$\text{or,} \quad x(t + t_c) = x_{cl}(t) + \eta(t) = x_{cl}(t) + \sum_{n\geq 0} c_n \, \psi_n(t). \quad (8.47)$$

(Since the trajectory is independent of the center of the instanton trajectory, the fluctuations must balance out the t_c dependence.) Multiplying Eq. (8.47) with $\psi_0(t)$ and integrating over time, we obtain

$$\int_{-\frac{T}{2}}^{\frac{T}{2}} \mathrm{d}t \, x(t + t_c) \psi_0(t)$$

$$= \int_{-\frac{T}{2}}^{\frac{T}{2}} \mathrm{d}t \, \left(x_{cl}(t) + \sum_{n \geq 0} c_n \, \psi_n(t) \right) \psi_0(t)$$

$$= \int_{-\frac{T}{2}}^{\frac{T}{2}} \mathrm{d}t \, \left(\left(\frac{S_0}{m} \right)^{-\frac{1}{2}} x_{cl}(t) \, \frac{\mathrm{d}x_{cl}(t)}{\mathrm{d}t} \right) + c_0$$

$$= \frac{1}{2} \left(\frac{S_0}{m} \right)^{-\frac{1}{2}} x_{cl}^2(t) \Bigg|_{-\frac{T}{2}}^{\frac{T}{2}} + c_0$$

$$= c_0 . \tag{8.48}$$

The first term vanishes because it has the same value at both the limits. (It is worth emphasizing here that we are only interested in large T limits when all these results hold.) This simple analysis shows that

$$c_0 = c_0(t_c) . \tag{8.49}$$

Therefore, we can easily change the c_0-integration to an integration over t_c. To obtain the Jacobian of this transformation to the leading order, let us consider an infinitesimal change in the path in Eq. (8.47) arising from a change in the coefficient of the zero mode. Namely, let

$$\delta \eta = \delta c_0 \, \psi_0(t) , \tag{8.50}$$

where we assume that δc_0 is infinitesimal. In this case,

$$\delta x(t + t_c) = \delta \eta = \delta c_0 \, \psi_0(t) = \delta c_0 \left(\frac{S_0}{m} \right)^{-\frac{1}{2}} \frac{\mathrm{d}x_{cl}(t)}{\mathrm{d}t} , \tag{8.51}$$

where we have used Eq. (8.25). However, we also note that this is precisely the change in the path that we would have obtained to leading order had we translated the center of the instanton as

$$t_c \rightarrow t_c + \delta t_c = t_c + \delta c_0 \left(\frac{S_0}{m}\right)^{-\frac{1}{2}}. \tag{8.52}$$

Namely, in this case,

$$\delta x(t + t_c) = x(t + t_c + \delta t_c) - x(t + t_c)$$

$$\simeq \delta t_c \frac{dx(t + t_c)}{dt} = \delta c_0 \left(\frac{S_0}{m}\right)^{-\frac{1}{2}} \frac{dx(t + t_c)}{dt}$$

$$\simeq \delta c_0 \left(\frac{S_0}{m}\right)^{-\frac{1}{2}} \frac{dx_{cl}(t)}{dt}. \tag{8.53}$$

Thus, from Eq. (8.52), we note that to leading order, we can determine the Jacobian of the transformation from the integration variable c_0 to t_c to be

$$\frac{dc_0(t_c)}{dt_c} \simeq \left(\frac{S_0}{m}\right)^{\frac{1}{2}}. \tag{8.54}$$

A more direct way to arrive at this result is to note from Eq. (8.48) that since

$$c_0(t_c) = \int_{-\frac{T}{2}}^{\frac{T}{2}} dt\, x(t + t_c)\psi_0(t)$$

$$\frac{dc_0}{dt_c} = \int_{-\frac{T}{2}}^{\frac{T}{2}} dt\, \frac{dx(t + t_c)}{dt_c}\, \psi_0(t)$$

$$= \int_{-\frac{T}{2}}^{\frac{T}{2}} dt\, \frac{dx(t + t_c)}{dt}\, \psi_0(t)$$

$$= \int_{-\frac{T}{2}}^{\frac{T}{2}} dt\, \left(\frac{dx_{cl}(t)}{dt} + \sum_{n \geq 0} c_n \frac{d\psi_n(t)}{dt}\right) \psi_0(t)$$

$$= \left(\frac{S_0}{m}\right)^{\frac{1}{2}} \int_{-\frac{T}{2}}^{\frac{T}{2}} dt\, \psi_0^2(t) + \sum_{n > 0} c_n \int_{-\frac{T}{2}}^{\frac{T}{2}} dt\, \frac{d\psi_n(t)}{dt}\, \psi_0(t), \tag{8.55}$$

where we have used Eq. (8.25). The $n = 0$ term drops out in the second term because for $n = 0$, the integrand is a total derivative of $\psi_0^2(t)$ which vanishes at both the limits. It is clear, therefore, that to leading order (since ψ_0 is normalized to unity),

$$\frac{\mathrm{d}c_0(t_c)}{\mathrm{d}t_c} = \left(\frac{S_0}{m}\right)^{\frac{1}{2}} + O(\hbar). \tag{8.56}$$

Namely, we are using here the fact that the higher moments of a Gaussian of the kind that we are dealing with in Eq. (8.46) are higher orders in $\hbar$.

Thus, we are ready to do the determinant calculation now. We substitute Eq. (8.54) or (8.56) into Eq. (8.46) to obtain

$$\int \mathcal{D}\eta \, e^{-\frac{1}{2\hbar} \iint \mathrm{d}t_1 \mathrm{d}t_2 \, \eta(t_1) \frac{\delta^2 S[x_{cl}]}{\delta x_{cl}(t_1) \delta x_{cl}(t_2)} \eta(t_2)}$$

$$= \left(\frac{S_0}{m}\right)^{\frac{1}{2}} \int_{-\frac{T}{2}}^{\frac{T}{2}} \mathrm{d}t_c \int \prod_{n>0} \mathrm{d}c_n \, e^{-\frac{1}{2\hbar} \sum_{n>0} \lambda_n c_n^2}$$

$$= \left(\frac{S_0}{m}\right)^{\frac{1}{2}} \frac{\int_{-\frac{T}{2}}^{\frac{T}{2}} \mathrm{d}t_c}{\sqrt{\det'(\frac{1}{\hbar}(-m\frac{\mathrm{d}^2}{\mathrm{d}t^2} + V''(x_{cl})))}} . \tag{8.57}$$

Here det′ stands for the value of the determinant of the operator without the zero mode. Let us also note here that even though the $\mathrm{d}t_c$ integral in Eq. (8.57) can be done trivially, we will leave it as it is for later purposes. Thus, from Eqs. (8.20) and (8.57) we obtain the form of the transition amplitude in the presence of an instanton to be

$$\langle a|e^{-\frac{1}{\hbar}HT}|-a\rangle_{\text{O.I.}} = \frac{N\left(\frac{S_0}{m}\right)^{\frac{1}{2}} e^{-\frac{1}{\hbar}S_0}}{\sqrt{\det'(\frac{1}{\hbar}(-m\frac{\mathrm{d}^2}{\mathrm{d}t^2} + V''(x_{cl})))}} \int_{-\frac{T}{2}}^{\frac{T}{2}} \mathrm{d}t_c . \tag{8.58}$$

8.4 Evaluating the Determinant

To evaluate $\det'$ in Eq. (8.58), let us define the quantity

$$\Delta(E) = \frac{\det(\frac{1}{\hbar}(-m\frac{d^2}{dt^2} + V''(x_{cl})) - E)}{\det(\frac{1}{\hbar}(-m\frac{d^2}{dt^2} + m\omega^2) - E)} = \frac{\prod_n(E_n - E)}{\prod_n(E_n^{(0)} - E)}, \quad (8.59)$$

where the determinant in the denominator corresponds to that of a free harmonic oscillator which we have already evaluated. It is easy to see again that both sides of Eq. (8.59) have the same analytic structure and, therefore must be equal. We note that

$$\Delta(E = 0) = 0, \quad (8.60)$$

since there is a zero eigenvalue for the determinant in the numerator and further,

$$\Delta(E = \infty) = 1. \quad (8.61)$$

If we eliminate the zero mode in Eq. (8.59) by dividing it out, then we obtain

$$-\frac{1}{E}\Delta(E)\mid_{E=0} = \frac{\prod_{n>0} E_n}{\prod_n E_n^{(0)}}$$

$$\text{or,} \quad -\frac{\partial\Delta(E)}{\partial E}\mid_{E=0} = \frac{\det'(\frac{1}{\hbar}(-m\frac{d^2}{dt^2} + V''(x_{cl})))}{\det(\frac{1}{\hbar}(-m\frac{d^2}{dt^2} + m\omega^2))}. \quad (8.62)$$

Clearly, if we can evaluate the left hand side of Eq. (8.62), then we would have evaluated $\det'$ since we already know the value of the determinant for the harmonic oscillator.

To evaluate this, let us consider the scattering problem for the Schrödinger equation

$$\frac{1}{\hbar}\left(-m\frac{d^2}{dt^2} + V''(x_{cl})\right)\psi = E\psi$$

$$\text{or,} \quad \left(-m\frac{d^2}{dt^2} + V''(x_{cl})\right)\psi = \hbar E\psi. \quad (8.63)$$

If we define the asymptotic solutions (Jost functions) as

$$\lim_{t \to \infty} f_+(t, E) \to e^{-ikt},$$

$$\lim_{t \to -\infty} f_-(t, E) \to e^{ikt}, \tag{8.64}$$

where we identify

$$k^2 = \frac{\hbar E}{m} - \omega^2,$$

$$\lim_{t \to \pm\infty} V''(x_{\text{cl}}) \to m\omega^2. \tag{8.65}$$

The Jost functions are two linearly independent solutions of the Schrödinger equation in Eq. (8.63) and consequently, any general solution can be written as a linear combination of the two. In particular, we can write

$$\lim_{t \to -\infty} f_+(t, E) \to A_+(E) \, e^{ikt} + B_+(E) \, e^{-ikt},$$

$$\lim_{t \to \infty} f_-(t, E) \to B_-(E) \, e^{ikt} + A_-(E) \, e^{-ikt}. \tag{8.66}$$

The linear independence of the Jost functions can be easily seen by calculating the Wronskian which has the value

$$W(f_+(t, E), f_-(t, E))$$

$$= f_+(t, E) \frac{\partial f_-(t, E)}{\partial t} - \frac{\partial f_+(t, E)}{\partial t} f_-(t, E)$$

$$= 2ik \, B_+(E) = 2ik \, B_-(E), \tag{8.67}$$

where the equality in the last step results from evaluating the Wronskian at the two different time limits $t \to \pm\infty$. This, in fact, shows that the two coefficients $B_+(E)$ and $B_-(E)$ are identical. With a bit more analysis, they can also be shown to be equal to $\Delta(E)$, namely,

$$B_+(E) = B_-(E) = \Delta(E). \tag{8.68}$$

Let us also note from Eq. (8.25) that the zero mode has the asymptotic form

$$\lim_{t \to \pm\infty} \psi_0(t) = \lim_{t \to \pm\infty} \left(\frac{S_0}{m}\right)^{-\frac{1}{2}} \frac{a\omega}{2} \text{sech}^2 \frac{\omega(t - t_c)}{2}$$

$$\to 2a\omega \left(\frac{S_0}{m}\right)^{-\frac{1}{2}} e^{\mp\omega t} = K\, e^{\mp\omega t}, \tag{8.69}$$

where we have defined

$$K = 2a\omega \left(\frac{S_0}{m}\right)^{-\frac{1}{2}}. \tag{8.70}$$

Thus, from Eqs. (8.64), (8.65), (8.66) and (8.69), we note that we can identify $(k(E = 0) = -i\omega)$

$$\lim_{|t| \to \infty} f_\pm(t, E = 0) = \lim_{|t| \to \infty} \frac{1}{K} \psi_0(t) \to e^{-\omega|t|}. \tag{8.71}$$

Thus, comparing Eqs. (8.71) and (8.72) with the asymptotic form of the Jost functions in Eqs. (8.64) and (8.66), we conclude that

$$A_+(E = 0) = 1, \qquad A_-(E = 0) = 1,$$

$$B_+(E = 0) = 0, \qquad B_-(E = 0) = 0. \tag{8.72}$$

Consequently, we obtain

$$\Delta(E = 0) = B_+(E = 0) = 0, \tag{8.73}$$

a result which we already know.

The asymptotic equations which the Jost functions satisfy (see Eqs. (8.63) and (8.65)) are

$$-m\frac{\partial^2 f_+(t, E)}{\partial t^2} - (\hbar E - m\omega^2)f_+(t, E) = 0,$$

$$-m\frac{\partial^2 f_-(t, E')}{\partial t^2} - (\hbar E' - m\omega^2)f_-(t, E') = 0. \tag{8.74}$$

Multiplying the first of these equations by $f_-(t, E')$ and the second by $f_+(t, E)$ and subtracting one from the other, we obtain

$$m \left(f_+(t, E) \frac{\partial^2 f_-(t, E')}{\partial t^2} - f_-(t, E') \frac{\partial^2 f_+(t, E)}{\partial t^2} \right)$$

$$= \hbar (E - E') f_+(t, E) f_-(t, E')$$

or, $\displaystyle \frac{\partial}{\partial t} \left(f_+(t, E) \frac{\partial f_-(t, E')}{\partial t} - f_-(t, E') \frac{\partial f_+(t, E)}{\partial t} \right)$

$$= \frac{\hbar}{m} (E - E') f_+(t, E) f_-(t, E')$$

or, $\displaystyle \frac{\partial}{\partial t} W(f_+(t, 0), f_-(t, E)) = -\frac{\hbar}{m} E f_+(t, 0) f_-(t, E)$

or, $\displaystyle \frac{\partial^2}{\partial E \partial t} W(f_+(t, 0), f_-(t, E)) \Big|_{E=0} = -\frac{\hbar}{m} f_+(t, 0) f_-(t, 0)$

$$= -\frac{\hbar}{mK^2} \psi_0^2(t). \tag{8.75}$$

Integrating this equation between $(-\frac{T}{2}, \frac{T}{2})$ with $T \to \infty$, we obtain

$$\lim_{E \to 0} \frac{\partial}{\partial E} W(f_+(t, 0), f_-(t, E)) \Big|_{t=-\infty}^{t=\infty} = -\frac{\hbar}{mK^2}. \tag{8.76}$$

On the other hand, from the asymptotic form of the Jost functions in Eq. (8.66), we see that

$$W(f_+(t, 0), f_-(t, E)) \big|_{t=-\infty}^{t=\infty}$$

$$= \left(e^{-\omega t} (ik \, B_-(E) e^{ikt} - ik \, A_-(E) e^{-ikt}) \right.$$

$$\left. + \omega \, e^{-\omega t} (B_-(E) e^{ikt} + A_-(E) e^{-ikt}) \right)_{t=\infty}$$

$$- \left(e^{\omega t} ik \, e^{ikt} - \omega e^{\omega t} e^{ikt} \right)_{t=-\infty}, \tag{8.77}$$

from which we determine

$$\lim_{E \to 0} \frac{\partial}{\partial E} W(f_+(t,0), f_-(t,E)) \Big|_{-\infty}^{\infty}$$

$$= 2\omega \frac{\partial B_-(E)}{\partial E} \Big|_{E=0} = 2\omega \frac{\partial \Delta(E)}{\partial E} \Big|_{E=0}. \tag{8.78}$$

Here we have used the identification in Eq. (8.68) as well as the results of Eq. (8.72). Comparing Eqs. (8.76) and (8.78), then, we obtain

$$2\omega \frac{\partial \Delta(E)}{\partial E} \Big|_{E=0} = -\frac{\hbar}{mK^2}$$

$$\text{or,} \quad -\frac{\partial \Delta(E)}{\partial E} \Big|_{E=0} = \frac{\hbar}{2m\omega K^2}. \tag{8.79}$$

We, therefore, determine the ratio in Eq. (8.62) to be

$$\frac{\det'(\frac{1}{\hbar}(-m\frac{d^2}{dt^2} + V''(x_{cl})))}{\det(\frac{1}{\hbar}(-m\frac{d^2}{dt^2} + m\omega^2))} = -\frac{\partial \Delta(E)}{\partial E} \Big|_{E=0} = \frac{\hbar}{2m\omega K^2}. \tag{8.80}$$

The one instanton contribution in Eq. (8.58) can now be explicitly determined and takes the form

$$\langle a | e^{-\frac{1}{\hbar}HT} | -a \rangle_{O.I.}$$

$$= \frac{N}{\sqrt{\det(\frac{1}{\hbar}(-m\frac{d^2}{dt^2} + \omega^2))}} \sqrt{\frac{\det(\frac{1}{\hbar}(-m\frac{d^2}{dt^2} + \omega^2))}{\det'(\frac{1}{\hbar}(-m\frac{d^2}{dt^2} + V''(x_{cl})))}}$$

$$\times \left(\frac{S_0}{m}\right)^{\frac{1}{2}} e^{-\frac{1}{\hbar}S_0} \int_{-\frac{T}{2}}^{\frac{T}{2}} dt_c$$

$$= \left(\frac{m\omega}{\pi\hbar}\right)^{\frac{1}{2}} e^{-\frac{\omega T}{2}} \sqrt{\frac{2m\omega}{\hbar}} K \left(\frac{S_0}{m}\right)^{\frac{1}{2}} e^{-\frac{1}{\hbar}S_0} \int_{-\frac{T}{2}}^{\frac{T}{2}} dt_c$$

$$= \left(\frac{m\omega}{\pi\hbar}\right)^{\frac{1}{2}} e^{-\frac{\omega T}{2}} \sqrt{\frac{2S_0\omega}{\hbar}} \, K \, e^{-\frac{1}{\hbar}S_0} \int_{-\frac{T}{2}}^{\frac{T}{2}} dt_c$$

$$= \left(\frac{m\omega}{\pi\hbar}\right)^{\frac{1}{2}} e^{-\frac{\omega T}{2}} \, r \int_{-\frac{T}{2}}^{\frac{T}{2}} dt_c \,, \tag{8.81}$$

where we have used Eq. (8.80) as well as the value of the path integral for the harmonic oscillator given in Eq. (3.28) (rotated to Euclidean space). We have also defined a new quantity, r, whose value using Eq. (8.70) is given by

$$r = \sqrt{\frac{2S_0\omega}{\hbar}} \, K \, e^{-\frac{1}{\hbar}S_0}$$

$$= 2\sqrt{\frac{2m}{\hbar}} \, \omega^{\frac{3}{2}} a \, e^{-\frac{1}{\hbar}S_0} \,. \tag{8.82}$$

The transition amplitude, in this case, separates into a product of two factors-one that of a simple harmonic oscillator arising from the trivial solution of the Euclidean equation of motion and the second giving the true contribution due to an instanton. We can, similarly, calculate the transition amplitude in the presence of an anti-instanton and it can be shown to be identical to the result obtained in Eq. (8.81).

8.5   Multi-Instanton Contributions

As we had discussed earlier, a string of widely separated instantons and anti-instantons also satisfies the Euclidean classical equation given in Eq. (8.5). The instanton density is small for weak coupling and in such a case these multi-instanton solutions will contribute to the transition amplitude as well and their contribution can be evaluated under an approximation commonly known as the dilute gas approximation. A typical example of a multi-instanton solution has the following form.

Let us consider a n-instanton solution with centers at $t_1, t_2, \ldots, t_n$ satisfying

$$-\frac{T}{2} \leq t_n \leq t_{n-1} \cdots \leq t_1 \leq \frac{T}{2}. \tag{8.83}$$

In such a case, the integral over the centers of the instantons gives

$$\int_{-\frac{T}{2}}^{\frac{T}{2}} dt_1 \int_{-\frac{T}{2}}^{t_1} dt_2 \cdots \int_{-\frac{T}{2}}^{t_{n-1}} dt_n = \frac{T^n}{n!}. \tag{8.84}$$

Furthermore, since the instantons and the anti-instantons are assumed to be noninteracting, their contributions to the transition amplitude will simply be multiplicative. Thus, a n-"instanton" solution will contribute an amount (see Eqs. (8.81) and (8.84))

$$\left(\frac{m\omega}{\pi\hbar}\right)^{\frac{1}{2}} e^{-\frac{\omega T}{2}} r^n \frac{T^n}{n!} = \left(\frac{m\omega}{\pi\hbar}\right)^{\frac{1}{2}} e^{-\frac{\omega T}{2}} \frac{(rT)^n}{n!}. \tag{8.85}$$

We have to recognize here that only an even number of instantons and anti-instantons can contribute to the transition amplitude of the form

$$\langle a|e^{-\frac{1}{\hbar}HT}|a\rangle, \qquad \langle -a|e^{-\frac{1}{\hbar}HT}|-a\rangle. \tag{8.86}$$

Similarly, only a total of odd number of instantons and anti-instantons can contribute to transition amplitudes of the form

$$\langle a|e^{-\frac{1}{\hbar}HT}|-a\rangle, \qquad \langle -a|e^{-\frac{1}{\hbar}HT}|a\rangle. \tag{8.87}$$

Adding all such "instanton" contributions, we see from Eq. (8.85) that we will have

$$\langle -a|e^{-\frac{1}{\hbar}HT}| - a\rangle$$

$$= \sum_n \left(\frac{m\omega}{\pi\hbar}\right)^{\frac{1}{2}} e^{-\frac{\omega T}{2}} \frac{(rT)^{2n}}{2n!}$$

$$= \left(\frac{m\omega}{\pi\hbar}\right)^{\frac{1}{2}} e^{-\frac{\omega T}{2}} \cosh(rT)$$

$$= \left(\frac{m\omega}{\pi\hbar}\right)^{\frac{1}{2}} e^{-\frac{\omega T}{2}} \frac{(e^{rT} + e^{-rT})}{2}$$

$$= \frac{1}{2}\left(\frac{m\omega}{\pi\hbar}\right)^{\frac{1}{2}} \left(e^{-(\frac{\omega}{2}-r)T} + e^{-(\frac{\omega}{2}+r)T}\right). \tag{8.88}$$

Similarly, we can show that for odd number of instanton contributions, we will obtain

$$\langle a|e^{-\frac{1}{\hbar}HT}| - a\rangle = \left(\frac{m\omega}{\pi\hbar}\right)^{\frac{1}{2}} e^{-\frac{\omega T}{2}} \sinh(rT)$$

$$= \frac{1}{2}\left(\frac{m\omega}{\pi\hbar}\right)^{\frac{1}{2}} \left(e^{-(\frac{\omega}{2}-r)T} - e^{-(\frac{\omega}{2}+r)T}\right). \tag{8.89}$$

If we identify the two low lying states of the Hamiltonian as $|\pm\rangle$ with energy eigenvalues $E_\pm$ respectively, then we note that by inserting a complete set of energy states we will obtain for large T

$$\langle -a|e^{-\frac{1}{\hbar}HT}| - a\rangle$$

$$\simeq \langle -a|e^{-\frac{1}{\hbar}HT}|-\rangle\langle -| - a\rangle + \langle -a|e^{-\frac{1}{\hbar}HT}|+\rangle\langle +| - a\rangle \tag{8.90}$$

$$= e^{-\frac{1}{\hbar}E_-T}\langle -a|-\rangle\langle -| - a\rangle + e^{-\frac{1}{\hbar}E_+T}\langle -a|+\rangle\langle +| - a\rangle.$$

Comparing this with Eq. (8.88), then, we obtain

$$E_\pm = \hbar\left(\frac{\omega}{2} \pm r\right). \tag{8.91}$$

Therefore, the splitting between the two energy levels is obtained to be

$$\Delta E = E_+ - E_- = 2\hbar r$$

$$= 2\hbar \times 2\sqrt{\frac{2m}{\hbar}}\omega^{\frac{3}{2}}a\, e^{-\frac{1}{\hbar}S_0}$$

$$= 4\sqrt{2m\hbar}\omega^{\frac{3}{2}}a\, e^{-\frac{1}{\hbar}S_0}. \tag{8.92}$$

This splitting of energy levels calculated in the path integral formalism can then be compared with the result obtained through the WKB approximation in Eq. (7.77).

8.6 References

Coleman, S., "The Uses of Instantons", Erice Lectures, 1977.

Sakita, B., "Quantum Theory of Many Variable Systems and Fields", World Scientific Publishing.

Shifman, M. et al., Sov. Phys. Usp. **25**, 195 (1982).

Zinn-Justin, J., "Quantum Field Theory and Critical Phenomena", Oxford Univ. Press.

Chapter 9

Path Integral for Relativistic Theories

9.1 Systems with Many Degrees of Freedom

Thus far, we have only discussed one particle systems. However, the method of path integrals generalizes readily to systems with many particles or systems with many degrees of freedom. Let us consider a system with n-degrees of freedom characterized by the coordinates $x^\alpha(t)$, $\alpha = 1, 2, \cdots, n$. These coordinates, for example, can denote the coordinates of n-particles in one dimension or the coordinates of a single particle in n-dimensions. If $S[x]$ denotes the appropriate action for the system (namely, if it describes the dynamics of the system), then the transition amplitude in Eq. (2.28) can be easily shown to generalize to

$$\langle x_f, t_f | x_i, t_i \rangle = N \int \mathcal{D}x^\alpha \, e^{\frac{i}{\hbar} S[x^\alpha]} \,. \tag{9.1}$$

The action generically has the form

$$S[x^\alpha] = \int_{t_i}^{t_f} \mathrm{d}t \, L(x^\alpha, \dot{x}^\alpha) \,, \tag{9.2}$$

and we are supposed to integrate over all paths starting at x_i^α at $t = t_i$ and ending at x_f^α at $t = t_f$. We can also introduce appropriate sources, in this case, through the couplings

$$S[x^\alpha, J^\alpha] = S[x^\alpha] + \int_{t_i}^{t_f} \mathrm{d}t \, J^\alpha(t) x^\alpha(t) \,, \tag{9.3}$$

to define the transition amplitude in the presence of these sources as (see Eq. (4.40))

$$\langle x_f, t_f | x_i, t_i \rangle^J = N \int \mathcal{D}x^\alpha \, e^{\frac{i}{\hbar} S[x^\alpha, J^\alpha]} . \qquad (9.4)$$

As we have seen earlier in Eqs. (4.38) and (4.50), this allows us to derive the various transition amplitudes or matrix elements in a simple manner. We can also define, as before, the vacuum to vacuum transition amplitude in the limit of infinite time interval as (see section 4.4)

$$Z[J] = \langle 0|0 \rangle^J = N \int \mathcal{D}x^\alpha \, e^{\frac{i}{\hbar} S[x^\alpha, J^\alpha]} , \qquad (9.5)$$

where in the infinite time interval limit, the action in Eq. (9.3) has the form

$$S[x^\alpha, J^\alpha] = \int_{-\infty}^{\infty} dt \left(L(x^\alpha, \dot{x}^\alpha) + J^\alpha(t) x^\alpha(t) \right) , \qquad (9.6)$$

and the integration over the paths in Eq. (9.5) has no end point restriction in the sense that the initial and the final coordinates of the paths can be chosen arbitrarily.

The path integrals can also be extended to continuum field theories once we recognize that these theories describe physical systems with an infinite number of degrees of freedom. Thus, if $\phi(x, t)$ is the basic variable of a $1 + 1$ dimensional field theory, then the vacuum to vacuum transition amplitude in the presence of an external source can be written as

$$Z[J] = \langle 0|0 \rangle^J = N \int \mathcal{D}\phi \, e^{\frac{i}{\hbar} S[\phi, J]} , \qquad (9.7)$$

where

$$S[\phi, J] = S[\phi] + \iint_{-\infty}^{\infty} dt \, dx \, J(x, t) \phi(x, t) . \qquad (9.8)$$

(Incidentally, in all these discussions, we are going to assume that the relation between the Lagrangian and the Hamiltonian of the system is the canonical one which would lead to path integrals of the form in Eq. (9.5) or (9.7). If this is not the case, then one should take as the starting point, the path integral in the phase space as obtained in Eq. (2.22).) Before going into the discussion about the functional integration in the present case, it is worth emphasizing what we have discussed earlier, namely, it is the time ordered Green's functions in the vacuum which play the most important role in a field theory because the scattering matrix or the S-matrix can be obtained from them. This is why it is the vacuum functional which is the quantity of fundamental significance in these studies. The second point to note is that we have left the specific form of $S[\phi]$ arbitrary. Depending on the particular form of the action, we will be dealing with different kinds of field theories - both non-relativistic and relativistic.

Returning now to the question of the functional integration, let us recall that in the $0 + 1$ dimensional case, we defined the path integral by dividing up the time interval into infinitesimal steps (see Eqs. (2.18) and (2.20)). Here, in addition, we have to divide up the space interval into infinitesimal steps as well. Thus, let us assume that

$$-\frac{L}{2} \le x \le \frac{L}{2}, \tag{9.9}$$

with the understanding that we will take the limit $L \to \infty$ at the end. Let us further divide the length interval into $\bar{N}$ equal steps of length ϵ such that

$$\bar{N}\epsilon = L. \tag{9.10}$$

$-\frac{L}{2}$ $\qquad\qquad\qquad\qquad\qquad\qquad\qquad\qquad$ $\frac{L}{2}$

If we now label the position as well as the value of the field variable at any intermediate point on the trajectory as

$$x_m = -\frac{L}{2} + m\epsilon,$$

$$\phi(x_m) = \phi_m, \qquad 0 \le m \le \bar{N}, \tag{9.11}$$

then, just as in the case of quantum mechanics, we can define (see Eq. (2.29), for example)

$$\int \mathcal{D}\phi = \lim_{L\to\infty} \lim_{\substack{\epsilon\to 0, \bar{N}\to\infty \\ \bar{N}\epsilon=L}} \int \prod_m \mathrm{d}\phi_m \,. \qquad (9.12)$$

Unlike the case of quantum mechanics, which we have extensively discussed, however, the path integral, in the present case, does not exist in the sense that the integrations defined in Eq. (9.12) are divergent in the continuum limit. However, if we absorb the divergence into the normalization constant, N, in Eq. (9.5) or (9.7), then the Green's functions can still be defined uniquely since they are defined as ratios for which the divergent constants simply drop out.

For field theories in higher dimensions where the basic variables are $\phi(\mathbf{x}, t)$, we can define the vacuum generating functional exactly in an analogous manner. Namely, we have

$$Z[J] = \langle 0|0\rangle^J = N \int \mathcal{D}\phi \, e^{\frac{i}{\hbar} S[\phi, J]} \,, \qquad (9.13)$$

where

$$S[\phi, J] = S[\phi] + \int \mathrm{d}^n x \, J(\mathbf{x}, t)\phi(\mathbf{x}, t) \,, \qquad (9.14)$$

and the integrations are over the entire space-time manifold in higher dimension with n denoting the dimensionality of the space-time manifold. The functional integral, in such a case, is defined by taking a hypercube divided into a lattice of infinitesimal spacing and then identifying the functional integral with a product of ordinary integrals of the field values at each of the lattice sites.

9.2 Relativistic Scalar Field Theory

With all these preliminaries, let us take a specific form of the action in Eq. (9.14). Namely, let us choose a relativistic scalar field theory in $3 + 1$ dimensions described by the Lagrangian density

$$\mathcal{L}(\phi, \partial_\mu \phi) = \frac{1}{2}\partial_\mu \phi \partial^\mu \phi - \frac{m^2}{2}\phi^2 - \frac{\lambda}{4!}\phi^4 \,, \qquad (9.15)$$

with $\lambda > 0$ (This condition merely corresponds to the fact that we would like the potential to be bounded from below so that the quantum theory will have a meaningful ground state.) so that

$$S[\phi] = \int d^4x \, \mathcal{L}(\phi, \partial_\mu \phi) , \qquad (9.16)$$

and

$$S[\phi, J] = S[\phi] + \int d^4x \, J(\mathbf{x}, t)\phi(\mathbf{x}, t) . \qquad (9.17)$$

It is worth noting here that the Lagrangian can be obtained from the Lagrangian density in Eq. (9.15) by integrating over the space variables as

$$L = \int d^3x \, \mathcal{L}(\phi, \partial_\mu \phi) . \qquad (9.18)$$

This theory is quite similar (see Eq. (4.63)) to the anharmonic oscillator which we discussed earlier except that it is a relativistic field theory invariant under global Poincare transformations. This is a self-interacting theory which can describe spin zero particles with mass m. The Euler-Lagrange equations following from the action in Eq. (9.17) take the form

$$\frac{\delta S[\phi, J]}{\delta \phi(x)} = \partial_\mu \partial^\mu \phi + m^2 \phi + \frac{\lambda}{3!}\phi^3 - J(x) = 0 . \qquad (9.19)$$

Here we have chosen to represent the space-time variables in a compact notation of x for simplicity. Commonly, this theory described by the action in Eq. (9.17) or by the dynamical equations in Eq. (9.19) is also known as the ϕ^4-theory.

In the absence of interaction, namely, when $\lambda = 0$, the action in Eq. (9.17) is quadratic in the field variables and hence the generating functional can be evaluated in much the same way as in the case of quantum mechanical systems (see chapters 2, 3 and in particular, 4). However, let us first define the Feynman Green's function associated with this theory. The equation satisfied by the Green's function is

$$\left(\partial_\mu \partial^\mu + m^2\right) G(x - x') = -\delta^4(x - x') . \qquad (9.20)$$

Defining the Fourier transforms as

$$G(x - x') = \int \frac{d^4k}{(2\pi)^2} \, G(k) \, e^{-ik \cdot (x - x')} \, ,$$

$$\delta^4(x - x') = \frac{1}{(2\pi)^4} \int d^4k \, e^{-ik \cdot (x - x')} \, , \qquad (9.21)$$

and substituting these back into the differential equation, Eq. (9.20), we obtain

$$\frac{1}{(2\pi)^2} \left(-k^2 + m^2 \right) G(k) = -\frac{1}{(2\pi)^4}$$

$$\text{or,} \quad G(k) = \frac{1}{(2\pi)^2} \frac{1}{k^2 - m^2} \, . \qquad (9.22)$$

Here we are using the scalar product for the four vectors with the metrics introduced in Eqs. (1.3) and (1.4) and k^2 represents the invariant length square of the conjugate four vector k^μ. The Feynman Green's function or the propagator is, then, defined following Eq. (3.75) as

$$G_F(x - x') = \lim_{\epsilon \to 0^+} \int \frac{d^4k}{(2\pi)^4} \frac{1}{k^2 - m^2 + i\epsilon} \, e^{-ik \cdot (x - x')} \, . \qquad (9.23)$$

We can also think of the Feynman Green's function as satisfying the differential equation (see Eq. (3.76))

$$\lim_{\epsilon \to 0^+} \left(\partial_\mu \partial^\mu + m^2 - i\epsilon \right) G_F(x - x') = -\delta^4(x - x') \, . \qquad (9.24)$$

Going back to the generating functional in Eq. (9.14), we note that for the present case, if $\lambda = 0$, then we can define

$$Z_0[J] = N \int \mathcal{D}\phi \, e^{\frac{i}{\hbar} S_0[x,J]}$$

$$= N \int \mathcal{D}\phi \, e^{\frac{i}{\hbar} \int d^4x \, \left(\frac{1}{2} \partial_\mu \phi \partial^\mu \phi - \frac{m^2}{2} \phi^2 + J\phi \right)}$$

$$= N \int \mathcal{D}\phi \, e^{-\frac{i}{\hbar} \int d^4x \, \left(\frac{1}{2} \phi (\partial_\mu \partial^\mu + m^2) \phi - J\phi \right)} \, , \qquad (9.25)$$

with $S_0[x, J]$ representing the free, quadratic part of the action. The field, $\phi(\mathbf{x}, t)$, is assumed to satisfy the asymptotic condition

$$\lim_{|\mathbf{x}| \to \infty} \phi(\mathbf{x}, t) \to 0.$$

Furthermore, the normalization constant, N, for the path integral is normally chosen such that

$$Z[0] = 1.$$

Let us note here once again that the integral in Eq. (9.25) should be properly evaluated by rotating to Euclidean space as discussed in section 4.1. Alternately, we can also define

$$Z_0[J] = \lim_{\epsilon \to 0^+} N \int \mathcal{D}\phi \, e^{-\frac{i}{\hbar} \int d^4 x \, (\frac{1}{2} \phi(\partial_\mu \partial^\mu + m^2 - i\epsilon)\phi - J\phi)}. \tag{9.26}$$

If we now redefine the variable of integration to be

$$\tilde{\phi}(x) = \phi(x) + \int d^4 x' \, G_F(x - x') J(x'), \tag{9.27}$$

with G_F defined in Eq. (9.23), then, we obtain

$$\lim_{\epsilon \to 0^+} \int d^4 x \, \frac{1}{2} \tilde{\phi}(\partial_\mu \partial^\mu + m^2 - i\epsilon)\tilde{\phi}$$

$$= \lim_{\epsilon \to 0^+} \int d^4 x \, \frac{1}{2} \left(\phi(x) + \int d^4 x' \, G_F(x - x') J(x') \right)$$

$$\times (\partial_\mu \partial^\mu + m^2 - i\epsilon) \left(\phi(x) + \int d^4 x'' \, G_F(x - x'') J(x'') \right)$$

$$= \lim_{\epsilon \to 0^+} \int d^4 x \, \left[\frac{1}{2} \phi(x)(\partial_\mu \partial^\mu + m^2 - i\epsilon)\phi(x) - J(x)\phi(x) \right.$$

$$\left. - \frac{1}{2} \int d^4 x' \, J(x) G_F(x - x') J(x') \right], \tag{9.28}$$

where we have used Eq. (9.24). Substituting Eq. (9.28) back into the generating functional in Eq. (9.26), we obtain (Note that the

Jacobian for the change of variable in Eq. (9.27) is trivial.)

$$Z_0[J] = \lim_{\epsilon \to 0^+} N \int \mathcal{D}\phi \, e^{-\frac{i}{\hbar}\int d^4x(\frac{1}{2}\phi(x)(\partial_\mu\partial^\mu + m^2 - i\epsilon)\phi(x) - J(x)\phi(x))}$$

$$= \lim_{\epsilon \to 0^+} e^{-\frac{i}{2\hbar}\int\int d^4x \, d^4x' \, J(x)G_F(x-x')J(x')}$$

$$\times N \int \mathcal{D}\tilde{\phi} \, e^{-\frac{i}{\hbar}\int d^4x \, \frac{1}{2}\tilde{\phi}(x)(\partial_\mu\partial^\mu + m^2 - i\epsilon)\tilde{\phi}(x)}$$

$$= N \left[\det(\partial_\mu\partial^\mu + m^2)\right]^{-\frac{1}{2}}$$

$$\times e^{-\frac{i}{2\hbar}\int\int d^4x \, d^4x' \, J(x)G_F(x-x')J(x')}$$

$$= Z_0[0] \, e^{-\frac{i}{2\hbar}\int\int d^4x \, d^4x' \, J(x)G_F(x-x')J(x')} . \tag{9.29}$$

Here we have used a generalization of the result in Eq. (4.2) for a field theory.

As in the case of the harmonic oscillator, we note that when $\lambda = 0$,

$$\langle 0|\phi(x)|0\rangle = \frac{(-i\hbar)}{Z_0[J]} \frac{\delta Z_0[J]}{\delta J(x)}\bigg|_{J=0}$$

$$= \frac{(-i\hbar)}{Z_0[J]} \left(-\frac{i}{\hbar}\int d^4x' \, G_F(x-x')J(x')\right) Z_0[J]\bigg|_{J=0}$$

$$= 0,$$

$$\langle 0|T(\phi(x)\phi(y))|0\rangle = \frac{(-i\hbar)^2}{Z_0[J]} \frac{\delta^2 Z_0[J]}{\delta J(x)\delta J(y)}\bigg|_{J=0}$$

$$= \frac{(-i\hbar)^2}{Z_0[J]} \left(-\frac{i}{\hbar}G_F(x-y)\right) Z_0[J]\bigg|_{J=0}$$

$$= i\hbar \, G_F(x-y) . \tag{9.30}$$

Namely, we obtain once again the result that the Feynman propagator is nothing other than the time ordered two point correlation function in the vacuum (see Eq. (4.62)).

Just as the path integral for the anharmonic oscillator cannot be evaluated in a closed form, the ϕ^4-theory does not also have a closed form expression for the generating functional. However, we can evaluate it perturbatively at least when the coupling is weak. We note that we can write (as we had also noted earlier in Eq. (4.68) in the case of the anharmonic oscillator)

$$\phi(x) \rightarrow \frac{\delta}{\delta J(x)}, \qquad (9.31)$$

when acting on the free, quadratic action $S_0[\phi, J]$. Therefore, we can rewrite the generating functional of Eq. (9.13), in the present case, also as

$$Z[J] = N \int \mathcal{D}\phi \, e^{\frac{i}{\hbar} \int d^4x \left(\frac{1}{2} \partial_\mu \phi \partial^\mu \phi - \frac{m^2}{2}\phi^2 - \frac{\lambda}{4!}\phi^4 + J\phi\right)}$$

$$= N \int \mathcal{D}\phi \left(e^{-\frac{i\lambda}{4!\hbar} \int d^4x \, \phi^4(x)}\right) e^{\frac{i}{\hbar} S_0[\phi, J]}$$

$$= N \int \mathcal{D}\phi \left(e^{-\frac{i\lambda}{4!\hbar} \int d^4x \, (-i\hbar \frac{\delta}{\delta J(x)})^4}\right) e^{\frac{i}{\hbar} S_0[\phi, J]}$$

$$= \left(e^{-\frac{i\lambda}{4!\hbar} \int d^4x \, (-i\hbar \frac{\delta}{\delta J(x)})^4}\right) N \int \mathcal{D}\phi \, e^{\frac{i}{\hbar} S_0[\phi, J]}$$

$$= \left(e^{-\frac{i\lambda}{4!\hbar} \int d^4x \, (-i\hbar \frac{\delta}{\delta J(x)})^4}\right) Z_0[J]. \qquad (9.32)$$

Once again, we note that this is very analogous to the result obtained in Eq. (4.69) for the anharmonic oscillator.

A power series expansion in λ for the generating functional in Eq. (9.32) follows by Taylor expanding the exponential involving the interaction terms. Thus, we have

$$Z[J] = \left[1 - \frac{i\lambda\hbar^3}{4!} \int d^4x \frac{\delta^4}{\delta J^4(x)} + \frac{1}{2!} \left(-\frac{i\lambda\hbar^3}{4!}\right)^2 \right.$$

$$\left. \times \left(\int d^4x \frac{\delta^4}{\delta J^4(x)}\right) \left(\int d^4y \frac{\delta^4}{\delta J^4(y)}\right) + \cdots \right] Z_0[J]$$

$$= Z_0[0] \left[1 - \frac{i\lambda\hbar^3}{4!} \int d^4x \frac{\delta^4}{\delta J^4(x)} + \frac{1}{2!} \left(-\frac{i\lambda\hbar^3}{4!} \right)^2 \right.$$

$$\left. \times \left(\int d^4x \frac{\delta^4}{\delta J^4(x)} \right) \left(\int d^4y \frac{\delta^4}{\delta J^4(y)} \right) + \cdots \right]$$

$$\times e^{-\frac{i}{2\hbar} \int\int d^4x_1 d^4x_2 J(x_1) G_F(x_1-x_2) J(x_2)} . \tag{9.33}$$

To obtain a feeling for how the actual calculations are carried out, let us derive some of the Green's functions to low orders in the coupling constant for the ϕ^4-theory. Let us recall from Eq. (4.50) that, by definition,

$$\langle 0|T(\phi(x_1)\phi(x_2) \cdots \phi(x_n))|0 \rangle$$

$$= \frac{(-i\hbar)^n}{Z[J]} \left. \frac{\delta^n Z[J]}{\delta J(x_1)\delta J(x_2) \cdots \delta J(x_n)} \right|_{J=0}. \tag{9.34}$$

Furthermore, from the symmetry of $Z[J]$ in Eqs. (9.29) and (9.32) (namely, from the fact that it is invariant under $J \leftrightarrow -J$), we conclude that the vacuum expectation value of the time ordered product of an odd number of fields must vanish in this theory. In other words,

$$\langle 0|T(\phi(x_1) \cdots \phi(x_{2n+1}))|0 \rangle$$

$$= \frac{(-i\hbar)^{2n+1}}{Z[J]} \left. \frac{\delta^{2n+1} Z[J]}{\delta J(x_1) \cdots \delta J(x_{2n+1})} \right|_{J=0} = 0. \tag{9.35}$$

Consequently, only the even order Green's functions will be nontrivial in this theory and let us calculate only the 2-point and the 4-point functions up to order λ. By definition,

$$\langle 0|T(\phi(x_1)\phi(x_2))|0 \rangle = \frac{(-i\hbar)^2}{Z[J]} \left. \frac{\delta^2 Z[J]}{\delta J(x_1)\delta J(x_2)} \right|_{J=0}. \tag{9.36}$$

Keeping terms up to order λ, we note from Eq. (9.33) that

$$Z[J] \simeq Z_0[0] \left(1 - \frac{i\lambda\hbar^3}{4!} \int d^4x \frac{\delta^4}{\delta J^4(x)} \right)$$

$$\times e^{-\frac{i}{2\hbar} \int\int d^4x_1 d^4x_2 J(x_1) G_F(x_1-x_2) J(x_2)} . \tag{9.37}$$

To evaluate this, let us note that

$$\frac{\delta^2}{\delta J^2(x)} e^{-\frac{i}{2\hbar} \iint d^4x_1 d^4x_2 \, J(x_1) G_F(x_1-x_2) J(x_2)}$$

$$= \frac{\delta}{\delta J(x)} \left[-\frac{i}{\hbar} \left(\int d^4x_3 G_F(x-x_3) J(x_3) \right) \right.$$

$$\left. \times e^{-\frac{i}{2\hbar} \iint d^4x_1 d^4x_2 J(x_1) G_F(x_1-x_2) J(x_2)} \right]$$

$$= \left[-\frac{i}{\hbar} G_F(0) \right.$$

$$\left. -\frac{1}{\hbar^2} \left(\int d^4x_3 G_F(x-x_3) J(x_3) \right) \left(\int d^4x_4 G_F(x-x_4) J(x_4) \right) \right]$$

$$\times e^{-\frac{i}{2\hbar} \iint d^4x_1 d^4x_2 J(x_1) G_F(x_1-x_2) J(x_2)} . \tag{9.38}$$

With some algebraic manipulations, this, then, leads to the result

$$\int d^4x \frac{\delta^4}{\delta J^4(x)} e^{-\frac{i}{2\hbar} \iint d^4x_1 d^4x_2 J(x_1) G_F(x_1-x_2) J(x_2)}$$

$$= \int d^4x \frac{\delta^2}{\delta J^2(x)} \left(\frac{\delta^2}{\delta J^2(x)} e^{-\frac{i}{2\hbar} \iint d^4x_1 d^4x_2 J(x_1) G_F(x_1-x_2) J(x_2)} \right)$$

$$= \int d^4x \left[-\frac{3}{\hbar^2} G_F(0) G_F(0) \right.$$

$$+ \frac{6i}{\hbar^3} G_F(0) \left(\int d^4x_3 G_F(x-x_3) J(x_3) \right)$$

$$\times \left(\int d^4x_4 G_F(x-x_4) J(x_4) \right)$$

$$+ \frac{1}{\hbar^4} \left(\int d^4x_3 G_F(x-x_3) J(x_3) \right) \left(\int d^4x_4 G_F(x-x_4) J(x_4) \right)$$

$$\left. \times \left(\int d^4x_5 G_F(x-x_5) J(x_5) \right) \left(\int d^4x_6 G_F(x-x_6) J(x_6) \right) \right]$$

$$\times e^{-\frac{i}{2\hbar} \iint d^4x_1 d^4x_2 J(x_1) G_F(x_1-x_2) J(x_2)} . \tag{9.39}$$

Putting this back into Eq. (9.37), we obtain the generating functional to linear power in λ to be

$$
Z[J] = Z_0[0] \left[1 + \frac{i\lambda\hbar}{8} G_F(0) G_F(0) \int d^4x \right.
$$

$$
+ \frac{\lambda}{4} G_F(0) \int d^4x \left(\int d^4x_3 G_F(x - x_3) J(x_3) \right)
$$

$$
\times \left(\int d^4x_4 G_F(x - x_4) J(x_4) \right)
$$

$$
- \frac{i\lambda}{4!\hbar} \int d^4x \left(\int d^4x_3 G_F(x - x_3) J(x_3) \right)
$$

$$
\times \left(\int d^4x_4 G_F(x - x_4) J(x_4) \right)
$$

$$
\left. \times \left(\int d^4x_5 G_F(x - x_5) J(x_5) \right) \left(\int d^4x_6 G_F(x - x_6) J(x_6) \right) \right]
$$

$$
\times e^{-\frac{i}{2\hbar} \iint d^4x_1 d^4x_2 J(x_1) G_F(x_1 - x_2) J(x_2)} . \tag{9.40}
$$

It now follows that

$$
Z[0] = Z_0[0] \left(1 + \frac{i\lambda\hbar}{8} G_F(0) G_F(0) \int d^4x \right) . \tag{9.41}
$$

This is clearly divergent and as we have argued earlier, the divergence can be absorbed into the normalization constant. From Eq. (9.40) we can also calculate to linear order in λ

$$
\frac{\delta^2 Z[J]}{\delta J(x_1) \delta J(x_2)} \bigg|_{J=0}
$$

$$
= Z_0[0] \left[\left(1 + \frac{i\lambda\hbar}{8} G_F(0) G_F(0) \int d^4x \right) \left(-\frac{i}{\hbar} G_F(x_1 - x_2) \right) \right.
$$

$$
\left. + \frac{\lambda}{2} G_F(0) \int d^4x\, G_F(x - x_1) G_F(x - x_2) \right] . \tag{9.42}
$$

Therefore, to linear order in λ, we obtain

$$\langle 0|T(\phi(x_1)\phi(x_2))|0\rangle$$

$$= \frac{(-i\hbar)^2}{Z[J]} \left.\frac{\delta^2 Z[J]}{\delta J(x_1)\delta J(x_2)}\right|_{J=0}$$

$$= -\frac{\hbar^2}{Z_0[0]\left(1 + \frac{i\lambda\hbar}{8}G_F(0)G_F(0)\int d^4x\right)}$$

$$\times Z_0[0]\left[\left(1 + \frac{i\lambda\hbar}{8}G_F(0)G_F(0)\int d^4x\right)\left(-\frac{i}{\hbar}G_F(x_1 - x_2)\right)\right.$$

$$\left. + \frac{\lambda}{2}G_F(0)\int d^4x\, G_F(x - x_1)G_F(x - x_2)\right]$$

$$\simeq -\hbar^2\left[-\frac{i}{\hbar}G_F(x_1 - x_2) + \frac{\lambda}{2}G_F(0)\int d^4x G_F(x - x_1)G_F(x - x_2)\right]$$

$$= i\hbar G_F(x_1 - x_2) - \frac{\lambda\hbar^2}{2}G_F(0)\int d^4x G_F(x - x_1)G_F(x - x_2).$$

$$(9.43)$$

Here in the second term, we have only kept the leading order term coming from the expansion of the denominator since we are interested in the 2-point function up to order λ. We note that the first term is, of course, the Feynman propagator for the free theory defined in Eq. (9.23). The second term, on the other hand, is a first order quantum correction. It is worth pointing it out here that $G_F(0)$ is a divergent quantity as we can readily check from the form of the propagator. Thus, we find that the first order correction to the propagator in this theory is divergent. This is, indeed, a general feature of quantum field theories, namely, the quantum corrections in a field theory lead to divergences which are then taken care of by what is commonly known as the process of renormalization.

Next, let us calculate the 4-point function up to order λ. To leading order, we note from Eq. (9.40) that

$$\frac{\delta^4 Z[J]}{\delta J(x_1)\delta J(x_2)\delta J(x_3)\delta J(x_4)}\bigg|_{J=0}$$

$$= Z_0[0]\left[-\frac{1}{\hbar^2}\left(1 + \frac{i\lambda\hbar}{8}G_F(0)G_F(0)\int d^4x\right)\right.$$

$$\times\left(G_F(x_1 - x_2)G_F(x_3 - x_4)\right.$$

$$\left. + G_F(x_1 - x_3)G_F(x_2 - x_4) + G_F(x_1 - x_4)G_F(x_2 - x_3)\right)$$

$$-\frac{i\lambda}{2\hbar}G_F(0)\int d^4x\left\{G_F(x_1 - x_2)G_F(x - x_3)G_F(x - x_4)\right.$$

$$+ G_F(x_1 - x_3)G_F(x - x_2)G_F(x - x_4)$$

$$+ G_F(x_1 - x_4)G_F(x - x_2)G_F(x - x_3)$$

$$+ G_F(x_2 - x_3)G_F(x - x_1)G_F(x - x_4)$$

$$+ G_F(x_2 - x_4)G_F(x - x_1)G_F(x - x_3)$$

$$\left. + G_F(x_3 - x_4)G_F(x - x_1)G_F(x - x_2)\right\}$$

$$\left. -\frac{i\lambda}{\hbar}\int d^4x G_F(x - x_1)G_F(x - x_2)G_F(x - x_3)G_F(x - x_4)\right].$$

$$(9.44)$$

Substituting this back into the definition of the 4-point function in Eq. (9.34) and keeping terms only up to order λ, we obtain

$$\langle 0|T(\phi(x_1)\phi(x_2)\phi(x_3)\phi(x_4))|0\rangle$$

$$= \frac{(-i\hbar)^4}{Z[J]}\frac{\delta^4 Z[J]}{\delta J(x_1)\delta J(x_2)\delta J(x_3)\delta J(x_4)}\bigg|_{J=0}$$

$$= -\hbar^2 \Big(G_F(x_1 - x_2)G_F(x_3 - x_4) + G_F(x_1 - x_3)G_F(x_2 - x_4)$$

$$+ G_F(x_1 - x_4)G_F(x_2 - x_3) \Big)$$

$$- \frac{i\lambda\hbar^3}{2} G_F(0) \int d^4x \Big\{ G_F(x_1 - x_2)G_F(x - x_3)G_F(x - x_4)$$

$$+ G_F(x_1 - x_3)G_F(x - x_2)G_F(x - x_4)$$

$$+ G_F(x_1 - x_4)G_F(x - x_2)G_F(x - x_3)$$

$$+ G_F(x_2 - x_3)G_F(x - x_1)G_F(x - x_4)$$

$$+ G_F(x_2 - x_4)G_F(x - x_1)G_F(x - x_3)$$

$$+ G_F(x_3 - x_4)G_F(x - x_1)G_F(x - x_2) \Big\}$$

$$- i\lambda\hbar^3 \int d^4x\, G_F(x - x_1)G_F(x - x_2)G_F(x - x_3)G_F(x - x_4).$$

$$(9.45)$$

9.3 Feynman Rules

These lowest order calculations are enough to convince any one interested in the subject that a systematic procedure needs to be developed to keep track of the perturbative expansion. The Feynman rules do precisely this. Let us note that the basic elements in our ϕ^4-theory are the Feynman propagator for the free theory and the interaction. Let us represent these diagrammatically as

$$x_1 \quad\rule[0.5ex]{3em}{0.4pt}\quad x_2 = i\hbar\, G_F(x_1 - x_2)\,,$$

$$= V(x_1, x_2, x_3, x_4)$$

$$= \frac{i}{\hbar} \frac{\delta^4 S[\phi]}{\delta\phi(x_1)\delta\phi(x_2)\delta\phi(x_3)\delta\phi(x_4)}\bigg|_{\phi=0}$$

$$= -\frac{i\lambda}{\hbar} \int d^4x \delta(x - x_1)\delta(x - x_2)\delta(x - x_3)\delta(x - x_4).$$

(9.46)

The interaction vertex is understood to be the part of the graph without the external lines or the propagators. It is clear that given these basic elements, we can construct various nontrivial graphs by joining the vertex to the propagators. Let us further use the rule that in evaluating such graphs, we must integrate over the intermediate points where a vertex connects with the propagators. With these rules, then, we can obtain the value for the following simple diagram to be

$$= \int d^4y_1 d^4y_2 d^4y_3 d^4y_4 i\hbar G_F(x_1 - y_1)i\hbar G_F(y_2 - x_2)$$

$$\times i\hbar G_F(y_3 - y_4)V(y_1, y_2, y_3, y_4)$$

$$= (i\hbar)^3 \int d^4y d^4y_1 d^4y_2 d^4y_3 d^4y_4 G_F(x_1 - y_1)G_F(y_2 - x_2)$$

$$\times G_F(y_3 - y_4)\left(-\frac{i\lambda}{\hbar}\delta(y - y_1)\delta(y - y_2)\delta(y - y_3)\delta(y - y_4)\right)$$

$$= -\lambda\hbar^2 G_F(0) \int d^4y \, G_F(x_1 - y)G_F(y - x_2).$$

(9.47)

There is one final rule. Namely, that if the internal part of a diagram has a symmetry, then the true value of the diagram is obtained by dividing with this symmetry factor. For the present case, the internal bubble in the diagram for Eq. (9.47) is invariant under a rotation by $180°$. The symmetry factor, in this case, is $2^1 = 2$. (The symmetry factor for a Feynman diagram is the most difficult to determine by naive inspection and should be obtained through a careful evaluation- when necessary, going back to the Wick expansion of field theory.) Dividing by this factor, we obtain the value of this diagram to be

$$x_1 \underset{}{\overline{Q}} x_2 = -\frac{\lambda \hbar^2}{2} G_F(0) \int \mathrm{d}^4 x\, G_F(x_1 - x) G_F(x - x_2), \quad (9.48)$$

which we recognize to be the first order (linear in λ) correction to the propagator in Eq. (9.43).

Each such diagram that can be constructed from the basic elements in Eq. (9.46) is known as a Feynman diagram of the theory and corresponds to a basic term in the perturbative expansion. Thus, for example, we now immediately recognize from Eqs. (9.43) and (9.45) that up to order λ, we can write

$$\langle 0|T(\phi(x_1)\phi(x_2))|0\rangle$$

$$= x_1 \overline{} x_2 + x_1 \overline{Q} x_2 \, ,$$

$$\langle 0|T(\phi(x_1)\phi(x_2)\phi(x_3)\phi(x_4))|0\rangle$$

$$= \begin{matrix} x_1 \\ x_3 \end{matrix} \overline{} \begin{matrix} x_2 \\ x_4 \end{matrix} + \begin{matrix} x_1 \\ x_2 \end{matrix} \overline{} \begin{matrix} x_3 \\ x_4 \end{matrix} + \begin{matrix} x_1 \\ x_2 \end{matrix} \overline{} \begin{matrix} x_4 \\ x_3 \end{matrix}$$

$$+ \begin{matrix} x_3 \\ x_1 \end{matrix} \overline{Q} \begin{matrix} x_4 \\ x_2 \end{matrix} + \begin{matrix} x_2 \\ x_1 \end{matrix} \overline{Q} \begin{matrix} x_4 \\ x_3 \end{matrix} + \begin{matrix} x_2 \\ x_1 \end{matrix} \overline{Q} \begin{matrix} x_3 \\ x_4 \end{matrix}$$

$$+ \begin{matrix} x_1 \\ x_2 \end{matrix} \overline{Q} \begin{matrix} x_4 \\ x_3 \end{matrix} + \begin{matrix} x_1 \\ x_2 \end{matrix} \overline{Q} \begin{matrix} x_3 \\ x_4 \end{matrix} + \begin{matrix} x_1 \\ x_3 \end{matrix} \overline{Q} \begin{matrix} x_2 \\ x_4 \end{matrix}$$

$$+ \begin{matrix} x_1 \\ x_2 \end{matrix} \times \begin{matrix} x_4 \\ x_3 \end{matrix} \, .$$

$$(9.49)$$

9.4 Connected Diagrams

The Feynman diagrams, as can be seen from the diagrams in Eq. (9.49), clearly consist of two classes of diagrams, one where each part of the diagram is connected to the rest of the diagram and another where parts of the diagram are disconnected. The Feynman diagrams, which consist of parts that are not connected to one another, are known as disconnected Feynman diagrams. It is clear from the above simple example that the generating functional $Z[J]$ generates Green's functions which contain disconnected diagrams as well. As we have discussed earlier (see the discussion following Eq. (4.51)), the logarithm of $Z[J]$ generates Green's functions which contain only the connected diagrams (otherwise known as the connected Green's functions) and these give rise to the physical scattering matrix elements. Namely,

$$W[J] = -i\hbar \ln Z[J] \,, \tag{9.50}$$

generates connected Green's functions. We note from Eqs. (9.50), (9.34) and (9.35) that

$$\left. \frac{\delta W[J]}{\delta J(x_1)} \right|_{J=0} = -\frac{i\hbar}{Z[J]} \left. \frac{\delta Z[J]}{\delta J(x_1)} \right|_{J=0} = \langle 0|\phi(x_1)|0\rangle. \tag{9.51}$$

Similarly, for the two point function, we obtain

$$-i\hbar \left. \frac{\delta^2 W[J]}{\delta J(x_1)\delta J(x_2)} \right|_{J=0}$$

$$= (-i\hbar)^2 \left[\frac{1}{Z[J]} \frac{\delta^2 Z[J]}{\delta J(x_1)\delta J(x_2)} - \frac{1}{Z^2[J]} \frac{\delta Z[J]}{\delta J(x_1)} \frac{\delta Z[J]}{\delta J(x_2)} \right]_{J=0}$$

$$= \langle 0|T(\phi(x_1)\phi(x_2))|0\rangle - \langle 0|\phi(x_1)|0\rangle \, \langle 0|\phi(x_2)|0\rangle$$

$$= \langle 0|T(\phi(x_1)\phi(x_2))|0\rangle_c \,. \tag{9.52}$$

In a similar manner, with some algebra, it can be shown that

$$(-i\hbar)^2 \left. \frac{\delta^3 W[J]}{\delta J(x_1)\delta J(x_2)\delta J(x_3)} \right|_{J=0}$$

$$= \langle 0|T(\phi(x_1)\phi(x_2)\phi(x_3))|0\rangle - \langle 0|\phi(x_1)|0\rangle \, \langle 0|T(\phi(x_2)\phi(x_3))|0\rangle$$

$$- \langle 0|\phi(x_2)|0\rangle \langle 0|T(\phi(x_3)\phi(x_1))|0\rangle - \langle 0|\phi(x_3)|0\rangle \, \langle 0|T(\phi(x_1)\phi(x_2))|0\rangle$$

$$+ 2\langle 0|\phi(x_1)|0\rangle \, \langle 0|\phi(x_2)|0\rangle \, \langle 0|\phi(x_3)|0\rangle$$

$$= \langle 0|T(\phi(x_1)\phi(x_2)\phi(x_3)|0\rangle_c \,, \tag{9.53}$$

and so on.

We can, of course, check explicitly that the connected Green's functions involve only connected Feynman diagrams as follows. Note that for the ϕ^4-theory, as we have discussed earlier in Eq. (9.35)

$$\langle 0|\phi(x)|0\rangle = 0 \,. \tag{9.54}$$

Consequently, up to order λ, we note from Eq. (9.52) that

$$\langle 0|T(\phi(x_1)\phi(x_2))|0\rangle = \langle 0|T(\phi(x_1)\phi(x_2))|0\rangle_c$$

$$= x_1 \underline{\hspace{1.5cm}} x_2 + x_1 \underline{\hspace{0.3cm}}^{\bigcirc}\underline{\hspace{0.3cm}} x_2 \,. \tag{9.55}$$

Similarly, for the 4-point function, we have

$$\langle 0|T(\phi(x_1)\phi(x_2)\phi(x_3)\phi(x_4))|0\rangle_c$$

$$= \langle 0|T(\phi(x_1)\phi(x_2)\phi(x_3)\phi(x_4))|0\rangle$$

$$- \langle 0|T(\phi(x_1)\phi(x_2))|0\rangle \, \langle 0|T(\phi(x_3)\phi(x_4))|0\rangle$$

$$- \langle 0|T(\phi(x_1)\phi(x_3))|0\rangle \, \langle 0|T(\phi(x_2)\phi(x_4))|0\rangle$$

$$- \langle 0|T(\phi(x_1)\phi(x_4))|0\rangle \, \langle 0|T(\phi(x_2)\phi(x_3))|0\rangle$$

$$= \begin{matrix} x_1 & & x_4 \\ & \diagdown\!\!\!\!\diagup & \\ x_2 & & x_3 \end{matrix} \,. \tag{9.56}$$

Thus, we see that $W[J]$ generates connected Green's functions. Given this, we can write down the diagrammatic expansion of the connected 2-point Green's function up to order λ^2 in this theory simply as

$$x_1 \,\text{—}\!\!\bigoplus\!\!\text{—}\, x_2$$

$$= x_1 \,\text{———}\, x_2 \; + \; x_1 \,\text{——}\overset{\text{\large$\mathcal{Q}$}}{\text{——}}\, x_2 \; + \; x_1 \,\text{—}\overset{\text{\large$\mathcal{QQ}$}}{\text{——}}\, x_2$$

$$+ \; x_1 \,\text{—}\!\!\bigominus\!\!\text{—}\, x_2 \; + \; x_1 \,\text{——}\overset{\text{\large$\mathcal{8}$}}{\text{——}}\, x_2 \,.$$

$$(9.57)$$

The organization of the perturbation series now becomes quite straightforward. Note that while $W[J]$ generates connected diagrams, it contains diagrams that are reducible to two connected diagrams upon cutting an internal line. Thus, in the 2-point function represented above, the third graph is reducible upon cutting the internal propagator. Such diagrams are called 1P (one particle) reducible. It is clear that the 1P irreducible diagrams are in some sense more fundamental since we can construct all the connected diagrams from them. We will take up the study of the 1PI (one particle irreducible) diagrams in the next chapter.

9.5   References

Huang, K., "Quarks, Leptons and Gauge Fields", World Scientific
        Publishing.
Jona-Lasinio, J., Nuovo Cimento **34**, 1790 (1964).
Schwinger, J., Proc. Natl. Sci. USA **37**, 452 (1951); *ibid* **37**, 455
        (1951).

Chapter 10

Effective Action

10.1 The Classical Field

As we saw in the last chapter, the generating functional for Green's functions in a scalar field theory is given by

$$Z[J] = e^{\frac{i}{\hbar}W[J]} = N \int \mathcal{D}\phi \, e^{\frac{i}{\hbar}S[\phi,J]}, \qquad (10.1)$$

where $W[J]$ generates connected Green's functions. We note that the one point function in the presence of an external source is given by

$$\frac{\delta W[J]}{\delta J(x)} = \frac{(-i\hbar)}{Z[J]} \frac{\delta Z[J]}{\delta J(x)} = \langle 0|\phi(x)|0\rangle^J. \qquad (10.2)$$

For the ϕ^4-theory, we have seen in Eq. (9.35) that this vacuum expectation value of the field operator vanishes in the absence of external sources (namely, when $J = 0$). In general, however, let us note that we can write

$$\langle 0|\phi(x)|0\rangle = \langle 0|e^{\frac{i}{\hbar}P\cdot x}\phi(0)e^{-\frac{i}{\hbar}P\cdot x}|0\rangle, \qquad (10.3)$$

where we have identified $e^{-\frac{i}{\hbar}P\cdot x}$ with the generator of space-time translations. Assuming that the vacuum state in our Hilbert space is unique and that it is Poincare invariant, namely, that it satisfies

$$e^{-\frac{i}{\hbar}P\cdot a}|0\rangle = |0\rangle$$

$$\text{or,} \quad P_\mu|0\rangle = 0, \qquad (10.4)$$

we obtain from Eqs. (10.3) and (10.4) that

$$\langle 0|\phi(x)|0\rangle = \langle 0|\phi(0)|0\rangle = \text{constant}. \tag{10.5}$$

Thus, from the symmetry arguments alone, we conclude that the one point function can, in general, be a constant, independent of space-time coordinates. For the ϕ^4-theory, this constant coincides with zero, namely, we have

$$\langle 0|\phi(x)|0\rangle = 0. \tag{10.6}$$

The value of the one point function is quite important in the study of symmetries. As we will see later, a nonvanishing value of this quantity signals the spontaneous break down of some symmetry in the theory.

In the presence of an external source, however, the vacuum expectation value of the field operator is a functional of the source and need not be zero. Let us denote this by

$$\frac{\delta W[J]}{\delta J(x)} = \frac{(-i\hbar)}{Z[J]}\frac{\delta Z[J]}{\delta J(x)} = \langle 0|\phi(x)|0\rangle^J = \phi_{\rm c}(x), \tag{10.7}$$

and note that it is indeed a functional of the external source. The field $\phi_{\rm c}(x)$ which is only a classical variable is known as the classical field. To understand the meaning of the classical field as well as the reason for its name, let us analyze the generating functional in Eq. (10.1)

$$Z[J] = N \int \mathcal{D}\phi \, e^{\frac{i}{\hbar}S[\phi,J]},$$

in some detail. Since $Z[J]$ is independent of $\phi(x)$, an arbitrary, infinitesimal change in $\phi(x)$ in the integrand on the right hand side will leave the generating functional invariant. Namely, under such a change,

$$\delta Z[J] = N \int \mathcal{D}\phi \, \frac{i}{\hbar}\delta S[\phi, J]e^{\frac{i}{\hbar}S[\phi,J]}$$

$$= \frac{iN}{\hbar} \int \mathcal{D}\phi \left(\int d^4x \delta\phi(x)\frac{\delta S[\phi, J]}{\delta\phi(x)} \right) e^{\frac{i}{\hbar}S[\phi,J]}$$

$$= 0. \tag{10.8}$$

Here we are assuming that the functional integration measure in Eq. (10.1) does not change under a redefinition of the field variable. (If it does change, then we will have an additional term coming from the change of the measure. Such a term plays an important role in the study of anomalies.) Since the relation in Eq. (10.8) must be true for any arbitrary variation $\delta\phi(x)$ of the field variable, it now follows that we must have

$$N \int \mathcal{D}\phi \, \frac{\delta S[\phi, J]}{\delta\phi(x)} \, e^{\frac{i}{\hbar} S[\phi, J]} = 0 \,. \tag{10.9}$$

This merely expresses what we already know from our study of quantum mechanics. Namely, that the Euler-Lagrange equations of a theory hold only as an expectation value equation (Ehrenfest's theorem) or more explicitly,

$$\langle 0| \frac{\delta S[\phi, J]}{\delta\phi(x)} |0\rangle^J = 0 \,. \tag{10.10}$$

The Euler-Lagrange equation (the classical equation) has the generic form (see, for example, Eq. (9.8) or (9.17))

$$-\frac{\delta S[\phi, J]}{\delta\phi(x)} = F(\phi(x)) - J(x) = 0 \,, \tag{10.11}$$

where the quantity $F(\phi(x))$ depends on the specific dynamics of the system and for the ϕ^4-theory, we note from, Eq. (9.15), that it has the particular form

$$F(\phi(x)) \equiv -\frac{\delta S[\phi]}{\delta\phi(x)} = (\partial_\mu \partial^\mu + m^2)\phi(x) + \frac{\lambda}{3!}\phi^3(x) \,. \tag{10.12}$$

Using Eq. (10.11) in Eq. (10.9), then, we obtain

$$-N \int \mathcal{D}\phi \, \frac{\delta S[\phi, J]}{\delta\phi(x)} \, e^{\frac{i}{\hbar} S[\phi, J]}$$

$$= N \int \mathcal{D}\phi \, (F(\phi(x)) - J(x)) \, e^{\frac{i}{\hbar} S[\phi, J]} = 0 \,. \tag{10.13}$$

Let us recall that

$$-i\hbar \frac{\delta Z[J]}{\delta J(x)} = N \int \mathcal{D}\phi \, \phi(x) e^{\frac{i}{\hbar} S[\phi, J]} \,, \tag{10.14}$$

which allows us to use the identification

$$\phi(x) \rightarrow -i\hbar \frac{\delta}{\delta J(x)},$$

using which, we can write Eq. (10.13) also as

$$\left[F\left(-i\hbar \frac{\delta}{\delta J(x)} \right) - J(x) \right] Z[J] = 0$$

$$\text{or,} \quad \left[F\left(-i\hbar \frac{\delta}{\delta J(x)} \right) - J(x) \right] e^{\frac{i}{\hbar} W[J]} = 0$$

$$\text{or,} \quad e^{-\frac{i}{\hbar} W[J]} \left[F\left(-i\hbar \frac{\delta}{\delta J(x)} \right) - J(x) \right] e^{\frac{i}{\hbar} W[J]} = 0$$

$$\text{or,} \quad F\left(\frac{\delta W[J]}{\delta J(x)} - i\hbar \frac{\delta}{\delta J(x)} \right) - J(x) = 0$$

$$\text{or,} \quad F\left(\phi_c(x) - i\hbar \frac{\delta}{\delta J(x)} \right) - J(x) = 0. \tag{10.15}$$

This is, of course, the full dynamical equation of the theory at the quantum level. It is quite different from the classical Euler-Lagrange equation in Eq. (10.11). But let us note that in the limit $\hbar \rightarrow 0$, the complete equation in Eq. (10.15) reduces to the form

$$F(\phi_c(x)) - J(x) = 0, \tag{10.16}$$

which is the familiar classical Euler-Lagrange equation in Eq. (10.11). It is for this reason that $\phi_c(x)$ is called the classical field. To get a better feeling for the quantum equation in Eq. (10.15) as well as Eq. (10.16) in the $\hbar \rightarrow 0$ limit, let us consider specifically the ϕ^4-theory. In this case, as we have noted earlier in Eq. (10.12), the Euler-Lagrange equation takes the form

$$F(\phi(x)) - J(x) = (\partial_\mu \partial^\mu + m^2)\phi(x) + \frac{\lambda}{3!} \phi^3(x) - J(x) = 0.$$

Consequently, the quantum equation takes the form

$$F\left(\phi_c(x) - i\hbar\frac{\delta}{\delta J(x)}\right) - J(x) = 0$$

or, $(\partial_\mu\partial^\mu + m^2)\left(\phi_c(x) - i\hbar\frac{\delta}{\delta J(x)}\right)$

$$+\frac{\lambda}{3!}\left(\phi_c(x) - i\hbar\frac{\delta}{\delta J(x)}\right)^3 - J(x) = 0$$

or, $(\partial_\mu\partial^\mu + m^2)\phi_c(x) + \frac{\lambda}{3!}\phi_c^3(x) - J(x)$

$$-\frac{i\lambda\hbar}{2!}\phi_c(x)\frac{\delta\phi_c(x)}{\delta J(x)} - \frac{\lambda\hbar^2}{3!}\frac{\delta^2\phi_c(x)}{\delta J(x)\delta J(x)} = 0. \quad (10.17)$$

In terms of $W[J]$, this can also be written as

$$(\partial_\mu\partial^\mu + m^2)\frac{\delta W[J]}{\delta J(x)} + \frac{\lambda}{3!}\left(\frac{\delta W[J]}{\delta J(x)}\right)^3 - J(x)$$

$$-\frac{i\lambda\hbar}{2!}\frac{\delta W[J]}{\delta J(x)}\frac{\delta^2 W[J]}{\delta J^2(x)} - \frac{\lambda\hbar^2}{3!}\frac{\delta^3 W[J]}{\delta J^3(x)} = 0. \quad (10.18)$$

We can think of Eq. (10.18) as the master equation governing the full dynamics of the quantum theory. By taking higher functional derivatives, we can determine from this, the dynamical equations satisfied by various Green's functions of the theory. These equations are also known as the Schwinger-Dyson equations and the Bethe-Salpeter equations are a special case of these equations. We should note here that in quantum field theory, whenever there are products of field operators at the same space-time point, the expressions become ill defined and have to be regularized (defined) in some manner. The manifestation of this problem is quite clear in Eq. (10.18). The equation involves second and third functional derivatives at the same space-time point which are not at all well defined. One needs to develop a systematic regularization procedure to handle these difficulties. The discussion of these topics lies outside the scope of these lectures.

Let us note, on the other hand, that in the limit $\hbar \to 0$, all such ill defined terms vanish and we have from Eq. (10.17)

$$(\partial_\mu \partial^\mu + m^2)\phi_c(x) + \frac{\lambda}{3!}\phi_c^3(x) - J(x) = 0, \tag{10.19}$$

which is the classical Euler-Lagrange equation. Furthermore, let us note that we can solve this equation iteratively through the use of the propagator defined in Eqs. (9.23) and (9.24) (Green's function) as follows. First, we note that the solution for $\phi_c(x)$ can be written as an integral equation

$$\phi_c(x) = -\int d^4x'\, G_F(x - x')(J(x') - \frac{\lambda}{3!}\phi_c^3(x'))$$

$$= -\int d^4x' G_F(x - x')J(x') + \frac{\lambda}{3!}\int d^4x'\, G_F(x - x')\phi_c^3(x'), \tag{10.20}$$

which can be solved iteratively. The iterative solution has the form

$$\phi_c(x) \simeq -\int d^4x'\, G_F(x - x')J(x') + \frac{\lambda}{3!}\int d^4x' G_F(x - x')$$

$$\times \left(\int d^4x'' G_F(x' - x'')(-J(x'') + \frac{\lambda}{3!}\phi_c^3(x''))\right)^3$$

$$= -\int d^4x' G_F(x - x')J(x') - \frac{\lambda}{3!}\int d^4x_1 d^4x_2 d^4x_3 d^4x_4 G_F(x - x_1)$$

$$\times G_F(x_1 - x_2)G_F(x_1 - x_3)G_F(x_1 - x_4)J(x_2)J(x_3)J(x_4) + \cdots \tag{10.21}$$

We can diagrammatically represent this if we introduce a vertex describing the interaction of the field with the external source as

$$\underset{x}{\rule{3cm}{0.4pt}}\!\!\times \;=\; \frac{i}{\hbar}\left.\frac{\delta S[\phi, J]}{\delta \phi(x)}\right|_{\phi=0} = \frac{i}{\hbar}J(x). \tag{10.22}$$

In this case, the iterative solution for the classical field can be written

diagrammatically as

$$\phi_{\rm c}(x) = \quad \cdots + \frac{1}{3!} \quad \cdots$$

$$+ \frac{1}{2!3!} \quad \cdots + \cdots . \tag{10.23}$$

In other words, the classical field in the limit $\hbar \to 0$ generates all the tree diagrams or the Born diagrams. It is, therefore, also known as the Born functional or the generating functional for tree diagrams. The combinatoric factors cancel out when taking functional derivatives of $\phi_{\rm c}$ with respect to sources and we obtain the n-point tree amplitudes.

10.2 Effective Action

The classical field, as we have seen in Eq. (10.7), is defined as

$$\frac{\delta W[J]}{\delta J(x)} = \phi_{\rm c}(x) .$$

This relation indicates that the variables $J(x)$ and $\phi_{\rm c}(x)$ are in some sense conjugate variables. As we have argued earlier, the classical field is a functional of the source $J(x)$. However, we can also invert the defining relation for the classical field and solve for the source $J(x)$ as a functional of $\phi_{\rm c}(x)$ at least perturbatively. In fact, let us define a new functional through a Legendre transformation as

$$\Gamma[\phi_{\rm c}] = W[J] - \int {\rm d}^4 x\, J(x)\phi_{\rm c}(x) . \tag{10.24}$$

It is clear, then, that

$$\frac{\delta\Gamma[\phi_c]}{\delta\phi_c(x)} = \frac{\delta W[J]}{\delta\phi_c(x)} - \int d^4y \frac{\delta J(y)}{\delta\phi_c(x)}\phi_c(y) - J(x)$$

$$= \int d^4y \frac{\delta W[J]}{\delta J(y)}\frac{\delta J(y)}{\delta\phi_c(x)} - \int d^4y \frac{\delta J(y)}{\delta\phi_c(x)}\frac{\delta W[J]}{\delta J(y)} - J(x)$$

or,
$$\frac{\delta\Gamma[\phi_c]}{\delta\phi_c(x)} = -J(x)\,. \tag{10.25}$$

In this derivation, we have used the chain rule for functional derivatives as well as the definition of the classical field. We note that Eq. (10.25), indeed, defines the source as a functional of the classical field and has the same structure as the classical Euler-Lagrange equation for a system in the presence of an external source. Let us recall from Eqs. (10.11) and (10.12) that it is given by

$$\frac{\delta S[\phi]}{\delta\phi(x)} = -J(x)\,. \tag{10.26}$$

It is for this reason that $\Gamma[\phi_c]$ is also known as the effective action functional. Note that as we have discussed earlier in Eq. (10.5), when

$$J \to 0, \quad \phi_c(x) \to \text{constant}.$$

In the framework of the effective action, we see from Eq. (10.25) that the value of this constant is determined from the equation

$$\frac{\delta\Gamma[\phi_c]}{\delta\phi_c(x)}\bigg|_{\phi_c(x)=\text{const.}} = 0\,. \tag{10.27}$$

This is an extremum equation and is much easier to analyze to determine whether a symmetry is spontaneously broken.

To understand the meaning of this new functional, $\Gamma[\phi_c]$, let us note that if we treat $\phi_c(x)$ as our independent variable, then we can write

$$\frac{\delta}{\delta J(x)} = \int d^4y \frac{\delta\phi_c(y)}{\delta J(x)}\frac{\delta}{\delta\phi_c(y)}$$

$$= \int d^4y \frac{\delta^2 W[J]}{\delta J(x)\delta J(y)}\frac{\delta}{\delta\phi_c(y)}\,. \tag{10.28}$$

Using this in Eq. (10.25), then, we obtain

$$\frac{\delta}{\delta J(y)}\left(\frac{\delta\Gamma[\phi_c]}{\delta\phi_c(x)}\right) = -\delta^4(x-y)$$

or, $$\int d^4z\,\frac{\delta^2 W[J]}{\delta J(y)\delta J(z)}\frac{\delta^2\Gamma[\phi_c]}{\delta\phi_c(z)\delta\phi_c(x)} = -\delta^4(x-y)\,. \quad (10.29)$$

Introducing the compact notation

$$W^{(n)} = \frac{\delta^n W[J]}{\delta J(x_1)\cdots\delta J(x_n)},$$

$$\Gamma^{(n)} = \frac{\delta^n \Gamma[\phi_c]}{\delta\phi_c(x_1)\cdots\delta\phi_c(x_n)}, \quad (10.30)$$

we can write Eq. (10.29) also in the compact form (We can view this as an operatorial equation where the appropriate coordinate dependences will arise by taking matrix elements in the coordinate basis.)

$$W^{(2)}\Gamma^{(2)} = -1\,. \quad (10.31)$$

We recall from Eqs. (9.30) and (9.52) that the full propagator of the theory is defined to be

$$W^{(2)}\Big|_{J=0} = -G\,. \quad (10.32)$$

Furthermore, recalling that when $J=0$, $\phi_c(x) = \phi_c = $ constant, we have from Eq. (10.31)

$$W^{(2)}\Big|_{J=0}\Gamma^{(2)}\Big|_{\phi_c} = -1$$

$$\text{or,} \quad G\Gamma^{(2)}\Big|_{\phi_c} = 1\,. \quad (10.33)$$

In other words, $\Gamma^{(2)}\big|_{\phi_c}$ is the inverse of the propagator at every order of the perturbation theory. Thus, writing

$$\Gamma^{(2)}\Big|_{\phi_c} = \Gamma_0^{(2)} + \Sigma = G_F^{-1} + \Sigma\,, \quad (10.34)$$

where Σ denotes the quantum corrections in $\Gamma^{(2)}|_{\phi_c}$, we have from Eq. (10.33)

$$G(G_F^{-1} + \Sigma) = 1$$

or, $G = \dfrac{1}{G_F^{-1} + \Sigma}$

$$= \frac{1}{G_F^{-1}} - \frac{1}{G_F^{-1}}\Sigma\frac{1}{G_F^{-1}} + \frac{1}{G_F^{-1}}\Sigma\frac{1}{G_F^{-1}}\Sigma\frac{1}{G_F^{-1}} \cdots$$

$$= G_F - G_F\Sigma G_F + G_F\Sigma G_F\Sigma G_F + \cdots . \tag{10.35}$$

Introducing the diagrammatic representation

$$\frac{i}{\hbar}\Sigma = -\!\!\bigcirc\!\!- \quad , \tag{10.36}$$

we have the diagrammatic relation for the propagator as

$$+ \underset{x}{\,} \bigcirc\!\bigcirc \underset{y}{\,} + \cdots . \tag{10.37}$$

It is clear from the above relation that Σ is nothing other than the 1P irreducible (1PI) 2-point vertex function. It is also known as the proper self energy diagram.

Given the relation in Eq. (10.29)

$$\int \mathrm{d}^4 z\, \frac{\delta^2 W[J]}{\delta J(y)\delta J(z)}\, \frac{\delta^2\Gamma[\phi_c]}{\delta\phi_c(z)\delta\phi_c(x)} = -\delta^4(x-y)\,,$$

we can differentiate this with respect to $\frac{\delta}{\delta J(\omega)}$ to obtain

$$\int \mathrm{d}^4 z\, \frac{\delta^3 W[J]}{\delta J(\omega)\delta J(y)\delta J(z)}\, \frac{\delta^2\Gamma[\phi_c]}{\delta\phi_c(z)\delta\phi_c(x)} \tag{10.38}$$

$$= -\int \mathrm{d}^4 z\mathrm{d}^4\sigma\, \frac{\delta^2 W[J]}{\delta J(y)\delta J(z)}\, \frac{\delta^3\Gamma[\phi_c]}{\delta\phi_c(z)\delta\phi_c(x)\delta\phi_c(\sigma)}\, \frac{\delta^2 W[J]}{\delta J(\sigma)\delta J(\omega)}\,.$$

Recalling that

$$W^{(2)}\Gamma^{(2)} = -1\,,$$

we can also rewrite this equation as

$$\frac{\delta^3 W[J]}{\delta J(x)\delta J(y)\delta J(z)}$$

$$= \int \mathrm{d}^4 x'\mathrm{d}^4 y'\mathrm{d}^4 z' \frac{\delta^2 W[J]}{\delta J(x)\delta J(x')} \frac{\delta^2 W[J]}{\delta J(y)\delta J(y')} \frac{\delta^2 W[J]}{\delta J(z)\delta J(z')}$$

$$\times \frac{\delta^3 \Gamma[\phi_\mathrm{c}]}{\delta\phi_\mathrm{c}(x')\delta\phi_\mathrm{c}(y')\delta\phi_\mathrm{c}(z')}\,. \tag{10.39}$$

In compact notation, we can write this equation also as

$$W^{(3)} = W^{(2)}W^{(2)}W^{(2)}\Gamma^{(3)}\,. \tag{10.40}$$

Furthermore, introducing the diagrammatic representations,

$$= -i\hbar\, W^{(2)}(x,y)\Big|_{J=0}\,,$$

$$= (-i\hbar)^2\, W^{(3)}(x,y,z)\Big|_{J=0}\,,$$

$$= \frac{i}{\hbar}\,\Gamma^{(3)}(x,y,z)\Big|_{\phi_\mathrm{c}}\,, \tag{10.41}$$

we note that Eq. (10.39) or (10.40) can also be diagrammatically represented as

$$\tag{10.42}$$

This shows that $\Gamma^{(3)}\big|_{\phi_c}$ gives the proper 3-point vertex function (in other words, it is 1PI).

A similar calculation for the 4-point connected Green's function leads to the diagrammatic relation

Thus, we see that we can expand the effective action functional as

$$\Gamma[\phi_c] = \sum_{n=1}^{\infty} \int d^4x_1 \cdots d^4x_n \frac{1}{n!}\, \Gamma^{(n)}(x_1, \cdots, x_n)\Big|_{\phi_c} \phi_c(x_1) \cdots \phi_c(x_n),$$

(10.43)

where (we are assuming $\phi_c = 0$ in the above expansion)

$$\Gamma^{(n)}(x_1, \cdots, x_n)\Big|_{\phi_c},$$

is the proper (1PI) n-point vertex function of the theory. It is for this reason that $\Gamma[\phi_c]$ is known as the 1PI generating functional. Let us note here that the 1PI vertex functions with suitable external wave functions lead to the scattering matrix of the theory.

Since $\Gamma[\phi_c]$ is an effective action, it can also be expanded alternately in powers of the derivative or momentum like the classical action. Thus, since the tree level action has the form

$$S[\phi] = \int d^4x \left(\frac{1}{2} \partial_\mu \phi \partial^\mu \phi - \frac{m^2}{2} \phi^2 - \frac{\lambda}{4!} \phi^4 \right)$$

$$= \int d^4x \left(\frac{1}{2} \partial_\mu \phi \partial^\mu \phi - V(\phi) \right), \tag{10.44}$$

we expect to be able to write $\Gamma[\phi_c]$ also in the form

$$\Gamma[\phi_c] = \int d^4x \left(-V_{\text{eff}}(\phi_c(x)) + \frac{1}{2} A(\phi_c(x)) \partial_\mu \phi_c(x) \partial^\mu \phi_c(x) + \cdots \right), \tag{10.45}$$

where the terms neglected are higher order in the derivatives. Let us recall that when the sources are set equal to zero, then the classical field takes a constant value $\phi_c(x) = \phi_c$. In this limit all the derivative terms in the expansion in Eq. (10.45) vanish, leading to

$$\Gamma[\phi_c] = - \int d^4x \, V_{\text{eff}}(\phi_c) = -V_{\text{eff}}(\phi_c) \int d^4x. \tag{10.46}$$

In other words, in this limit, the effective action simply picks out the effective potential including quantum corrections to all orders. The quantity $\int d^4x$ which represents the space-time volume is also conventionally written as

$$\int d^4x = (2\pi)^4 \delta^4(0). \tag{10.47}$$

Let us also note that the constant value of $\phi_c = \langle \phi \rangle$ when the sources are turned off is obtained from the extremum equation in Eq. (10.27)

$$\left. \frac{\delta \Gamma[\phi_c]}{\delta \phi_c(x)} \right|_{\phi_c(x)=\phi_c} = 0. \tag{10.48}$$

In terms of the effective potential, this condition becomes equivalent to

$$\left. \frac{\partial V_{\text{eff}}(\phi_c)}{\partial \phi_c} \right|_{\phi_c=\langle \phi \rangle} = 0, \tag{10.49}$$

which is a familiar extremization condition from the study of classical mechanics. Note here that since ϕ_c is an ordinary variable (namely, it does not depend on space-time coordinates), only ordinary partial derivatives are involved in the above extremum condition. In this sense, this equation is much easier to analyze than a functional equation. We also note that the renormalized values of the masses and the coupling constants (including all quantum corrections) can be obtained (in this theory) from the effective potential simply as

$$\left. \frac{\partial^2 V_{\text{eff}}}{\partial \phi_c^2} \right|_{\phi_c = \langle \phi \rangle} = m_R^2 \,,$$

$$\left. \frac{\partial^4 V_{\text{eff}}}{\partial \phi_c^4} \right|_{\phi_c = \langle \phi \rangle} = \lambda_R \,. \tag{10.50}$$

The study of the effective potential is, therefore, quite important. It is particularly useful in analyzing when the quantum corrections can change the qualitative tree level or classical behavior of a theory.

10.3 Loop Expansion

We have already described the Feynman rules in the coordinate space. However, for most practical calculations, it is quite useful to work in the momentum space. The Feynman rules can be readily generalized to the momentum space given the rules in the coordinate space. (This simply involves taking Fourier transforms.) For the ϕ^4-theory, for example, the momentum space Feynman rules take the form

$$\xrightarrow{\quad p \quad} = i\hbar G_F(p) = \lim_{\epsilon \to 0} \frac{i\hbar}{p^2 - m^2 + i\epsilon} \,,$$

$$
\begin{array}{c} p_1 \\ p_2 \end{array} \!\!\!\!\!\!\!\!\times\!\!\!\!\!\!\!\! \begin{array}{c} p_4 \\ p_3 \end{array} = -\frac{i\lambda}{\hbar} \, \delta^4(p_1 + p_2 + p_3 + p_4) \,. \tag{10.51}
$$

In evaluating a Feynman diagram, we should integrate over the intermediate momenta, namely, the momenta of the internal propagators. Thus, for example, let us evaluate the 1PI 2-point vertex function at order λ.

$$I = \quad \underset{p_1 \qquad\qquad\qquad p_2}{\underrightarrow{\qquad\qquad}} \quad .$$

According to our rules, we obtain

$$I = \frac{1}{2}\left(-\frac{i\lambda}{\hbar}\right)\int \frac{d^4k}{(2\pi)^4}\frac{i\hbar}{k^2-m^2}\,\delta^4(p_1-p_2+k-k)$$

$$= \frac{\lambda}{2}\,\delta^4(p_1-p_2)\int \frac{d^4k}{(2\pi)^4}\frac{1}{k^2-m^2}\,. \tag{10.52}$$

Note that the factor of $\frac{1}{2}$ in front of the integral in Eq. (10.52) simply corresponds to the symmetry factor of the diagram which we discussed earlier. In writing the propagator, we have not explicitly included the $i\epsilon$ term although it should always be kept in mind in evaluating the integral. We will discuss the actual evaluation of the integrals later. For the present, let us simply note that the calculations indeed take a simpler form in the momentum space.

Let us next recognize from the form of the exponent in the path integral that the quantity which determines the dynamics of the system is

$$\frac{1}{\hbar}\mathcal{L}(\phi,\partial_\mu\phi) = \frac{1}{\hbar}\left(\frac{1}{2}\partial_\mu\phi\partial^\mu\phi - \frac{m^2}{2}\phi^2 - \frac{\lambda}{4!}\phi^4\right). \tag{10.53}$$

The Planck's constant which measures the quantum nature of an amplitude comes as a multiplicative factor in the exponent. As we have seen in Eqs. (9.46) and (10.51), the consequence of this is that each vertex has a factor of $\frac{1}{\hbar}$ associated with it. On the other hand, the propagator which is the inverse of the operator in the quadratic part of the Lagrangian, comes multiplied with a factor of $\hbar$. Thus, suppose we are considering a proper vertex diagram (1PI diagram) with V vertices and I internal lines or propagators, then the total number of $\hbar$ factors associated with such a diagram is given by

$$P = I - V. \tag{10.54}$$

In other words, such a diagram will behave like $\sim \hbar^P$.

Let us also calculate the number of independent momentum integrations associated with such a diagram. First, let us note that in a proper vertex diagram, there are no external propagators or legs. Second, all the momenta associated with the internal lines must be integrated. Since there are I internal lines, there must, therefore, be I momentum integrations. Of course, not all such momenta will be independent since at each vertex there are momentum conserving δ-functions. Each such δ-function will reduce the number of momentum integration by one. Since there are V vertices, there will be as many momentum conserving δ-functions. However, we will need to have an overall momentum conserving δ-function for the amplitude. Hence, the δ-functions will effectively reduce the number of momentum integration by $V - 1$. Therefore, the number of independent internal momentum integrations will be given by

$$L = I - (V - 1) = I - V + 1 = P + 1. \tag{10.55}$$

But the number of independent momentum integrations precisely measures the number of loops in a diagram and from the above relation we note that the number of loops associated with a diagram is related to the power of $\hbar$ associated with a diagram. In fact, the number of loops exceeds the power of $\hbar$ by one. Therefore, expanding an amplitude in powers of $\hbar$ is also equivalent to an expansion in the number of loops.

The loop expansion provides a valid perturbative expansion simply because $\hbar$ is a small quantity. This expansion is quite useful and is very different from expanding in powers of the coupling constant. This follows mainly from the fact that the expansion parameter, $\hbar$, multiplies the entire Lagrangian. Consequently, it is insensitive to how we divide the Lagrangian into a free part and an interaction part. The loop expansion is, therefore, unaffected by any such separation. This is particularly useful if the theory exhibits spontaneous symmetry breakdown in which case, as we will see later, the vacuum expectation value $\phi_c = \langle \phi \rangle$ becomes dependent on the coupling constants of the theory. Shifting the fields around such a value complicates perturbation in powers of the coupling constants. However, as we have argued, the loop expansion is unaffected by such a shift.

10.4 Effective Potential at One Loop

Let us next calculate the effective potential for the ϕ^4- theory at one loop. In this case,

$$S[\phi] = \int d^4x \left(\frac{1}{2} \partial_\mu \phi \partial^\mu \phi - \frac{m^2}{2} \phi^2 - \frac{\lambda}{4!} \phi^4 \right).$$

As we have seen earlier in Eq. (10.26), in this case, the classical dynamical equations are given by

$$-\frac{\delta S[\phi]}{\delta \phi(x)} = (\partial_\mu \partial^\mu + m^2)\phi(x) + \frac{\lambda}{3!}\phi^3(x) = J(x).$$

Furthermore, we note from Eq. (10.17) that the classical field, $\phi_c(x)$, satisfies the equation

$$(\partial_\mu \partial^\mu + m^2)\phi_c(x) + \frac{\lambda}{3!}\phi_c^3(x)$$

$$-\frac{i\lambda\hbar}{2!}\phi_c(x)\frac{\delta\phi_c(x)}{\delta J(x)} - \frac{\lambda\hbar^2}{3!}\frac{\delta^2\phi_c(x)}{\delta J^2(x)} = J(x). \tag{10.56}$$

If we now use the relations (see Eqs. (10.7) and (10.25))

$$\frac{\delta\Gamma[\phi_c]}{\delta\phi_c(x)} = -J(x),$$

$$\frac{\delta\phi_c(x)}{\delta J(y)} = \frac{\delta^2 W[J]}{\delta J(x)\delta J(y)} = -G(x - y, \phi_c), \tag{10.57}$$

and remember that we are interested only in one loop effects, then keeping terms up to linear power in $\hbar$, we obtain from Eq. (10.56)

$$\frac{\delta\Gamma[\phi_c]}{\delta\phi_c(x)} = -(\partial_\mu\partial^\mu + m^2)\phi_c(x) - \frac{\lambda}{3!}\phi_c^3(x) + \frac{i\lambda\hbar}{2!}\phi_c(x)\frac{\delta^2 W[J]}{\delta J^2(x)} + O(\hbar^2)$$

$$= \frac{\delta S[\phi_c]}{\delta\phi_c(x)} - \frac{i\lambda\hbar}{2}\phi_c(x)G(x-x,\phi_c) + O(\hbar^2)$$

or, $\quad \dfrac{\delta}{\delta\phi_c(x)}(\Gamma[\phi_c] - S[\phi_c]) = -\dfrac{i\lambda\hbar}{2}\phi_c(x)G(0,\phi_c) + O(\hbar^2)$. $\quad$ (10.58)

Now, if we expand the effective action as

$$\Gamma[\phi_c] = S[\phi_c] + \hbar S_1[\phi_c] + O(\hbar^2), \tag{10.59}$$

then, Eq. (10.58) gives

$$\frac{\delta}{\delta\phi_c(x)}(\hbar S_1[\phi_c]) + O(\hbar^2) = -\frac{i\lambda\hbar}{2}\phi_c(x)G(0,\phi_c) + O(\hbar^2). \tag{10.60}$$

Therefore, to linear order in $\hbar$, we can consistently write

$$\frac{\delta S_1[\phi_c]}{\delta\phi_c(x)} = -\frac{i\lambda}{2}\phi_c(x)G(0,\phi_c). \tag{10.61}$$

From the structure of the action (see Eq. (10.45))

$$S_1[\phi_c] = \int d^4x\,(-V_1(\phi_c(x)) + \cdots), \tag{10.62}$$

we obtain

$$\left.\frac{\delta S_1[\phi_c]}{\delta\phi_c(x)}\right|_{\phi_c(x)=\phi_c} = -\frac{\partial V_1}{\partial\phi_c}. \tag{10.63}$$

Thus, if we restrict to $\phi_c(x) = \phi_c = $ constant, then Eq. (10.61) takes the form

$$\frac{\partial V_1(\phi_c)}{\partial\phi_c} = \frac{i\lambda}{2}\,\phi_c G(0,\phi_c). \tag{10.64}$$

Let us note that although the Green's function $G(x-y,\phi_c)$ can itself have a power series expansion in $\hbar$, consistency requires that we only use the lowest order expression for the Green's function in the above equation. Furthermore, noting from Eq. (10.29) that

$$\int d^4z \frac{\delta^2\Gamma[\phi_c]}{\delta\phi_c(x)\delta\phi_c(z)} \frac{\delta^2 W[J]}{\delta J(z)\delta J(y)} = -\delta^4(x-y),$$

to the lowest order of the Green's function that we are interested in, this relation gives

$$\int d^4z \frac{\delta^2 S[\phi_c]}{\delta\phi_c(x)\delta\phi_c(z)}\bigg|_{\phi_c} (-G(z-y,\phi_c)) = -\delta^4(x-y)$$

or, $\int d^4z((\partial_\mu\partial^\mu + m^2 + \frac{\lambda}{2}\phi_c^2)\delta^4(x-z))G(z-y,\phi_c) = -\delta^4(x-y)$

or, $(\partial_\mu\partial^\mu + m^2 + \frac{\lambda}{2}\phi_c^2)G(x-y,\phi_c) = -\delta^4(x-y)$

or, $(\partial_\mu\partial^\mu + m_{\text{eff}}^2)G(x-y,\phi_c) = -\delta^4(x-y).$ (10.65)

Here we have used the form of the action from Eq. (9.15) as well as Eq. (10.32) and have defined

$$m_{\text{eff}}^2 = m^2 + \frac{\lambda}{2}\phi_c^2.$$ (10.66)

The Green's function can now be trivially determined and as we have seen before in Eq. (9.23) has the form

$$G(x-y,\phi_c) = \int \frac{d^4k}{(2\pi)^4} \frac{1}{k^2 - m_{\text{eff}}^2} e^{-ik\cdot(x-y)}.$$ (10.67)

Substituting this back into Eq. (10.64), we obtain

$$\frac{\partial V_1(\phi_c)}{\partial\phi_c} = \frac{i\lambda}{2}\phi_c G(0,\phi_c) = \frac{i\lambda}{2}\phi_c \int \frac{d^4k}{(2\pi)^4} \frac{1}{k^2 - m_{\text{eff}}^2}$$

or, $V_1(\phi_c) = \frac{i\lambda}{2}\int_0^{\phi_c} d\phi_c\, \phi_c \int \frac{d^4k}{(2\pi)^4} \frac{1}{k^2 - m^2 - \frac{\lambda}{2}\phi_c^2}.$ (10.68)

Here, we are assuming that $V_1(\phi_c = 0) = 0$. If that is not true, then we have to add a constant term to the above expression. The one loop correction to the potential can now be determined by going over to Euclidean space and doing the integral. First, we note that if we interchange the orders of integration in Eq. (10.68), we obtain

$$V_1(\phi_c) = \frac{i\lambda}{2} \int \frac{d^4k}{(2\pi)^4} \int_0^{\phi_c} d\phi_c \frac{\phi_c}{k^2 - m^2 - \frac{\lambda}{2}\phi_c^2}$$

$$= -\frac{i}{2} \int \frac{d^4k}{(2\pi)^4} \int_0^{\phi_c} \frac{d\left(\frac{\lambda}{2}\phi_c^2\right)}{\frac{\lambda}{2}\phi_c^2 + m^2 - k^2}$$

$$= -\frac{i}{2} \int \frac{d^4k}{(2\pi)^4} \ln\left(\frac{\lambda}{2}\phi_c^2 + m^2 - k^2\right)\Big|_0^{\phi_c}$$

$$= -\frac{i}{2} \int \frac{d^4k}{(2\pi)^4} \ln\left(\frac{\frac{\lambda}{2}\phi_c^2 + m^2 - k^2}{m^2 - k^2}\right). \tag{10.69}$$

Now, rotating to Euclidean space (see section 4.1), we obtain

$$V_1(\phi_c) = -\frac{i}{2} \int \frac{i d^4k_E}{(2\pi)^4} \ln\left(\frac{\frac{\lambda}{2}\phi_c^2 + m^2 + k_E^2}{m^2 + k_E^2}\right)$$

$$= \frac{1}{2} \int \frac{d^3\Omega}{(2\pi)^4} k_E^3 dk_E \ln\left(\frac{\frac{\lambda}{2}\phi_c^2 + m^2 + k_E^2}{m^2 + k_E^2}\right). \tag{10.70}$$

Since the integrand does not depend on the angular variables, the angular integration can be done trivially and has the value

$$\int d^3\Omega = 2\pi^2, \tag{10.71}$$

so that we have

$$V_1(\phi_c) = \frac{1}{2}\frac{1}{(2\pi)^4}2\pi^2 \int_0^\infty d\left(\frac{1}{2}k_E^2\right) k_E^2 \ln\left(\frac{m_{\text{eff}}^2 + k_E^2}{m^2 + k_E^2}\right)$$

$$= \frac{1}{32\pi^2} \int_0^\infty dk_E^2 \, k_E^2 \ln\left(\frac{m_{\text{eff}}^2 + k_E^2}{m^2 + k_E^2}\right). \tag{10.72}$$

Defining

$$x = k_E^2, \qquad (10.73)$$

we note that the effective potential at one loop takes the form

$$V_1(\phi_c) = \frac{1}{32\pi^2} \int_0^\infty \mathrm{d}x \, x \left(\ln(x + m_{\text{eff}}^2) - \ln(x + m^2) \right). \qquad (10.74)$$

Clearly, the integrals are divergent and, therefore, we have to cut off the integral at some high momentum scale to obtain

$$V_1(\phi_c) = \frac{1}{32\pi^2} \int_0^{\Lambda^2} \mathrm{d}x \, x \left(\ln(x + m_{\text{eff}}^2) - \ln(x + m^2) \right)$$

$$\simeq \frac{1}{32\pi^2} \left\{ \left[\frac{\Lambda^4}{2} \left(\ln \Lambda^2 - \frac{1}{2} \right) + m_{\text{eff}}^2 \Lambda^2 \right. \right.$$

$$\left. - \frac{m_{\text{eff}}^4}{2} \ln \Lambda^2 + \frac{m_{\text{eff}}^4}{2} \left(\ln m_{\text{eff}}^2 - \frac{1}{2} \right) \right]$$

$$\left. - \left[\frac{\Lambda^4}{2} \left(\ln \Lambda^2 - \frac{1}{2} \right) + m^2 \Lambda^2 - \frac{m^4}{2} \ln \Lambda^2 + \frac{m^4}{2} \left(\ln m^2 - \frac{1}{2} \right) \right] \right\}$$

$$= \frac{1}{32\pi^2} \left[(m_{\text{eff}}^2 - m^2) \Lambda^2 - \frac{m_{\text{eff}}^4}{2} \ln \frac{\Lambda^2}{\mu^2} + \frac{m^4}{2} \ln \frac{\Lambda^2}{\mu^2} \right.$$

$$\left. + \frac{m_{\text{eff}}^4}{2} \left(\ln \frac{m_{\text{eff}}^2}{\mu^2} - \frac{1}{2} \right) - \frac{m^4}{2} \left(\ln \frac{m^2}{\mu^2} - \frac{1}{2} \right) \right]$$

$$= \frac{1}{32\pi^2} \left[\frac{\lambda}{2} \phi_c^2 \Lambda^2 - \frac{\lambda}{4} \phi_c^2 \left(2m^2 + \frac{\lambda}{2} \phi_c^2 \right) \ln \frac{\Lambda^2}{\mu^2} - \frac{m^4}{2} \left(\ln \frac{m^2}{\mu^2} - \frac{1}{2} \right) \right.$$

$$\left. + \frac{1}{2} (m^2 + \frac{\lambda}{2} \phi_c^2)^2 \left(\ln \frac{(m^2 + \frac{\lambda}{2} \phi_c^2)}{\mu^2} - \frac{1}{2} \right) \right]. \qquad (10.75)$$

Here we have introduced an arbitrary mass scale, μ, to write the expression in a meaningful manner. Note that the one loop potential, as it stands, diverges in the limit $\Lambda \to \infty$ which is the physical limit for the true value. This brings out one of the essential features of quantum field theory. Namely, point-like interactions necessarily induce divergences simply because the Heisenberg uncertainty principle, in this case, allows for an infinite uncertainty in the momentum being exchanged. This necessitates a systematic procedure for eliminating divergences in such theories. This is known as the renormalization theory which we will not go into. Let us simply note here that up to one loop, then, we can write the effective potential of the ϕ^4-theory to be

$$V_{\text{eff}}^{(1)} = V + V_1. \tag{10.76}$$

10.5 References

Goldstone, J., A. Salam and S. Weinberg, Phys. Rev. **127**, 965 (1962).

Jona-Lasinio, J., Nuovo Cimento **34**, 1790 (1964).

Nambu, Y., Phys. Lett. **26B**, 626 (1966).

Zinn-Justin, J., "Quantum Field Theory and Critical Phenomena", Oxford Univ. Press.

Chapter 11

Invariances and Their Consequences

11.1 Symmetries of the Action

Let us continue with the ϕ^4-theory and note that, in this case, we have

$$S[\phi] = \int \mathrm{d}^4 x\, \mathcal{L}(\phi, \partial_\mu \phi)\,, \qquad (11.1)$$

where the Lagrangian density has the form

$$\mathcal{L}(\phi, \partial_\mu \phi) = \frac{1}{2} \partial_\mu \phi \partial^\mu \phi - \frac{m^2}{2} \phi^2 - \frac{\lambda}{4!} \phi^4\,. \qquad (11.2)$$

We can write the action in Eq. (11.1) also in terms of the Lagrangian in the form

$$S[\phi] = \int \mathrm{d}t\, L\,, \qquad (11.3)$$

where

$$L = \int \mathrm{d}^3 x\, \mathcal{L}(\phi, \partial_\mu \phi)\,. \qquad (11.4)$$

Given this theory, where the basic variables are the fields $\phi(x)$, we can define the momentum conjugate to the field variables in a straightforward manner as

$$\Pi(x) = \frac{\partial \mathcal{L}}{\partial \dot{\phi}(x)} = \dot{\phi}(x)\,. \qquad (11.5)$$

This is the analogue of the relation between the momentum and velocity in classical mechanics, namely, $p = \dot{x}$ (for $m = 1$). In quantum field theory, in operator language, this then is the starting point for quantization. However, in the path integral formalism, we treat all variables classically. Therefore, let us analyze various concepts in the classical language. First, let us note that given the Lagrangian in Eq. (11.3), we can obtain the Hamiltonian through a Legendre transformation as

$$H = \int \mathrm{d}^3x\, \Pi(x)\dot{\phi}(x) - L\,. \tag{11.6}$$

In the present case, we can write this out in detail as

$$H = \int \mathrm{d}^3x \left(\Pi(x)\dot{\phi}(x) - \frac{1}{2}\dot{\phi}^2(x) + \frac{1}{2}\boldsymbol{\nabla}\phi(x)\cdot\boldsymbol{\nabla}\phi(x) \right.$$
$$\left. + \frac{m^2}{2}\phi^2(x) + \frac{\lambda}{4!}\phi^4(x) \right)$$
$$= \int \mathrm{d}^3x \left(\frac{1}{2}\Pi^2(x) + \frac{1}{2}\boldsymbol{\nabla}\phi\cdot\boldsymbol{\nabla}\phi(x) + \frac{m^2}{2}\phi^2(x) + \frac{\lambda}{4!}\phi^4(x) \right)\,.$$
$$\tag{11.7}$$

Sometimes, this is also written as

$$H = \int \mathrm{d}^3x \left(\frac{1}{2}\dot{\phi}^2(x) + \frac{1}{2}\boldsymbol{\nabla}\phi(x)\cdot\boldsymbol{\nabla}\phi(x) + \frac{m^2}{2}\phi^2(x) + \frac{\lambda}{4!}\phi^4(x) \right)\,.$$
$$\tag{11.8}$$

Given a Lagrangian density, which depends only on $\phi(x)$ and $\partial_\mu\phi(x)$, the Euler-Lagrange equation is obtained to be (This is simply the generalization of Eq. (1.28) to the case of a field theory.)

$$\partial_\mu \frac{\partial\mathcal{L}}{\partial\partial_\mu\phi(x)} - \frac{\partial\mathcal{L}}{\partial\phi(x)} = 0\,, \tag{11.9}$$

which gives the dynamics of the system. Given the dynamical equations, we can ask how unique is the Lagrangian density for the system. The answer, not surprisingly, turns out to be that the Lagrangian density is unique only up to total derivatives. Namely, both

$$\mathcal{L}(\phi,\partial_\mu\phi), \quad \text{and} \quad \mathcal{L}(\phi,\partial_\mu\phi) + \partial_\mu K^\mu(\phi,\partial_\lambda\phi), \tag{11.10}$$

give the same Euler-Lagrange equation. We can, of course, check this directly. But a more intuitive way to understand this is to note that with the usual assumptions about the asymptotic fall off of the field variables, we have

$$S_K = \int d^4x\, \partial_\mu K^\mu(\phi, \partial_\lambda \phi) = 0. \qquad (11.11)$$

In other words, a total divergence in the Lagrangian density does not contribute to the action. Consequently, the variation of S_K cannot contribute to the dynamical equations. (We note here that even when the asymptotic fall off of the fields is not fast enough, this statement remains true.) We can, of course, check this for specific examples explicitly. Thus, choosing

$$\mathcal{L}_K = \partial_\mu K^\mu = \partial_\mu(\phi \partial^\mu \phi) = \partial_\mu \phi \partial^\mu \phi + \phi \partial_\mu \partial^\mu \phi, \qquad (11.12)$$

the Euler-Lagrange equation gives

$$\partial_\mu \partial_\nu \frac{\partial \mathcal{L}_K}{\partial \partial_\mu \partial_\nu \phi} - \partial_\mu \frac{\partial \mathcal{L}}{\partial \partial_\mu \phi} + \frac{\partial \mathcal{L}}{\partial \phi} = \partial_\mu \partial^\mu \phi - 2\partial_\mu \partial^\mu \phi + \partial_\mu \partial^\mu \phi \equiv 0. \qquad (11.13)$$

With this analysis, therefore, it is clear that a given system of dynamical equations will remain invariant under a set of infinitesimal transformations of the field variables of the form

$$\phi \rightarrow \phi + \delta\phi, \qquad (11.14)$$

if and only if the corresponding Lagrangian density changes, at the most, by a total divergence under the same transformations. Namely, if

$$\mathcal{L} \rightarrow \mathcal{L} + \partial_\mu K^\mu, \qquad (11.15)$$

under a field transformation, then it defines an invariance of the dynamical equations. Note that in the special case when $K^\mu = 0$, then the Lagrangian density itself is invariant under the set of field transformations in Eq. (11.14) and, therefore, also defines a symmetry of the system. However, this is a very special case. In general, if under

$$\phi \rightarrow \phi + \delta\phi \,,$$

$$S[\phi] \rightarrow S[\phi + \delta\phi] = S[\phi] \,, \tag{11.16}$$

then we say that the field transformations define an invariance or a symmetry of the system, namely, the dynamical equations.

Continuous transformations, by definition, depend on a parameter of transformation continuously. This parameter can be a space-time independent parameter or it can depend on the coordinates of the field variables. In the first case, the transformations would change the field variables by the same amount at every space-time point. On the other hand, the change in the field variables, in the second case, will be different at different space-time points depending on the value of the parameter. Accordingly, these transformations are called global and local transformations respectively. The basic symmetries in gauge theories are local symmetries.

11.2   Noether's Theorem

Noether's theorem, very simply, says that for every continuous global symmetry of a system, there exists a current density which is conserved. More specifically, it says that for a system described by a Lagrangian density $\mathcal{L}(\phi, \partial_\mu \phi)$, if the infinitesimal global transformations

$$\phi(x) \rightarrow \phi(x) + \delta_\epsilon \phi(x) \,, \tag{11.17}$$

where ϵ is the constant parameter of transformation, define a symmetry of the system, in the sense that under these transformations

$$\mathcal{L} \rightarrow \mathcal{L} + \partial_\mu K^\mu(\phi, \partial_\lambda \phi, \delta_\epsilon \phi) \,, \tag{11.18}$$

then,

$$j_\epsilon^\mu = \frac{\partial \mathcal{L}}{\partial \partial_\mu \phi(x)} \delta_\epsilon \phi - K^\mu \,, \tag{11.19}$$

defines a current density which is conserved.

To see that j_ϵ^μ is indeed conserved, let us note that

$$\partial_\mu j_\epsilon^\mu = \partial_\mu \left(\frac{\partial \mathcal{L}}{\partial \partial_\mu \phi(x)} \delta_\epsilon \phi(x) \right) - \partial_\mu K^\mu$$

$$= \partial_\mu \frac{\partial \mathcal{L}}{\partial \partial_\mu \phi(x)} \delta_\epsilon \phi(x) + \frac{\partial \mathcal{L}}{\partial \partial_\mu \phi(x)} \partial_\mu \delta_\epsilon \phi(x) - \partial_\mu K^\mu$$

$$= \frac{\partial \mathcal{L}}{\partial \phi(x)} \delta_\epsilon \phi(x) + \frac{\partial \mathcal{L}}{\partial \partial_\mu \phi(x)} \delta_\epsilon (\partial_\mu \phi(x)) - \partial_\mu K^\mu . \quad (11.20)$$

Here, we have used the Euler-Lagrange equation in Eq. (11.9) as well as the fact that $\partial_\mu \delta_\epsilon \phi(x) = \delta_\epsilon \partial_\mu \phi(x)$. We note next that the first two terms in Eq. (11.20) simply give the change in the Lagrangian density under the transformations. Therefore, we can also write using Eq. (11.18)

$$\partial_\mu j_\epsilon^\mu = \delta_\epsilon \mathcal{L} - \partial_\mu K^\mu = 0 . \quad (11.21)$$

This shows that the current density given in Eq. (11.19) is indeed conserved. The current density defined in Eq. (11.19) depends on the parameter of transformation as well. A more fundamental quantity is the current without the parameter of the transformation and let us denote this symbolically as

$$j_\epsilon^\mu = \epsilon j^\mu . \quad (11.22)$$

We have to remember that this is only a symbolic relation simply because the parameter ϵ may, itself, have a tensorial structure in which case the current without the parameter will have a more complicated tensor structure as we will see shortly.

As in classical electrodynamics, we know that given a conserved current density, we can define a charge which is a constant of motion as

$$Q = \int d^3x \, j^0(\mathbf{x}, t) . \quad (11.23)$$

The fact that this charge is a constant, independent of time, can be seen simply as follows.

$$\frac{dQ}{dt} = \frac{d}{dt} \int d^3x \, j^0(\mathbf{x}, t)$$

$$= \int d^3x \, \partial_0 j^0(\mathbf{x}, t)$$

$$= \int d^3x \left(\partial_0 j^0(\mathbf{x}, t) + \boldsymbol{\nabla} \cdot \mathbf{j}(\mathbf{x}, t) \right) . \tag{11.24}$$

Here, we have added a total divergence which vanishes under our assumptions on the asymptotic behavior of the field variables. Thus, we have

$$\frac{dQ}{dt} = \int d^3x \, \partial_\mu j^\mu = 0 , \tag{11.25}$$

which follows from Eqs. (11.21) and (11.22), namely, the conservation of the current density. This shows that the charge is a constant of motion.

Another way to understand this result is to note that this implies classically that the Poisson bracket of Q with H vanishes. Quantum mechanically, the commutator of the two operators must vanish.

$$[Q, H] = 0 . \tag{11.26}$$

But this is precisely a symmetry condition in quantum mechanics. Namely, we know from our studies in quantum mechanics that a transformation is a symmetry if the generator of infinitesimal symmetry transformations commutes with the Hamiltonian. Conversely, any operator which commutes with the Hamiltonian is the generator of a symmetry transformation which leaves the system invariant. Thus, we recognize Q to be the generator of the infinitesimal symmetry transformations in the present case. This simply means that the infinitesimal change in any variable can be obtained from

$$\delta_\epsilon \phi = -i[\epsilon Q, \phi] . \tag{11.27}$$

(Classically, we should use appropriate Poisson bracket relations.) It is now clear that the vanishing of the commutator between Q and

H simply corresponds to the Hamiltonian being invariant under the symmetry transformations-which we expect.

In quantum field theory, the operator implementing finite symmetry transformations can be written in terms of the generator of infinitesimal transformations as

$$U(\alpha) = e^{-i\alpha Q}, \tag{11.28}$$

where α is the parameter of finite transformation. A field variable, under such a transformation, is supposed to change as

$$\phi(x) \rightarrow U(\alpha)\phi(x)U^{-1}(\alpha) = e^{-i\alpha Q}\phi(x)e^{i\alpha Q}. \tag{11.29}$$

And, furthermore, a true symmetry is supposed to leave the ground state or the vacuum invariant, namely,

$$U(\alpha)|0\rangle = e^{-i\alpha Q}|0\rangle = |0\rangle. \tag{11.30}$$

Equivalently, it follows from Eq. (11.30) that

$$Q|0\rangle = 0. \tag{11.31}$$

In other words, for a true symmetry, the conserved charge annihilates the vacuum. Therefore, in such a case, we note from Eq. (11.27) that

$$\langle 0|\delta_\epsilon \phi(x)|0\rangle = -i\langle 0|[\epsilon Q, \phi(x)]|0\rangle = 0, \tag{11.32}$$

where we have used Eq. (11.31). As we will see later, if there is a spontaneous breakdown of a symmetry, then, the conserved charge, Q, does not annihilate the vacuum and that the vacuum expectation value of the change in some operator in the theory becomes nonzero.

11.2.1 *Example*

As an example of Noether's theorem, let us study global space-time translations as a symmetry of quantum field theories. Let us continue to use the ϕ^4-theory for this discussion.

Let us define the infinitesimal translations

$$x^\mu \rightarrow x^\mu + \epsilon^\mu$$

$$\text{or,} \quad \delta_\epsilon x^\mu = \epsilon^\mu, \tag{11.33}$$

as the global transformations, where ϵ^μ is the constant parameter of transformation. In such a case,

$$\delta_\epsilon \phi(x) = \phi(x + \epsilon) - \phi(x) = \epsilon^\mu \partial_\mu \phi(x) \,,$$

$$\delta_\epsilon \partial_\mu \phi(x) = \partial_\mu (\delta_\epsilon \phi(x)) = \epsilon^\nu \partial_\mu \partial_\nu \phi(x) \,. \qquad (11.34)$$

Given this, we can, of course, obtain the infinitesimal change in the Lagrangian density in Eq. (11.2)

$$\mathcal{L} = \frac{1}{2} \partial_\mu \phi(x) \partial^\mu \phi(x) - \frac{m^2}{2} \phi^2(x) - \frac{\lambda}{4!} \phi^4(x) \,,$$

in a straightforward manner. A much simpler way to evaluate this, however, is to note that the Lagrangian density is effectively a function of x, namely, $\mathcal{L} = \mathcal{L}(x)$. Thus,

$$\delta_\epsilon \mathcal{L} = \mathcal{L}(x + \epsilon) - \mathcal{L}(x) = \epsilon^\mu \partial_\mu \mathcal{L}(x) = \partial_\mu K^\mu \,. \qquad (11.35)$$

Therefore, we readily identify

$$K^\mu = \epsilon^\mu \mathcal{L}(x) = \epsilon^\mu \mathcal{L}(\phi, \partial_\mu \phi) \,. \qquad (11.36)$$

On the other hand, we see from Eq. (11.2) that for this theory

$$\frac{\partial \mathcal{L}}{\partial \partial_\mu \phi(x)} = \partial^\mu \phi(x) \,. \qquad (11.37)$$

As a result, we see from Eqs. (11.34), (11.36) and (11.19), that the Noether current defined in Eq. (11.19), in this case, follows to be

$$\begin{aligned}
j_\epsilon^\mu(x) &= \frac{\partial \mathcal{L}}{\partial \partial_\mu \phi(x)} \delta_\epsilon \phi(x) - K^\mu \\
&= \partial^\mu \phi(x) \left(\epsilon^\nu \partial_\nu \phi(x) \right) - \epsilon^\mu \mathcal{L} \\
&= \epsilon^\nu \left(\partial^\mu \phi(x) \partial_\nu \phi(x) - \delta_\nu^\mu \mathcal{L} \right) \\
&= \epsilon_\nu \left(\partial^\mu \phi(x) \partial^\nu \phi(x) - \eta^{\mu\nu} \mathcal{L} \right) \,.
\end{aligned} \qquad (11.38)$$

This is, of course, the conserved current density and the current without the parameter of transformation has the form (see Eq. (11.22))

$$j_\epsilon^\mu(x) = \epsilon_\nu T^{\mu\nu}, \qquad (11.39)$$

where we see from Eq. (11.38) that

$$T^{\mu\nu} = \partial^\mu \phi(x) \partial^\nu \phi(x) - \eta^{\mu\nu} \mathcal{L}. \qquad (11.40)$$

There are several comments in order here. First, let us note that the fundamental conserved quantity (in this case $T^{\mu\nu}$) is not necessarily a vector. Its tensorial character depends completely on the parameter of transformation. Second, we note from Eq. (11.40) that the conserved quantity, in this case, is a symmetric second rank tensor, namely,

$$T^{\mu\nu} = T^{\nu\mu}. \qquad (11.41)$$

This is known as the stress tensor of the theory.

Let us also note from Eqs. (11.23) and (11.40) that the conserved charge, in this case, has a vectorial character and has the form

$$P^\mu = \int d^3x \, T^{0\mu}. \qquad (11.42)$$

To understand the meaning of the charges in Eq. (11.42) (there are, in fact, four of them), let us write them out explicitly. We note from Eqs. (11.40) and (11.42) that

$$P^0 = \int d^3x \, T^{00}$$

$$= \int d^3x \, ((\dot{\phi}(x))^2 - \mathcal{L})$$

$$= \int d^3x \left((\dot{\phi}(x))^2 - \frac{1}{2}\dot{\phi}^2(x) + \frac{1}{2}\boldsymbol{\nabla}\phi \cdot \boldsymbol{\nabla}\phi \right.$$

$$\left. + \frac{m^2}{2}\phi^2(x) + \frac{\lambda}{4!}\phi^4(x) \right)$$

$$= \int d^3x \left(\frac{1}{2}\dot{\phi}^2(x) + \frac{1}{2}\nabla\phi \cdot \nabla\phi + \frac{m^2}{2}\phi^2(x) + \frac{\lambda}{4!}\phi^4(x) \right)$$

$$= H,$$

$$P^i = \int d^3x \, T^{0i} = \int d^3x \, \dot{\phi}\partial^i\phi$$

$$\text{or,} \quad \mathbf{P} = -\int d^3x \, \dot{\phi}(x)\nabla\phi(x). \tag{11.43}$$

We recognize the first quantity (namely, P^0) as the Hamiltonian of the system obtained in Eq. (11.8) and from relativistic invariance we conclude that $\mathbf{P}$ must represent the total momentum of the system. Thus, we recover the familiar result that the space-time translations are generated by the energy-momentum operators of the theory.

11.3   Complex Scalar Field

So far we have discussed the ϕ^4-theory where the basic field variable is real. Such a theory, as we have mentioned before, can describe spin zero mesons which are charge neutral. Let us next consider a scalar field theory where the basic field variable is complex. Namely, in this case,

$$\phi^*(x) \neq \phi(x). \tag{11.44}$$

One way to study such a theory is to expand the complex field in terms of two real fields as, say

$$\phi(x) = \frac{1}{\sqrt{2}}(\phi_1(x) + i\phi_2(x)). \tag{11.45}$$

However, let us continue with the complex field, $\phi(x)$, as the basic variable. The real Lagrangian density describing quartic interactions can be generalized from Eq. (11.2) and written as

$$\mathcal{L}(\phi, \phi^*) = \partial_\mu\phi^*\partial^\mu\phi - m^2\phi^*\phi - \frac{\lambda}{4}(\phi^*\phi)^2, \tag{11.46}$$

with $\lambda > 0$. We can treat ϕ and ϕ^* as independent dynamical variables. Correspondingly, the two Euler-Lagrange equations following from Eq. (11.46) are given by

$$\partial_\mu \frac{\partial \mathcal{L}}{\partial \partial_\mu \phi^*} - \frac{\partial \mathcal{L}}{\partial \phi^*} = 0$$

$$\text{or, } (\partial_\mu \partial^\mu + m^2)\phi + \frac{\lambda}{2}(\phi^* \phi)\phi = 0, \tag{11.47}$$

and

$$\partial_\mu \frac{\partial \mathcal{L}}{\partial \partial_\mu \phi} - \frac{\partial \mathcal{L}}{\partial \phi} = 0$$

$$\text{or, } (\partial_\mu \partial^\mu + m^2)\phi^* + \frac{\lambda}{2}(\phi^* \phi)\phi^* = 0. \tag{11.48}$$

Thus, the two dynamical equations in Eqs. (11.47) and (11.48) correspond to two coupled scalar field equations. We should have expected this since having a complex field doubles the number of degrees of freedom.

Let us next note that if we make a phase transformation of the form

$$\phi(x) \rightarrow e^{-i\alpha}\phi(x),$$

$$\phi^*(x) \rightarrow e^{i\alpha}\phi^*(x), \tag{11.49}$$

where α is a real, constant (global) parameter of transformation or equivalently, an infinitesimal transformation of the form (ϵ is infinitesimal)

$$\delta_\epsilon \phi(x) = -i\epsilon\phi(x),$$

$$\delta_\epsilon \phi^*(x) = i\epsilon\phi^*(x), \tag{11.50}$$

then, we note that under such a transformation

$$\phi^* \phi \rightarrow e^{i\alpha}\phi^* e^{-i\alpha}\phi = \phi^* \phi. \tag{11.51}$$

Equivalently, under the infinitesimal transformations of Eq. (11.50), we note that

$$\delta_\epsilon(\phi^*\phi) = (\delta_\epsilon\phi^*)\phi + \phi^*(\delta_\epsilon\phi) = i\epsilon\phi^*\phi - i\epsilon\phi^*\phi = 0\,. \qquad (11.52)$$

Namely, under the transformation in Eq. (11.49) or (11.50) $\phi^*\phi$ remains unchanged. Similarly, we note that under the transformation of Eq. (11.49)

$$\partial_\mu\phi^*\partial^\mu\phi \rightarrow \partial_\mu(e^{i\alpha}\phi^*)\partial^\mu(e^{-i\alpha}\phi) = \partial_\mu\phi^*\partial^\mu\phi\,. \qquad (11.53)$$

Alternatively, from the form of the infinitesimal transformations in Eq. (11.50), we obtain

$$\delta_\epsilon(\partial_\mu\phi^*\partial^\mu\phi) = (\delta_\epsilon(\partial_\mu\phi^*))\partial^\mu\phi + \partial_\mu\phi^*(\delta_\epsilon(\partial^\mu\phi))$$

$$= \partial_\mu(\delta_\epsilon\phi^*)\partial^\mu\phi + \partial_\mu\phi^*(\partial^\mu(\delta_\epsilon\phi))$$

$$= i\epsilon\partial_\mu\phi^*\partial^\mu\phi - i\epsilon\partial_\mu\phi^*\partial^\mu\phi = 0\,. \qquad (11.54)$$

(It is important to recognize that the invariance in Eqs. (11.53) and (11.54) results because the parameter of transformation is assumed to be independent of space-time coordinates.)

In this case, therefore, we see that the constant phase transformations define a symmetry of the theory, in the sense that,

$$\mathcal{L} = \partial_\mu\phi^*\partial^\mu\phi - m^2\phi^*\phi - \frac{\lambda}{4}(\phi^*\phi)^2 \rightarrow \mathcal{L}\,. \qquad (11.55)$$

Equivalently,

$$\delta_\epsilon\mathcal{L} = 0\,. \qquad (11.56)$$

Such a symmetry is called an internal symmetry since the transformations do not change the space-time points. For such an invariance, we note that

$$K^\mu = 0\,. \qquad (11.57)$$

Therefore, the conserved current constructed through the Noether procedure has the form (see Eq. (11.19))

$$j_\epsilon^\mu = \frac{\partial \mathcal{L}}{\partial \partial_\mu \phi^*} \delta_\epsilon \phi^* + \frac{\partial \mathcal{L}}{\partial \partial_\mu \phi} \delta_\epsilon \phi$$

$$= \partial^\mu \phi(i\epsilon \phi^*) + \partial^\mu \phi^*(-i\epsilon \phi)$$

$$= i\epsilon \left(\phi^* \partial^\mu \phi - (\partial^\mu \phi^*) \phi \right)$$

$$= \epsilon(i\phi^* \overleftrightarrow{\partial^\mu} \phi) = \epsilon \, j^\mu, \tag{11.58}$$

where we have defined

$$j^\mu = i\phi^* \overleftrightarrow{\partial^\mu} \phi \equiv i \left(\phi^* \partial^\mu \phi - (\partial^\mu \phi^*) \phi \right). \tag{11.59}$$

The conserved current, j^μ, in this case, has a vectorial character very much like the electromagnetic current density. Therefore, it can be identified with the electromagnetic current associated with this system. This theory, therefore, can describe charged spin zero mesons. The conserved charge, for the present case, can be written as

$$Q = \int d^3x \, j^0$$

$$= \int d^3x \, i(\phi^* \dot{\phi} - \dot{\phi}^* \phi)$$

$$= i \int d^3x \left(\phi^* \dot{\phi} - \dot{\phi}^* \phi \right). \tag{11.60}$$

In quantum field theory, Q would represent the electric charge operator. As we have mentioned earlier, if the phase transformations in Eq. (11.49) or (11.50) define a true symmetry of the system, then the charge operator in Eq. (11.60) must annihilate the vacuum. In other words, we must have, in such a case,

$$Q|0\rangle = 0. \tag{11.61}$$

11.4   Ward Identities

Symmetries are quite important in the study of physical theories for various reasons. First of all, they lead to conserved quantities and conserved quantum numbers. But more importantly, they give rise to relations between various Green's functions and, therefore, between the transition amplitudes. Thus, as an example, let us consider the generating functional for the complex scalar field.

$$Z[J, J^*] = e^{iW[J,J^*]} = N \int \mathcal{D}\phi \mathcal{D}\phi^* \, e^{iS[\phi,\phi^*,J,J^*]} . \qquad (11.62)$$

Let us note here that we have now set $\hbar = 1$ for simplicity and that we have defined

$$S[\phi, \phi^*, J, J^*] = S[\phi, \phi^*] + \int d^4x (J^*\phi + J\phi^*) , \qquad (11.63)$$

with $S[\phi, \phi^*]$ representing the dynamical action for the system. Here, we note that J^* is the source for the field ϕ whereas J corresponds to the source for ϕ^*. Note also that even though the action $S[\phi, \phi^*]$ is invariant under the global phase transformations of Eq. (11.50), the complete action $S[\phi, \phi^*, J, J^*]$ is not unless we simultaneously change J and J^* also. In fact, let us note that infinitesimally,

$$\delta_\epsilon S[\phi, \phi^*, J, J^*] = \delta_\epsilon S[\phi, \phi^*] + \delta_\epsilon \left(\int d^4x \, (J^*\phi + J\phi^*) \right)$$

$$= \int d^4x \, (J^*\delta_\epsilon\phi + J\delta_\epsilon\phi^*)$$

$$= -i\epsilon \int d^4x \, (J^*\phi - J\phi^*) . \qquad (11.64)$$

Since the generating functional does not depend on the field variables (that is, the fields are all integrated out), making a field redefinition in the integrand of the path integral should not change the generating functional. In particular, if the redefinition corresponds to the infinitesimal phase transformations defined in Eq. (11.50), then we will have

$$
\delta_\epsilon Z[J, J^*] = 0 = N \int \mathcal{D}\phi \mathcal{D}\phi^* \, i\delta_\epsilon S \, e^{iS[\phi, \phi^*, J, J^*]}
$$

$$
= N \int \mathcal{D}\phi \mathcal{D}\phi^* \left(\epsilon \int \mathrm{d}^4 x \, (J^*\phi - J\phi^*) \right) e^{iS[\phi, \phi^*, J, J^*]} \,,
$$

$$(11.65)$$

where we have used Eq. (11.64). In general, one should also worry about the change coming from the Jacobian under a field redefinition. In the present case, it does not contribute.

Let us recall that by definition,

$$
\frac{\delta Z[J, J^*]}{\delta J(x)} = iN \int \mathcal{D}\phi \mathcal{D}\phi^* \, \phi^*(x) \, e^{iS[\phi, \phi^*, J, J^*]} \,,
$$

$$
\frac{\delta Z[J, J^*]}{\delta J^*(x)} = iN \int \mathcal{D}\phi \mathcal{D}\phi^* \, \phi(x) \, e^{iS[\phi, \phi^*, J, J^*]} \,.
$$

$$(11.66)$$

Using this then, Eq. (11.65) becomes

$$
\epsilon \int \mathrm{d}^4 x \left(J^*(x) \left(-i \frac{\delta Z}{\delta J^*(x)} \right) - J(x) \left(-i \frac{\delta Z}{\delta J(x)} \right) \right) = 0 \,. \quad (11.67)
$$

This must hold for any arbitrary value of the parameter ϵ and, therefore, we conclude that

$$
\int \mathrm{d}^4 x \left(J^*(x) \frac{\delta}{\delta J^*(x)} - J(x) \frac{\delta}{\delta J(x)} \right) Z[J, J^*] = 0
$$

or, $\quad \displaystyle \int \mathrm{d}^4 x \left(J^*(x) \frac{\delta}{\delta J^*(x)} - J(x) \frac{\delta}{\delta J(x)} \right) e^{iW[J, J^*]} = 0$

or, $\quad \displaystyle \int \mathrm{d}^4 x \left(J^*(x) \frac{\delta W}{\delta J^*(x)} - J(x) \frac{\delta W}{\delta J(x)} \right) = 0 \,. \quad (11.68)$

This is the master equation for defining symmetry relations. By taking higher functional derivatives of Eq. (11.68), we can obtain relations between various connected Green's functions as a result of the symmetry in the problem. In this case, the symmetry relations are quite simple (simply because the symmetry transformations in Eq. (11.50) are simple), but in the case of more complicated symmetries such as gauge symmetries, such relations are extremely useful and go under the name of Ward Identities of the theory (also known as Slavnov-Taylor identities particularly in the case of gauge symmetries). It is interesting to note that we could also have obtained the Ward identities from a combined set of transformations of the form

$$\delta_\epsilon \phi = -i\epsilon\phi, \qquad \delta_\epsilon \phi^* = i\epsilon\phi^*,$$

$$\delta_\epsilon J = -i\epsilon J, \qquad \delta_\epsilon J^* = i\epsilon J^*. \tag{11.69}$$

In such a case, it is easy to see that the complete action in Eq. (11.63) is invariant. Namely,

$$\delta_\epsilon S[\phi, \phi^*, J, J^*] = 0. \tag{11.70}$$

Therefore, from

$$Z[J, J^*] = e^{iW[J,J^*]} = N \int \mathcal{D}\phi \mathcal{D}\phi^* \, e^{iS[\phi,\phi^*,J,J^*]},$$

we obtain

$$\delta_\epsilon Z[J, J^*] = N \int \mathcal{D}\phi \mathcal{D}\phi^* \, (i\delta_\epsilon S) \, e^{iS[\phi,\phi^*,J,J^*]}$$

or, $\quad i\delta_\epsilon W[J, J^*] e^{iW[J,J^*]} = 0$

or, $\quad \delta_\epsilon W[J, J^*] = 0$

or, $\quad \int \mathrm{d}^4x \left(\frac{\delta W}{\delta J^*(x)} \delta_\epsilon J^*(x) + \frac{\delta W}{\delta J(x)} \delta_\epsilon J(x) \right) = 0$

or, $\quad i\epsilon \int \mathrm{d}^4x \left(J^*(x) \frac{\delta W}{\delta J^*(x)} - J(x) \frac{\delta W}{\delta J(x)} \right) = 0$

or, $\quad \int \mathrm{d}^4x \left(J^*(x) \frac{\delta W}{\delta J^*(x)} - J(x) \frac{\delta W}{\delta J(x)} \right) = 0. \tag{11.71}$

This is, of course, the same relation as in Eq. (11.68).

Let us note that in the case of a complex scalar field, we will have a complex classical field defined by (see Eq. (10.7))

$$\phi_c(x) = \frac{\delta W}{\delta J^*(x)} = \langle 0|\phi(x)|0\rangle^{J,J^*},$$

$$\phi_c^*(x) = \frac{\delta W}{\delta J(x)} = \langle 0|\phi^*(x)|0\rangle^{J,J^*}. \tag{11.72}$$

From the transformation properties of the fields $\phi(x)$ and $\phi^*(x)$ in Eq. (11.50), we can immediately determine the transformation properties of the vacuum expectation values in Eq. (11.72). Namely, we obtain (This corresponds to asking, by how much would $\phi_c(x)$ change if we change $\phi(x)$ according to Eq. (11.50).)

$$\delta_\epsilon \phi_c(x) = \langle 0|\delta_\epsilon \phi(x)|0\rangle^{J,J^*} = -i\epsilon \langle 0|\phi(x)|0\rangle^{J,J^*} = -i\epsilon \phi_c(x),$$

$$\delta_\epsilon \phi_c^*(x) = \langle 0|\delta_\epsilon \phi^*(x)|0\rangle^{J,J^*} = i\epsilon \langle 0|\phi^*(x)|0\rangle^{J,J^*} = i\epsilon \phi_c^*(x), \tag{11.73}$$

where we have assumed the invariance of the ground state under such a transformation. From the transformation properties of the classical fields in Eq. (11.73), we can now work out the Ward Identities for the 1PI vertex functions. In the present case, we note that we have

$$\Gamma[\phi_c, \phi_c^*] = W[J, J^*] - \int d^4x \left(J^* \phi_c(x) + J(x)\phi_c^*(x) \right), \tag{11.74}$$

from which it follows that

$$\delta_\epsilon \Gamma[\phi_c, \phi_c^*] = -\int d^4x \left(J^* \delta_\epsilon \phi_c(x) + J(x)\delta_\epsilon \phi_c^*(x) \right)$$

$$= i\epsilon \int d^4x \left(J^*(x)\phi_c(x) - J(x)\phi_c^*(x) \right)$$

$$= i\epsilon \int d^4x \left(J^*(x)\frac{\delta W}{\delta J^*(x)} - J(x)\frac{\delta W}{\delta J(x)} \right)$$

$$= 0. \tag{11.75}$$

Here in the last step, we have used the relation in Eq. (11.68). On the other hand, using Eq. (11.73), we obtain

$$\delta_\epsilon \Gamma[\phi_c, \phi_c^*] = \int d^4x \left(\frac{\delta \Gamma}{\delta \phi_c(x)} \delta_\epsilon \phi_c(x) + \frac{\delta \Gamma}{\delta \phi_c^*(x)} \delta_\epsilon \phi_c^*(x) \right)$$

$$= -i\epsilon \int d^4x \left(\frac{\delta \Gamma}{\delta \phi_c(x)} \phi_c(x) - \frac{\delta \Gamma}{\delta \phi_c^*(x)} \phi_c^*(x) \right). \quad (11.76)$$

Therefore, following Eq. (11.75), we can set this to zero and noting that the parameter ϵ is arbitrary, we obtain

$$\int d^4x \left(\frac{\delta \Gamma}{\delta \phi_c(x)} \phi_c(x) - \frac{\delta \Gamma}{\delta \phi_c^*(x)} \phi_c^*(x) \right) = 0. \quad (11.77)$$

This is the master equation from which we can derive relations between various 1PI vertex functions, as a consequence of the symmetry in the theory, by taking higher order functional derivatives.

11.5　Spontaneous Symmetry Breaking

Let us next consider the complex scalar field theory defined by the following Lagrangian density.

$$\mathcal{L} = \partial_\mu \phi^* \partial^\mu \phi + m^2 \phi^* \phi - \frac{\lambda}{4} (\phi^* \phi)^2, \qquad \lambda > 0. \quad (11.78)$$

This is the same Lagrangian density as in Eq. (11.2) except for the sign in the mass term which is opposite. It is clear that this Lagrangian density is also invariant under the global phase transformations in Eq. (11.49) or (11.50) since each of the terms is. Therefore, the phase transformations define a symmetry of this theory as well and according to Noether's theorem, there exists a conserved charge which is the same as given in Eq. (11.59).

However, if we look at the potential of this theory, namely,

$$V(\phi, \phi^*) = -m^2 \phi^* \phi + \frac{\lambda}{4} (\phi^* \phi)^2, \quad (11.79)$$

then, we note that for constant field configurations, the extrema of the potential occur at

$$\frac{\partial V}{\partial \phi^*} = \left(-m^2 + \frac{\lambda}{2} \phi^* \phi \right) \phi = 0 \,,$$

$$\frac{\partial V}{\partial \phi} = \left(-m^2 + \frac{\lambda}{2} \phi^* \phi \right) \phi^* = 0 \,. \tag{11.80}$$

There are two solutions of these extremum conditions which are easily obtained to be

$$\phi_c = \phi_c^* = 0 \,,$$

$$\text{or,} \quad \phi_c^* \phi_c = \frac{2m^2}{\lambda} \,. \tag{11.81}$$

However, it is quite easy to see that

$$\left. \frac{\partial^2 V}{\partial \phi^* \partial \phi} \right|_{\phi=\phi^*=0} = -m^2 \,, \tag{11.82}$$

whereas

$$\left. \frac{\partial^2 V}{\partial \phi^* \partial \phi} \right|_{\phi^* \phi = \frac{2m^2}{\lambda}} = \left. \frac{\lambda}{2} \phi^* \phi \right|_{\phi^* \phi = \frac{2m^2}{\lambda}} = m^2 \,. \tag{11.83}$$

Consequently, the extremum at $\phi_c = \phi_c^* = 0$ is really a local maximum of the potential energy whereas the true minimum occurs at

$$\phi_c^* \phi_c = \frac{2m^2}{\lambda} \,. \tag{11.84}$$

Note that since for constant field configurations the derivative terms vanish, this also defines the true minimum of energy or the true ground state of this theory.

To better understand what is involved here, let us rewrite the complex field in terms of two real scalar fields. Namely, let us write

$$\phi = \frac{1}{\sqrt{2}} (\sigma + i\rho) \,, \tag{11.85}$$

where we assume that σ and ρ are real (Hermitian) scalar fields. In terms of these variables, then, the minimum of the potential occurs at

$$\phi_c^*\phi_c = \frac{1}{2}(\sigma_c^2 + \rho_c^2) = \frac{2m^2}{\lambda}$$

$$\text{or,} \quad \sigma_c^2 + \rho_c^2 = \frac{4m^2}{\lambda}. \tag{11.86}$$

It is clear that, in this case, there is an infinite number of degenerate minima lying on a circle in the $\sigma - \rho$ plane. For simplicity, let us choose $\rho_c = 0$. Then, the minimum of the potential can be chosen to be at

$$\sigma_c^2 = \frac{4m^2}{\lambda}$$

$$\text{or,} \quad \sigma_c = \pm\frac{2m}{\sqrt{\lambda}}. \tag{11.87}$$

Let us, in fact choose the minimum to be at

$$\sigma_c = \frac{2m}{\sqrt{\lambda}}, \qquad \rho_c = 0. \tag{11.88}$$

In this case, therefore, we see that one of the fields develops a vacuum expectation value, namely,

$$\sigma_c = \langle 0|\sigma(x)|0\rangle = \frac{2m}{\sqrt{\lambda}},$$

$$\rho_c = \langle 0|\rho(x)|0\rangle = 0. \tag{11.89}$$

To understand further what is involved, let us plot the potential in Eq. (11.79) as a function of σ and ρ for constant values of the fields.

$$V(\sigma,\rho) = -\frac{m^2}{2}(\sigma^2 + \rho^2) + \frac{\lambda}{16}(\sigma^2 + \rho^2)^2, \qquad \lambda > 0. \tag{11.90}$$

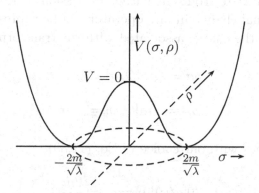

Thus, the potential, in the present case, is very much like the instanton potential in Eq. (7.45), but the minima are infinitely degenerate. Popularly, such a potential is also known as the Mexican hat potential.

Let us also note that since

$$\phi = \frac{1}{\sqrt{2}}(\sigma + i\rho),$$

we can deduce from the transformation rule in Eq. (11.50)

$$\delta\phi = -i\epsilon\phi,$$

that

$$\frac{1}{\sqrt{2}}(\delta\sigma + i\delta\rho) = -i\epsilon\,\frac{1}{\sqrt{2}}(\sigma + i\rho)$$

$$\text{or,}\quad \delta\sigma + i\delta\rho = \epsilon(\rho - i\sigma). \tag{11.91}$$

From this, we conclude that under the global phase transformations, the real scalar fields transform as

$$\delta\sigma = \epsilon\rho,$$

$$\delta\rho = -\epsilon\sigma. \tag{11.92}$$

In other words, the global phase transformations correspond to a rotation in the σ - ρ plane.

Let us also note from our earlier discussion in Eq. (11.27) that the infinitesimal change in any operator can be expressed as a commutator with the charge associated with the transformation as

$$\delta\sigma = -i\epsilon\,[Q,\sigma] = \epsilon\rho\,,$$

$$\delta\rho = -i\epsilon\,[Q,\rho] = -\epsilon\sigma\,. \tag{11.93}$$

Therefore, since with our choice in Eq. (11.89)

$$\langle 0|\sigma(x)|0\rangle = \sigma_{\rm c} = \frac{2m}{\sqrt{\lambda}}\,,$$

we conclude using Eq. (11.92) that

$$\langle 0|\delta\rho|0\rangle = -\epsilon\langle 0|\sigma|0\rangle = -\epsilon\,\frac{2m}{\sqrt{\lambda}}$$

$$\text{or,}\quad -i\epsilon\langle 0|[Q,\rho]|0\rangle = -\epsilon\,\frac{2m}{\sqrt{\lambda}}\,. \tag{11.94}$$

It is clear, therefore, that in the present case, we must have

$$Q|0\rangle \neq 0\,, \tag{11.95}$$

in order that the relation in Eq. (11.94) is consistent. In such a case, we say that the symmetry of the Hamiltonian (or the theory) is spontaneously broken.

Since Q does not annihilate the vacuum of the present theory, let

$$Q|0\rangle = |\chi\rangle\,. \tag{11.96}$$

We know from Eq. (11.26) that the symmetry of the Hamiltonian implies that

$$[Q,H] = 0\,.$$

Assuming that the vacuum state has zero energy (i.e. $H|0\rangle = 0$), we then obtain using Eq. (11.96)

$$[Q, H]|0\rangle = 0$$

$$\text{or, } (QH - HQ)|0\rangle = 0$$

$$\text{or, } HQ|0\rangle = 0$$

$$\text{or, } H|\chi\rangle = 0. \tag{11.97}$$

In other words, the state $|\chi\rangle$ defined in Eq. (11.96) would appear to be degenerate with the vacuum in energy. We can, therefore, think of this as another vacuum. The problem with this interpretation is that this state is not normalizable. This can be easily seen from Eq. (11.96) and (11.23) as follows. (Q is seen from Eq. (11.60) to be hermitian.)

$$\langle\chi|\chi\rangle = \langle 0|QQ|0\rangle$$

$$= \langle 0| \int d^3x \, j^0(\mathbf{x}, t)Q|0\rangle$$

$$= \int d^3x \, \langle 0|e^{iP\cdot x}j^0(0)e^{-iP\cdot x}Q|0\rangle. \tag{11.98}$$

We have already seen that the Hamiltonian commutes with Q expressing the fact that it is independent of time. Since Q does not depend on spatial coordinates, it follows that the momentum operator also commutes with Q. In fact, in general, we can write

$$[P_\mu, Q] = 0. \tag{11.99}$$

There are many ways of obtaining this result besides the argument given above. The most intuitive way is to note that P_μ generates space-time translations whereas Q generates a phase transformation in the internal Hilbert space. Both these transformations are independent of each other and, therefore, their order should not matter which is equivalent to saying that the generators must commute. A consequence of their commutativity is that we have

$$e^{-iP\cdot x}Q = Qe^{-iP\cdot x}. \tag{11.100}$$

Using this in Eq. (11.98), then, we obtain

$$\langle \chi | \chi \rangle = \int d^3x \, \langle 0 | e^{iP \cdot x} j^0(0) Q e^{-iP \cdot x} | 0 \rangle$$

$$= \int d^3x \, \langle 0 | j^0(0) Q | 0 \rangle$$

$$= \langle 0 | j^0(0) Q | 0 \rangle \int d^3x \to \infty \,, \tag{11.101}$$

where we have used the property of the ground state, namely,

$$P_\mu | 0 \rangle = 0 \,. \tag{11.102}$$

In other words, the state $|\chi\rangle$ is not normalizable and hence cannot be thought of as another vacuum. This analysis also shows that the finite transformation operator

$$U(\alpha) = e^{-i\alpha Q} \,, \tag{11.103}$$

does not act unitarily on the Hilbert space. In fact, it is straightforward to show that the charge Q does not exist when there is spontaneous breakdown of the symmetry. Let us note, however, that even though Q may not exist, commutators such as

$$[Q, \phi(x)] \,,$$

are well defined in such a theory and as a result expressions such as

$$U(\alpha)\phi(x)U^{-1}(\alpha) = e^{-i\alpha Q} \phi(x) e^{i\alpha Q} \,,$$

are also well defined. Another way of saying this is to note that while the operator $U(\alpha)$ defines unitary transformations for the field variables, it does not act unitarily on the Hilbert space. This is another manifestation of spontaneous symmetry breaking.

To analyze further the consequences of spontaneous symmetry breaking, let us note that even classically, if the potential has a nontrivial minimum, then a stable perturbation would require us to expand the theory about the stable minimum. Thus, let us expand

$$\sigma \to \langle \sigma \rangle + \sigma = \frac{2m}{\sqrt{\lambda}} + \sigma,$$

$$\rho \to \rho. \tag{11.104}$$

Then, the Lagrangian density of the theory in Eq. (11.78) would become

$$
\begin{aligned}
\mathcal{L} &= \partial_\mu \phi^* \partial^\mu \phi + m^2 \phi^* \phi - \frac{\lambda}{4}(\phi^* \phi)^2 \\
&= \frac{1}{2}\partial_\mu \sigma \partial^\mu \sigma + \frac{1}{2}\partial_\mu \rho \partial^\mu \rho + \frac{m^2}{2}(\sigma^2 + \rho^2) - \frac{\lambda}{16}(\sigma^2 + \rho^2)^2 \\
&\to \frac{1}{2}\partial_\mu \left(\sigma + \frac{2m}{\sqrt{\lambda}} \right) \partial^\mu \left(\sigma + \frac{2m}{\sqrt{\lambda}} \right) + \frac{1}{2}\partial_\mu \rho \partial^\mu \rho \\
&\quad + \frac{m^2}{2}\left(\left(\sigma + \frac{2m}{\sqrt{\lambda}} \right)^2 + \rho^2 \right) - \frac{\lambda}{16}\left(\left(\sigma + \frac{2m}{\sqrt{\lambda}} \right)^2 + \rho^2 \right)^2 \\
&= \frac{1}{2}\partial_\mu \sigma \partial^\mu \sigma + \frac{1}{2}\partial_\mu \rho \partial^\mu \rho + \frac{m^2}{2}\left(\sigma^2 + \rho^2 + \frac{4m}{\sqrt{\lambda}}\sigma + \frac{4m^2}{\lambda} \right) \\
&\quad - \frac{\lambda}{16}\left(\sigma^2 + \rho^2 + \frac{4m}{\sqrt{\lambda}}\sigma + \frac{4m^2}{\lambda} \right)^2 \\
&= \frac{1}{2}\partial_\mu \sigma \partial^\mu \sigma + \frac{1}{2}\partial_\mu \rho \partial^\mu \rho + \sigma^2 \left(\frac{m^2}{2} - m^2 - \frac{m^2}{2} \right) \\
&\quad + \rho^2 \left(\frac{m^2}{2} - \frac{m^2}{2} \right) + \sigma \left(\frac{2m^3}{\sqrt{\lambda}} - \frac{2m^3}{\sqrt{\lambda}} \right) \\
&\quad + \left(\frac{2m^4}{\lambda} - \frac{m^4}{\lambda} \right) - \frac{m\sqrt{\lambda}}{2}\sigma(\sigma^2 + \rho^2) - \frac{\lambda}{16}(\sigma^2 + \rho^2)^2 \\
&= \frac{1}{2}\partial_\mu \sigma \partial^\mu \sigma + \frac{1}{2}\partial_\mu \rho \partial^\mu \rho - m^2\sigma^2 + \frac{m^4}{\lambda} \\
&\quad - \frac{m\sqrt{\lambda}}{2}\sigma(\sigma^2 + \rho^2) - \frac{\lambda}{16}(\sigma^2 + \rho^2)^2 . \tag{11.105}
\end{aligned}
$$

Thus, we see the interesting fact that while the field σ remains massive with the right sign for the mass term, the field ρ indeed has become massless. This is a general feature of spontaneous symmetry breaking, namely, whenever a continuous symmetry is spontaneously broken in a manifestly Lorentz invariant theory, there necessarily arise massless fields (particles). These are known as the Goldstone fields or Goldstone modes (particles). In the present case, we note that ρ corresponds to the Goldstone field and let us recall our earlier result, namely,

$$\langle 0|\delta\rho|0\rangle = -\epsilon \frac{2m}{\sqrt{\lambda}}. \tag{11.106}$$

This is also a general feature of theories with spontaneously broken symmetries. Namely, in such theories, the change in the Goldstone field under the symmetry transformation acquires a nonzero vacuum expectation value.

In terms of the potential, it is easier to understand the Goldstone mode intuitively. The minimum of the potential occurs along a valley and the Goldstone mode simply reflects the motion along the valley of the potential. In particle physics, one does not know of elementary spin zero particles which are massless. The closest that comes to being massless is the pi-meson. The Goldstone particles were, therefore, not received well by the particle physics community. However, in the presence of gauge fields like the photon field, the Goldstone modes get absorbed into the longitudinal modes of the gauge bosons effectively making them massive. This is known as the Higgs mechanism and is widely used in the physical models of fundamental interactions.

A massless field or a particle, of course, has associated with it an infinite characteristic length (Compton length). The most familiar massless field is the photon field and we know that as a consequence of the photon being massless, the Coulomb force has an infinite range. In fact, we recognize that the two point function in such a theory will have an infinite correlation. Namely, two particles at infinite separation will still feel the presence of each other. Therefore, we conclude that when Goldstone modes are present, certain correlation lengths will become infinite.

11.6 Goldstone Theorem

In a manifestly Lorentz invariant quantum theory with a positive metric for the Hilbert space, the Goldstone theorem states that if there is spontaneous breakdown of a continuous symmetry, then there must exist massless particles (Goldstone particles) in the theory.

To see a general proof of this theorem, let us assume that we have a theory of n-scalar fields described by the Lagrangian density

$$\mathcal{L} = \mathcal{L}(\phi_i, \partial_\mu \phi_i), \qquad i = 1, 2, \cdots, n. \tag{11.107}$$

Furthermore, let us assume that the global transformations

$$\delta_\epsilon \phi_i = T_{ij}(\epsilon)\phi_j, \tag{11.108}$$

where we assume summation over repeated indices and where the global parameter of transformation, ϵ, may itself have an index, define a symmetry of the Lagrangian density in Eq. (11.107). In this case, we can define the generating functional with appropriate sources as

$$Z[J_i] = e^{iW[J_i]} = N \int \mathcal{D}\phi_i \, e^{iS[\phi_i, J_i]}. \tag{11.109}$$

Furthermore, the classical fields are defined to be

$$\phi_{ic}(x) = \frac{\delta W}{\delta J_i(x)} = \langle 0|\phi_i(x)|0\rangle^{J_k}. \tag{11.110}$$

The case of spontaneous symmetry breaking, of course, corresponds to having nontrivial ϕ_{ic}'s when the sources are turned off. Namely, even if for one of the values of i,

$$\phi_{ic} = \phi_{ic}(x)|_{J_k=0} = \left.\frac{\delta W}{\delta J_i(x)}\right|_{J_k=0} \neq 0, \tag{11.111}$$

then, we will have spontaneous breakdown of the symmetry.

Let us note from Eqs. (11.73) and (11.108) that the classical fields would transform under the symmetry transformations as

$$\delta_\epsilon \phi_{ic}(x) = T_{ij}(\epsilon)\phi_{jc}(x). \tag{11.112}$$

We also know from Eq. (10.25) that the 1PI vertex functional satisfies the defining relation

$$\frac{\delta\Gamma[\phi_{ic}]}{\delta\phi_{ic}(x)} = -J_i(x)\,. \tag{11.113}$$

When the source is turned off, this defines an extremum equation whose solutions, ϕ_{ic}, will have at least one nonzero value if the symmetry is spontaneously broken. Given the above relation, we can also obtain

$$-\delta_\epsilon J_i(x) = \int d^4y\, \frac{\delta^2\Gamma[\phi_{ic}]}{\delta\phi_{ic}(x)\delta\phi_{jc}(y)}\, \delta_\epsilon\phi_{jc}(y)$$

$$= \int d^4y\, \frac{\delta^2\Gamma[\phi_{ic}]}{\delta\phi_{ic}(x)\delta\phi_{jc}(y)}\, T_{jk}(\epsilon)\phi_{kc}(y)\,. \tag{11.114}$$

When we switch off the sources, consistency of Eq. (11.114) will lead to (in this case $\phi_{ic}(x) = \phi_{ic} = $ constant)

$$\int d^4y\, \frac{\delta^2\Gamma}{\delta\phi_{ic}(x)\delta\phi_{jc}(y)}\bigg|_{\phi_{ic}}\, T_{jk}(\epsilon)\phi_{kc} = 0$$

$$\text{or, } \int d^4y\, \left(G_F^{-1}(x-y)\right)_{ij}\, T_{jk}(\epsilon)\phi_{kc} = 0$$

$$\text{or, } \left(G_F^{-1}(p_\mu = 0)\right)_{ij}\, T_{jk}(\epsilon)\phi_{kc} = 0\,. \tag{11.115}$$

This system of equations will have a nontrivial solution (namely, there will be spontaneous breaking of the symmetry) only if

$$\det\left(G_F^{-1}(p_\mu = 0)\right)_{ij} = 0\,. \tag{11.116}$$

In other words, there must exist massless particles in the theory. This proves the Goldstone theorem.

11.7   References

Goldstone, J., Nuovo Cimento, **19**, 154 (1961).

Guralnik, G. S. et al, in "Advances in Particle Physics", Ed. R. L. Cool and R. E. Marshak, John-Wiley Publishing.

Hill, E. L., Rev. Mod. Phys., **23**, 253 (1957).

Itzykson, C. and J.-B. Zuber, "Quantum Field Theory", McGraw-Hill Publishing.

Nambu, Y. and G. Jona-Lasinio, Phys. Rev., **122**, 345 (1961).

Chapter 12

Gauge Theories

Gauge theories are very fundamental in our present understanding
of physical forces and, in this chapter, we will study how such theories are described in the path integral formalism. Gauge theories are
defined to be theories with a local symmetry and the symmetry is
based on some relevant symmetry group. As we have seen in the last
chapter symmetries, in general, have important consequences and in
the case of a local symmetry, the consequences are even more power-
ful. However, along with beautiful structures, local invariances also
bring difficulties. For theories with simple local symmetries such as
Maxwell's theory, these difficulties can be handled with ease, but
problems become quite severe when the relevant local symmetries
are more complex such as a non-Abelian symmetry. The path integral description in this case needs to be analyzed carefully. (Similar
difficulties also manifest in the canonical quantization of such theories.) In this chapter, we will start with Maxwell's theory where the
difficulties can be simply handled and then we will go into a detailed
discussion of non-Abelian gauge theories within the context of the
path integral formalism.

12.1 Maxwell Theory

Let us recall that the Lagrangian density for Maxwell's theory is
given by

$$\mathcal{L} = -\frac{1}{4} F_{\mu\nu} F^{\mu\nu}, \tag{12.1}$$

where the field strength tensor describing the electric and the magnetic fields is defined in terms of the vector potential A_μ as

$$F_{\mu\nu} = \partial_\mu A_\nu - \partial_\nu A_\mu = -F_{\nu\mu}. \tag{12.2}$$

As we know, Maxwell's theory is invariant under gauge transformations which are local transformations. Explicitly, under a local change of the vector potential by a gradient,

$$A_\mu(x) \rightarrow A_\mu(x) + \partial_\mu \alpha(x), \tag{12.3}$$

the field strength tensor remains invariant

$$
\begin{aligned}
F_{\mu\nu} &= \partial_\mu A_\nu - \partial_\nu A_\mu \\
&\rightarrow \partial_\mu \left(A_\nu + \partial_\nu \alpha(x) \right) - \partial_\nu \left(A_\mu + \partial_\mu \alpha(x) \right) \\
&= \partial_\mu A_\nu - \partial_\nu A_\mu = F_{\mu\nu}.
\end{aligned} \tag{12.4}
$$

Namely, the physically observable electric and magnetic fields are not sensitive to a redefinition of the vector potential by a gauge transformations. Since Maxwell's equations can be described in terms of the field strength tensors, the gauge transformation of Eq. (12.3) defines a symmetry (invariance) of the Maxwell theory. These are simple local gauge transformations belonging to the Abelian $U(1)$ group and correspondingly, Maxwell's theory is known as an Abelian gauge theory. For completeness, we note here that the gauge transformation of Eq. (12.3) can be written in terms of local $U(1)$ phase transformations as

$$
\begin{aligned}
A_\mu(x) &\rightarrow A_\mu(x) + i e^{i\alpha(x)} \partial_\mu e^{-i\alpha(x)} \\
&= U^{-1}(x) A_\mu(x) U(x) + i U^{-1}(x) \partial_\mu U(x),
\end{aligned} \tag{12.5}
$$

where $U(x) = e^{-i\alpha(x)} \in U(1)$.

The gauge invariance, of course, restricts the structure of the theory (Lagrangian density) and leads to difficulties that can be seen as follows. Let us note that the canonical momenta conjugate to the field variables A_μ can be calculated from the Lagrangian density and take the forms

$$\Pi^\mu = \frac{\partial \mathcal{L}}{\partial \dot{A}_\mu} = -F^{0\mu},$$ (12.6)

which implies the constraint

$$\Pi^0 = -F^{00} = 0.$$ (12.7)

This is, in fact, a general feature of gauge theories, namely, the local invariance leads to constraints. As a result, the canonical quantization of such theories is nontrivial. However, we note that since Maxwell's theory as well as physical quantities such as the field strength tensor are invariant under the gauge transformation in Eq. (12.3), we can choose a gauge (a particular form of the vector potential defined by the gauge choice) to work with. If we choose the gauge $\nabla \cdot \mathbf{A} = 0$ (Coulomb gauge), then the equations of motion

$$\partial_\mu F^{\mu\nu} = 0,$$ (12.8)

lead to (for $\nu = 0$)

$$\partial_\mu F^{\mu 0} = \partial_i F^{i0} = 0$$

$$\text{or, } \partial_i \left(\partial^i A^0 - \partial^0 A^i \right) = 0$$

$$\text{or, } A^0 = A_0 = 0.$$ (12.9)

On the other hand, if sources (charges and currents) are present, the gauge

$$\nabla \cdot \mathbf{A} = 0,$$ (12.10)

would lead to

$$\nabla^2 A_0 = -j_0$$

$$\text{or, } A_0 = -\frac{1}{\nabla^2} j_0.$$ (12.11)

In either case, the true dynamics of the theory is, therefore, contained in the transverse physical degrees of freedom and the longitudinal degrees of freedom can be expressed in terms of these. The canonical quantization can now be carried out, but we lose manifest Lorentz invariance in the process. Let us emphasize here that the final result for the calculation of any amplitude in the canonical formalism remains manifestly Lorentz invariant. However, there is no manifest Lorentz invariance in the intermediate steps.

We can ask what would happen if we were to treat the theory in the path integral formalism as opposed to the canonical formalism. Here we note that we can write the Lagrangian density for the Maxwell theory also as

$$\mathcal{L} = -\frac{1}{4}F_{\mu\nu}F^{\mu\nu} = \frac{1}{2}A_\mu P^{\mu\nu} A_\nu + \text{ total derivatives}, \qquad (12.12)$$

where

$$P^{\mu\nu} = \eta^{\mu\nu}\Box - \partial^\mu\partial^\nu \,,$$

$$P^{\mu\nu}P_\nu^\lambda = (\eta^{\mu\nu}\Box - \partial^\mu\partial^\nu)\left(\delta_\nu^\lambda\Box - \partial_\nu\partial^\lambda\right)$$

$$= \eta^{\mu\lambda}\Box^2 - \partial^\mu\partial^\lambda\Box - \partial^\mu\partial^\lambda\Box + \partial^\mu\partial^\lambda\Box$$

$$= \Box\left(\eta^{\mu\lambda}\Box - \partial^\mu\partial^\lambda\right)$$

$$= \Box P^{\mu\lambda}. \qquad (12.13)$$

With a suitable normalization $P^{\mu\nu}$ can be thought of as a projection operator. ($\overline{P}^{\mu\nu} = \frac{1}{\Box}P^{\mu\nu}$ is the normalized projection operator.) In fact, we note that

$$\partial_\mu P^{\mu\nu} = \partial_\mu \left(\eta^{\mu\nu}\Box - \partial^\mu\partial^\nu\right)$$

$$= (\partial^\nu\Box - \Box\partial^\nu) = 0 = \partial_\nu P^{\mu\nu}. \qquad (12.14)$$

Therefore, this is the transverse projection operator, namely, it projects the components of any vector transverse (perpendicular) to the gradient operator ∂^μ. As a result, the inverse of $P^{\mu\nu}$ does not exist and the Green's function and, therefore, the Feynman propagator cannot be defined. This implies that if we were to apply the path integral formalism naively, neither the generating functional will exist nor can we carry out perturbation theory.

We note here that whenever the determinant of the matrix of highest derivatives in the Lagrangian density vanishes, the system is singular and contains constraints among the field variables. In such a case, without any further input, the Cauchy initial value problem cannot be uniquely solved, simply because the Green's function does not exist. As a consequence, we see that the naive canonical quantization has unpleasant features in the case of Maxwell's theory since the fields are constrained and the momentum corresponding to A_0 vanishes. In a physical gauge such as the Coulomb gauge, we can solve for the constraints and quantize only the true dynamical degrees of freedom. However, in this process, we give up manifest Lorentz invariance since we single out the transverse degrees of freedom.

We can, of course, take an alternative approach. Namely, since we realize that the difficulties in quantization arise because of the singular nature of the Lagrangian density, we could modify the theory to make it nonsingular. Let us consider, for example, the Lagrangian density (This formulation of the theory is due to Fermi and this gauge is known as the Feynman-Fermi gauge.)

$$\mathcal{L} = -\frac{1}{4} F_{\mu\nu} F^{\mu\nu} - \frac{1}{2} \left(\partial_\mu A^\mu \right)^2 - j_\mu A^\mu \,, \qquad (12.15)$$

where j^μ represents a conserved current (the sources in Maxwell's theory)

$$\partial_\mu j^\mu = 0 \,. \qquad (12.16)$$

Here we have generalized Maxwell's theory to include a conserved current. But more than that we have also added a term $-\frac{1}{2} \left(\partial_\mu A^\mu \right)^2$ to Maxwell's Lagrangian density. This term breaks gauge invariance and consequently leads to a nonsingular theory. But clearly this would appear to be different from Maxwell's theory. Therefore, at this point there is no justification for adding this new term to the Lagrangian density. But to understand the issue better, let us look at the equation of motion following from the action in the present case

$$\partial_\mu \frac{\partial \mathcal{L}}{\partial \partial_\mu A_\nu} - \frac{\partial \mathcal{L}}{\partial A_\nu} = 0$$

$$\text{or, } \partial_\mu F^{\mu\nu} + \partial^\nu (\partial \cdot A) - j^\nu = 0$$

$$\text{or, } \partial_\mu F^{\mu\nu} + \partial^\nu (\partial \cdot A) = j^\nu . \tag{12.17}$$

Without the second term on the left hand side in Eq. (12.17), this is just Maxwell's equations in the presence of conserved sources. If we now write out the left hand side, it takes the form

$$\partial_\mu \left(\partial^\mu A^\nu - \partial^\nu A^\mu \right) + \partial^\nu (\partial \cdot A) = j^\nu$$

$$\text{or, } \Box A^\nu = j^\nu$$

$$\text{or, } \Box \partial \cdot A = \partial \cdot j = 0 . \tag{12.18}$$

An alternative way to see this is to note that if we take the divergence of the equations of motion in (12.17), we have

$$\partial_\nu \partial_\mu F^{\mu\nu} + \Box (\partial \cdot A) = \partial_\nu j^\nu$$

$$\text{or, } \Box (\partial \cdot A) = 0 , \tag{12.19}$$

where we have used the anti-symmetry of the field strength tensor as well as the conservation of j^μ.

Thus we see that although the presence of the term $-\frac{1}{2}\chi^2$ in the Lagrangian density where

$$\chi = \partial \cdot A , \tag{12.20}$$

seems to modify the theory, χ is in reality a free field and, therefore, the presence of this additional term in the Lagrangian density would not change the physics of Maxwell's theory. Furthermore, we realize that if we restrict classically

$$\chi = 0, \quad \text{at} \quad t = 0 ,$$

and

$$\frac{\partial \chi}{\partial t} = 0, \quad \text{at} \quad t = 0 , \tag{12.21}$$

then Eq. (12.19) would determine $\chi = 0$ at all times and we get back our familiar Maxwell's theory. Thus, classically we can think of Maxwell's theory as described by the modified Lagrangian density of Eq. (12.15) with the supplementary condition

$$\partial \cdot A = 0 \,. \tag{12.22}$$

In the quantum theory, however, the field variables, A_μ, are operators and Eq. (12.22) is hard to impose on the theory as an operator relation. One can think of imposing such a condition on the vector space of states to select out the physical Hilbert space, namely,

$$\partial_\mu A^\mu |\text{phys}\rangle = 0 \,. \tag{12.23}$$

However, this, too, turns out to be too stringent a condition. This not only demands that certain kinds of photons are not present in the physical state, but it also requires that those photons cannot be emitted either. Gupta and Bleuler weakened the supplementary condition on the physical states to have the form

$$\partial_\mu A^{\mu\,(+)}(x) |\text{phys}\rangle = 0 \,, \tag{12.24}$$

where $A^{\mu\,(+)}(x)$ is the positive frequency part of the Maxwell's field and contains only the destruction operator. (This is commonly known as the Gupta-Bleuler quantization.) We remark here that since (see Eq. (12.19))

$$\Box \partial_\mu A^\mu(x) = 0 \,, \tag{12.25}$$

$\partial_\mu A^\mu(x)$ is like a free scalar field. Therefore, it can be decomposed into positive and negative frequency parts uniquely in a relativistically invariant manner and this decomposition is preserved under time evolution. Furthermore, we note that the theory has four degrees of freedom resulting from the four components of A_μ and since the components can be time-like or space-like, the resulting vector space of states has the problem of a negative metric. On the other hand, the Gupta-Bleuler condition (12.24) selects out three kind of photon states as being physical, two transverse photon states with a positive norm and a linear combination of the time-like and the longitudinal photon states with zero norm. The state with zero norm, however, is orthogonal to every other state (including itself) and, therefore, decouples from the theory and the physical Hilbert space effectively consist of the physical transverse photon states. We can think of the Gupta-Bleuler supplementary condition as imposing the Lorentz condition

$$\partial_\mu A^\mu(x) = 0 \,,$$

on the physical quantum states since

$$\langle \psi | \partial_\mu A^\mu | \psi \rangle = 0 \,, \tag{12.26}$$

where $|\psi\rangle$ represents a physical state.

Thus, we see that the physical subspace of the theory selected by the supplementary condition contains states with positive semi-definite norm (negative norm states are eliminated by the supplementary condition or the physical state condition and, consequently, there is no problem with a probabilistic interpretation). Since the zero norm states are orthogonal to all the states including themselves, if we further mod out the states by the zero norm states, we have the true physical subspace of the theory where the norm of states is positive definite, namely,

$$\overline{V}_{\text{phys}} = \frac{V_{\text{phys}}}{V_0} \,, \tag{12.27}$$

where V_0 represents the set of states with zero norm.

Since the modified theory in Eq. (12.15) (which is equivalent to Maxwell's theory) is nonsingular, it can be described in the path integral formalism in the standard manner. In this simple gauge theory, therefore, there is an easy solution to the problem associated with the local invariance of the theory. As we will see in the next section, the difficulties become more severe in a non-Abelian gauge theory.

12.2   Non-Abelian Gauge Theory

As we have mentioned earlier, non-Abelian gauge theories are based on nontrivial symmetry groups where the generators of the group satisfy a non-commutative algebra. Let us take brief digression into the structure of $SU(n)$. If we assume that $T^a, a = 1, 2, \cdots, n^2 - 1(\dim SU(n))$ define the generators of the group, then the generators satisfy the Lie algebra of the form (we take the generators to be Hermitian)

$$\left[T^a, T^b\right] = i f^{abc} T^c, \tag{12.28}$$

and the Jacobi identity for the algebra is given by

$$\left[\left[T^a, T^b\right], T^c\right] + \left[\left[T^c, T^a\right], T^b\right] + \left[\left[T^b, T^c\right], T^a\right] = 0. \tag{12.29}$$

This imposes a restriction on the structure constants of the form

$$i f^{abp} \left[T^p, T^c\right] + i f^{cap} \left[T^p, T^b\right] + i f^{bcp} \left[T^p, T^a\right] = 0$$

$$\text{or,} \quad f^{abp} f^{pcq} T^q + f^{cap} f^{pbq} T^q + f^{bcp} f^{paq} T^q = 0$$

$$\text{or,} \quad f^{abp} f^{pcq} + f^{cap} f^{pbq} + f^{bcp} f^{paq} = 0. \tag{12.30}$$

We can find the generators in various representations of the group much like in the case of angular momentum. However, a particular representation that is very important as well as useful in the study of gauge theories is given by

$$(T^a)_{bc} = -i f^{abc}. \tag{12.31}$$

This is consistent with the hermiticity requirement for the generators

$$\left(T^{a\dagger}\right)_{bc} = ((T^a)_{cb})^* = \left(-i f^{acb}\right)^* = i f^{acb} = -i f^{abc} = (T^a)_{bc}. \tag{12.32}$$

Furthermore, we can easily check that this representation satisfies the Lie algebra,

$$\left[T^a, T^b\right]_{cq} = \left(T^a T^b - T^b T^a\right)_{cq}$$

$$= (T^a)_{cp} \left(T^b\right)_{pq} - \left(T^b\right)_{cp} (T^a)_{pq}$$

$$= \left(-i f^{acp}\right)\left(-i f^{bpq}\right) - \left(-i f^{bcp}\right)\left(-i f^{apq}\right)$$

$$= -f^{acp}f^{bpq} + f^{bcp}f^{apq}$$

$$= -f^{cap}f^{pbq} - f^{bcp}f^{paq}$$

$$= f^{abp}f^{pcq} \qquad \text{(by Jacobi identity)}$$

$$= if^{abp}\left(-if^{pcq}\right)$$

$$= if^{abp}\left(T^p\right)_{cq}, \tag{12.33}$$

so that the identification

$$\left(T^a_{(\text{adj})}\right)_{bc} = -if^{abc}, \tag{12.34}$$

indeed forms a representation of the Lie algebra known as the adjoint representation.

For any representation of a simple Lie group we can write

$$\text{Tr } T^a T^b = C_2 \delta^{ab}. \tag{12.35}$$

We note that C_2 is a normalization constant which determines the values of the structure constants. It depends on the representation but not on the indices a and b. To prove this let us note that we can always diagonalize the tensor $\text{Tr}\left(T^a T^b\right)$ such that (this is a symmetric real matrix)

$$\text{Tr}\left(T^a T^b\right) = \begin{cases} 0 & \text{if } a \neq b, \\ K_a & \text{if } a = b. \end{cases} \tag{12.36}$$

Let us next note that the quantity

$$h^{abc} = \text{Tr}\left(\left[T^a, T^b\right] T^c\right)$$

$$= \text{Tr}\left(T^a T^b T^c\right) - \text{Tr}\left(T^b T^a T^c\right), \tag{12.37}$$

is completely antisymmetric in all its indices. Furthermore, using the commutation relations we have

$$h^{abc} = \text{Tr}\left(if^{abp}T^p T^c\right)$$

$$= if^{abp}\,\text{Tr}\left(T^p T^c\right)$$

$$= if^{abp}K_p \delta^{pc}$$

$$= iK_c f^{abc}, \qquad \text{(no sum on } c). \tag{12.38}$$

On the other hand, we note that

$$
\begin{aligned}
h^{acb} &= \operatorname{Tr}\left([T^a, T^c]\, T^b\right) \\
&= if^{acp} \operatorname{Tr}\left(T^p T^b\right) \\
&= if^{acp} K_p \delta^{pb} \\
&= iK_b f^{acb} = -iK_b f^{abc}, \qquad \text{(no sum on } b\text{)}.
\end{aligned} \tag{12.39}
$$

However, since h^{abc} is completely antisymmetric

$$
h^{acb} = -h^{abc}. \tag{12.40}
$$

Comparing the two results we see that

$$
K_b = K_c = K. \tag{12.41}
$$

(Alternatively, note that

$$
\operatorname{Tr} T^a T^b = K_a \delta^{ab},
$$
$$
\operatorname{Tr} T^b T^a = K_b \delta^{ab}. \tag{12.42}
$$

By cyclicity of the trace, the two must equal and hence we have

$$
K_a = K_b, \tag{12.43}
$$

which proves Eq. (12.41).) In other words, we see that we can write

$$
\operatorname{Tr}\left(T^a T^b\right) = C_2 \delta^{ab}, \tag{12.44}
$$

where the constant C_2 depends only on the representation. (It is chosen to be $\frac{1}{2}$ for the fundamental representation to which the fermions belong in $SU(n)$.) We note that if we write

$$
\operatorname{Tr} T^a T^b = T(R)\, \delta^{ab}, \tag{12.45}
$$

then $T(R)$ is known as the index of the representation R. Similarly, we have

$$
(T^a T^a)_{mn} = C(R)\, \delta_{mn}, \tag{12.46}
$$

where $C(R)$ is known as the Casimir of the representation R. The two are clearly related as

$$T(R)\dim G = C(R)\dim R,\qquad(12.47)$$

where $\dim G, \dim R$ denote respectively the dimensionalities of the group G and the representation R.

To construct a gauge theory with $SU(n)$ symmetry, we note that the gauge potentials must belong to some representation of the symmetry group and, therefore, can be written as matrices. Furthermore, these can be expanded in terms of the basic generators of the Lie algebra as

$$A_\mu(x) = T^a A_\mu^a(x),\qquad(12.48)$$

where summation over repeated indices is understood. The gauge transformation in Eq. (12.5) can now be generalized to

$$A_\mu(x) \rightarrow U^{-1}(x)A_\mu(x)U(x) + iU^{-1}(x)\partial_\mu U(x),\qquad(12.49)$$

where, in the present case,

$$U(x) = e^{-i\alpha(x)} = e^{-i\alpha^a(x)T^a},\qquad(12.50)$$

belongs to the non-Abelian group $SU(n)$. Furthermore, since U and A_μ are now matrices, they do not commute in general and the actual gauge transformation in Eq. (12.49) is much more complicated than (12.3). In fact, for the simple case of an infinitesimal gauge transformation, Eq. (12.49) takes the form (here the parameter of transformation $\theta(x)$ is assumed to be infinitesimal)

$$A_\mu(x) \rightarrow A_\mu(x)+\partial_\mu\theta(x)-i\left[A_\mu(x),\theta(x)\right] = A_\mu(x)+D_\mu\theta(x),\quad(12.51)$$

where we have defined

$$D_\mu\theta(x) = \partial_\mu\theta(x) - i\left[A_\mu(x),\theta(x)\right],\qquad(12.52)$$

also known as the covariant derivative.

To construct the Lagrangian density for the gauge field in a gauge invariant manner, we note that under a gauge transformation (12.49), the tensor representing the Abelian field strength (see (12.2)) would transform as

$$f_{\mu\nu} = \partial_\mu A_\nu - \partial_\nu A_\mu$$
$$\rightarrow \partial_\mu \left[U^{-1} A_\nu U + i U^{-1} \left(\partial_\nu U \right) \right]$$
$$- \partial_\nu \left[U^{-1} A_\mu U + i U^{-1} \left(\partial_\mu U \right) \right]$$
$$= i \left(\partial_\mu U^{-1} \right) \left(\partial_\nu U \right) - i \left(\partial_\nu U^{-1} \right) \left(\partial_\mu U \right)$$
$$+ \left(\partial_\mu U^{-1} \right) A_\nu U + U^{-1} A_\nu \left(\partial_\mu U \right)$$
$$- \left(\partial_\nu U^{-1} \right) A_\mu U - U^{-1} A_\mu \left(\partial_\nu U \right)$$
$$+ U^{-1} \left(\partial_\mu A_\nu - \partial_\nu A_\mu \right) U . \tag{12.53}$$

Namely, we see that, unlike QED, in the present case $f_{\mu\nu} = \partial_\mu A_\nu - \partial_\nu A_\mu$ is not invariant under a gauge transformation. Let us also note that under a gauge transformation

$$-i \left[A_\mu, A_\nu \right] = -i \left(A_\mu A_\nu - A_\nu A_\mu \right)$$
$$\rightarrow -i \left[\left(U^{-1} A_\mu U + i U^{-1} \left(\partial_\mu U \right) \right), \left(U^{-1} A_\nu U + i U^{-1} (\partial_\nu U) \right) \right]$$
$$= -i U^{-1} \left[A_\mu, A_\nu \right] U + i \left(\left(\partial_\nu U^{-1} \right) \left(\partial_\mu U \right) - \left(\partial_\mu U^{-1} \right) \left(\partial_\nu U \right) \right)$$
$$- \left(\left(\partial_\mu U^{-1} \right) A_\nu U + U^{-1} A_\nu \left(\partial_\mu U \right) \right.$$
$$\left. - \left(\partial_\nu U^{-1} \right) A_\mu U - U^{-1} A_\mu \left(\partial_\nu U \right) \right) . \tag{12.54}$$

Thus, it is clear that, in the present case, if we define the field strength tensor as

$$F_{\mu\nu} = \partial_\mu A_\nu - \partial_\nu A_\mu - ig \left[A_\mu, A_\nu \right] , \tag{12.55}$$

then under a gauge transformation, $F_{\mu\nu}$ will transform covariantly, namely,

$$F_{\mu\nu} \rightarrow U^{-1} F_{\mu\nu} U . \tag{12.56}$$

It is now easy to construct the Lagrangian density for the kinetic part of the gauge field as (quadratic in derivatives)

$$\mathcal{L}_{\text{gauge}} = -\frac{1}{2} \text{Tr} \, F_{\mu\nu} F^{\mu\nu} = -\frac{1}{4} F^a_{\mu\nu} F^{\mu\nu \, a} , \tag{12.57}$$

where a particular normalization for the trace is assumed (namely, T^a's in the fundamental representation). This Lagrangian density is easily seen, from the cyclicity of trace, to be invariant under the gauge transformation (12.49). In components, the field strength tensor takes the form, (we have set the coupling constant to unity for simplicity)

$$F^a_{\mu\nu} = \partial_\mu A^a_\nu - \partial_\nu A^a_\mu + f^{abc} A^b_\mu A^c_\nu = -F^a_{\nu\mu}. \qquad (12.58)$$

In components, the infinitesimal form of the transformation for the gauge fields in Eq. (12.51) take the form

$$
\begin{aligned}
\delta A^a_\mu &= A'^a_\mu - A^a_\mu \\
&= \left(\partial_\mu \theta^a + f^{abc} A^b_\mu \theta^c \right) \\
&= \left(\partial_\mu \theta^a - f^{bac} A^b_\mu \theta^c \right) \\
&= \left(\partial_\mu \theta^a - i \left(-i f^{bac} \right) A^b_\mu \theta^c \right) \\
&= \left(\partial_\mu \theta^a - i \left(T^b_{(\text{adj})} \right)_{ac} A^b_\mu \theta^c \right),
\end{aligned}
\qquad (12.59)
$$

so that we can write

$$\delta A^a_\mu = \left(D^{(\text{adj})}_\mu \theta \right)^a, \qquad (12.60)$$

which shows that the gauge field, A_μ, transforms according to the adjoint representation of the group. The infinitesimal transformation of the field strength tensor can be obtained from Eq. (12.56) and leads to

$$
\begin{aligned}
\delta F^a_{\mu\nu} &= f^{abc} F^b_{\mu\nu} \theta^c \\
&= i \left(-i f^{cab} \right) \theta^c F^b_{\mu\nu} \\
&= i \left(T^c_{(\text{adj})} \right)_{ab} \theta^c F^b_{\mu\nu} \\
&= i \theta^c \left(T^c_{(\text{adj})} \right)_{ab} F^b_{\mu\nu}.
\end{aligned}
\qquad (12.61)
$$

Thus, the field strength tensor $F_{\mu\nu}$ as well as the gauge field A_μ transform according to the adjoint representation of the group. (It does not matter what representationt the matter fields belong to, the gauge field must transform in the adjoint representation.) We note here that unlike the photon field, here the gauge field has self interaction. Physically we understand this in the following way. In the present case, the gauge field carries the charge of the non-Abelian symmetry group (they have a nontrivial symmetry index) in contrast to the photon field which is charge neutral. Since gauge fields couple to any particle carrying charge of the symmetry group, in the case of non-Abelian symmetry they must possess self interactions.

Let us next briefly examine the difficulties that arise in trying to canonically quantize the theory along the lines of Maxwell theory. The Euler-Lagrange equations following from Eq. (12.57) take the form

$$\partial_\nu \frac{\partial \mathcal{L}}{\partial \partial_\nu A_\mu^a} - \frac{\partial \mathcal{L}}{\partial A_\mu^a} = 0$$

$$\text{or, } (D_\nu F^{\mu\nu})^a = 0, \tag{12.62}$$

where the covariant derivative is defined to be in the adjoint representation of the group and

$$F_{\mu\nu}^a = \partial_\mu A_\nu^a - \partial_\nu A_\mu^a + f^{abc} A_\mu^b A_\nu^c. \tag{12.63}$$

For $\mu = 0$, we see that the Euler-Lagrange equations lead to a constraint of the form

$$D_i F^{0i\,a} = 0, \tag{12.64}$$

which is a reflection of the gauge invariance of the theory. In fact, it can be explicitly checked the coefficient matrix of highest derivatives is the transverse projection operator just like in the Maxwell theory.

Let us define the momenta canonically conjugate to the field variables A_μ^a

$$\Pi^{\mu\,a}(x) = \frac{\partial \mathcal{L}}{\partial \dot{A}_\mu^a(x)} = -F^{0\mu\,a}(x). \tag{12.65}$$

Noting that the field strength $F_{\mu\nu}^a$ is antisymmetric in the indices μ, ν, we have

$$\Pi^{0\,a}(x) = F^{00\,a} = 0\,, \tag{12.66}$$

and

$$\Pi^{i\,a}(x) = -F^{0i\,a} = E_i^a(x)\,, \tag{12.67}$$

where we can think of $E_i^a(x)$ as the non-Abelian electric field strength.

We note that the momentum conjugate to A_0^a does not exist much like in the Maxwell theory. This implies that A_0^a is like a c-number quantity which commutes with every other operator in the theory. Thus we can choose a physical gauge condition to suitably set it equal to zero, namely,

$$A_0^a(x) = 0\,. \tag{12.68}$$

The analysis is parallel to what we have discussed in the case of Maxwell theory, although it is much more complicated. Barring technical issues such as the Gribov ambiguity (related to the existence of large gauge transformations), the theory can be defined in terms of only physical transverse gauge fields in this case. Therefore, canonical quantization leads to a lack of manifest Lorentz invariance.

We can try to quantize a non-Abelian gauge theory covariantly, very much along the lines of the Abelian theory, namely, by modifying the theory. Thus, for example, let us look at the theory

$$\mathcal{L} = -\frac{1}{4} F_{\mu\nu}^a F^{\mu\nu\,a} - \frac{1}{2} \left(\partial_\mu A^{\mu a} \right)^2\,. \tag{12.69}$$

The additional term in this theory clearly breaks gauge invariance and, consequently, makes the theory nonsingular, much like in the Maxwell theory. However, in the present case, there are serious differences from Maxwell's theory. For example, let us note that in the Abelian theory,

$$\Box(\partial \cdot A) = 0\,, \tag{12.70}$$

and hence classically we can impose the condition

$$\partial \cdot A = 0\,, \tag{12.71}$$

which then translates to the Gupta-Bleuler condition on the physical states

$$\partial \cdot A^{(+)}|\text{phys}\rangle = 0. \tag{12.72}$$

In the case of the non-Abelian theory, however, even classically the equations of motion are given by

$$\partial_\mu F^{\mu\nu\ a} + f^{abc} A^b_\mu F^{\mu\nu\ c} + \partial^\nu \left(\partial_\mu A^{\mu a}\right) = 0. \tag{12.73}$$

Contracting with ∂_ν, we obtain

$$\Box \left(\partial_\mu A^{\mu a}\right) = -f^{abc} \partial_\nu \left(A^b_\mu F^{\mu\nu\ c}\right) \neq 0. \tag{12.74}$$

Thus, in contrast to the Abelian theory, we note that $(\partial_\mu A^{\mu a})$ does not behave like a free field and, consequently, the additional term has truly modified the theory. Furthermore, since $(\partial_\mu A^{\mu a})$ is not a free field, it cannot be uniquely decomposed into a positive and a negative frequency part (in a time invariant manner), nor can we think of a supplementary condition such as

$$\partial_\mu A^{\mu a(+)}|\text{phys}\rangle = 0, \tag{12.75}$$

in a physically meaningful manner, since it is not invariant under time evolution. (Namely, the physical subspace would keep changing with time which is not desirable.) Correspondingly the analog of the Gupta-Bleuler condition for non-Abelian gauge theories does not appear to exist. Therefore, we need to analyze the question of modifying the theory in a more systematic and detailed manner. We would see next how we can derive intuition on this important question from the path integral quantization.

12.3   Path Integral for Gauge Theories

To understand better how gauge theories can be handled in the path integral formalism, let us go back to the Maxwell theory described the Lagrangian density

$$\mathcal{L}^{(J)} = -\frac{1}{4} F_{\mu\nu} F^{\mu\nu} + J^\mu A_\mu, \tag{12.76}$$

where J^μ represents a conserved current (source). The generating functional in the path integral formalism is given by

$$Z\left[J_\mu\right] = e^{iW[J_\mu]} = N \int \mathcal{D}A_\mu e^{iS^{(J)}[A_\mu]}$$

$$= N \int \mathcal{D}A_\mu \, e^{i\left[\frac{1}{2}(A_\mu, O^{\mu\nu} A_\nu) + (J^\mu, A_\mu)\right]},$$

where N is a normalization constant and

$$O^{\mu\nu}(x - y) = (\Box \eta^{\mu\nu} - \partial^\mu \partial^\nu)\, \delta^4(x - y),$$

$$(J^\mu, A_\mu) = \int \mathrm{d}^4x \, J^\mu(x) A_\mu(x),$$

$$(A_\mu, O^{\mu\nu} A_\nu) = \int \mathrm{d}^4x \mathrm{d}^4y \, A_\mu(x) O^{\mu\nu}(x - y) A_\nu(y).$$

$$(12.77)$$

Here we have introduced a compact notation to describe integrations as described above. The functional integral is a Gaussian integral which we have worked out in earlier chapters and leads to

$$Z\left[J_\mu\right] = e^{iW[J_\mu]} = N \left(\det\left(-O^{\mu\nu}\right)\right)^{-\frac{1}{2}} e^{-\frac{i}{2}\left(J^\mu, O_{\mu\nu}^{-1} J^\nu\right)}. \qquad (12.78)$$

However, as we have seen before, the operator $O^{\mu\nu}$ is a projection operator for transverse photons. The longitudinal vectors k_μ (or $\partial_\mu F$) are its eigenvectors with zero eigenvalue. Clearly therefore, the determinant of $O^{\mu\nu}$ vanishes. This implies that the generating functional does not exist. (The operator possesses zero modes and, consequently, the inverse of the matrix cannot be defined either.) Going over to the Euclidean space does not help either. In fact, as we can see even in the limit of vanishing sources, the generating functional does not exist.

The source of the difficulty is not hard to see. The Lagrangian density for Maxwell's theory is invariant under the gauge transformation

$$A_\mu \to A_\mu^{(\alpha)} = U A_\mu U^{-1} + iU^{-1}(\partial_\mu U), \qquad (12.79)$$

where

$$U(\alpha) = e^{-i\alpha(x)}. \qquad (12.80)$$

For a fixed A_μ, all the $A_\mu^{(\alpha)}$'s that are obtained by making a gauge transformations with all possible $\alpha(x)$ are said to lie on an "orbit" in the group space. The action S, on the other hand, is constant (invariant) on such orbits. Therefore, the generating functional, even in the absence of any sources, is proportional to the "volume" of the orbits denoted by

$$\int \prod_x d\alpha(x) \,. \tag{12.81}$$

(In the non-Abelian case, this should be replaced by the group invariant Haar measure $\prod_x dU(x)$.) This is an infinite factor (which is the reason why the Gaussian functional integral does not exist) and must be extracted out before doing any calculations. The method for extracting this factor out of the path integral is due to Faddeev and Popov and relies on the method of gauge fixing. We recognize that we should not integrate over all gauge field configurations because they are not really distinct. Rather we should integrate over each orbit only once.

The way this is carried out is by choosing a hypersurface which intersects each orbit only once, i.e., if

$$F(A_\mu) = 0 \,, \tag{12.82}$$

defines the hypersurface which intersects the orbits once, then even if A_μ does not satisfy the condition, we can find a gauge transformed $A_\mu^{(\alpha)}$ which does, namely,

$$F\left(A_\mu^{(\alpha)}\right) = 0 \,, \tag{12.83}$$

has a unique solution for $\alpha(x)$. This procedure is known as gauge fixing and the condition

$$F(A_\mu) = 0 \,, \tag{12.84}$$

is known as the gauge condition. Physical quantities are, of course, gauge independent and do not depend on the choice of the hypersurface (gauge). Thus, for example,

$$F\left(A_{\mu}\right) = \partial_{\mu}A^{\mu} = 0 \quad \text{is the Lorentz/Landau gauge,}$$

$$\boldsymbol{\nabla} \cdot \mathbf{A} = 0 \quad \text{is the Coulomb gauge,}$$

$$A_0 = 0 \quad \text{is the temporal gauge,}$$

$$A_3 = 0 \quad \text{is the axial gauge,} \tag{12.85}$$

and so on. We can already see the need for gauge fixing from the fact that because the action is gauge invariant so is the generating functional (if sources are transformed appropriately in the non-Abelian case). Therefore, it would lead only to gauge invariant Green's functions. On the other hand, we know from ordinary perturbation theory that the Green's functions are, in general, gauge dependent although the S-matrix (the scattering matrix) elements are gauge independent. Thus one has to fix a gauge without which even the Cauchy initial value problem cannot be solved. (Only physical quantities need to be gauge independent.)

To extract out the infinite gauge volume factor, let us do the following trick due to Faddeev and Popov. Let us define

$$\Delta_{\text{FP}}\left[A_{\mu}\right] \int \prod_x \mathrm{d}\alpha(x)\, \delta\left(F\left(A_{\mu}^{(\alpha)}(x)\right)\right) = 1\,. \tag{12.86}$$

(The integration measure should be $\mathrm{d}U(x)$ which is essential in the case of non-Abelian theories.) Note that the quantity $\Delta_{\text{FP}}\left[A_{\mu}\right]$ is gauge invariant which can be seen from the fact that

$$\Delta_{\text{FP}}^{-1}\left[A_{\mu}\right] = \int \prod_x \mathrm{d}\alpha(x)\, \delta\left(F\left(A_{\mu}^{(\alpha)}(x)\right)\right)\,. \tag{12.87}$$

Let us make a gauge transformation $A_{\mu} \to A_{\mu}^{(\alpha')}$. Then

$$\Delta_{\text{FP}}^{-1}\left[A_{\mu}^{(\alpha')}\right] = \int \prod_x \mathrm{d}\alpha(x)\, \delta\left(F\left(A_{\mu}^{(\alpha+\alpha')}(x)\right)\right)$$

$$= \int \prod_x \mathrm{d}\alpha(x)\, \delta\left(F\left(A_{\mu}^{(\alpha)}(x)\right)\right)$$

$$= \Delta_{\text{FP}}^{-1}\left[A_{\mu}\right]\,. \tag{12.88}$$

This follows from the fact that the measure in the group space is invariant under a gauge transformation. That is

$$\int d\alpha(x) = \int d\left(\alpha(x) + \alpha'(x)\right). \tag{12.89}$$

In the non-Abelian case, we should have the Haar measure which is gauge invariant, namely,

$$\int d(UU') = \int dU. \tag{12.90}$$

Remembering that $\Delta_{\text{FP}}[A_\mu]$ is gauge invariant we can now insert this identity factor into the generating functional to write

$$Z[J] = N \int \mathcal{D}A_\mu \left(\Delta_{\text{FP}}[A_\mu] \int \prod_x d\alpha(x) \delta\left(F\left(A_\mu^{(\alpha)}\right)\right)\right) e^{iS^{(J)}[A_\mu]}. \tag{12.91}$$

Furthermore, let us make an inverse gauge transformation

$$A_\mu \to A_\mu^{(-\alpha)}, \tag{12.92}$$

under which the generating functional takes the form

$$Z[J] = N \int \mathcal{D}A_\mu \Delta_{\text{FP}}[A_\mu] \int \prod_x d\alpha(x) \, \delta\left(F\left(A_\mu\right)\right) e^{iS^{(J)}[A_\mu]}$$

$$= N \left(\int \prod_x d\alpha(x)\right) \int \mathcal{D}A_\mu \Delta_{\text{FP}}[A_\mu] \, \delta\left(F\left(A_\mu\right)\right) e^{iS^{(J)}[A_\mu]}$$

$$= N \int \mathcal{D}A_\mu \Delta_{\text{FP}}[A_\mu] \, \delta\left(F\left(A_\mu\right)\right) e^{iS^{(J)}[A_\mu]}, \tag{12.93}$$

where the gauge volume has been factored and absorbed into the normalization constant N of the path integral.

This, therefore, gives the correct functional form for the generating functional. However, we still have to determine what $\Delta_{\text{FP}}[A_\mu]$

is. To do this let us note that

$$\Delta_{\text{FP}}^{-1}[A_\mu] = \int \prod_x d\alpha(x) \, \delta\left(F\left(A_\mu^{(\alpha)}\right)\right)$$

$$= \int \prod_x dF \, \delta\left(F\left(A_\mu^{(\alpha)}\right)\right) \left(\det \frac{\delta\alpha}{\delta F}\right)$$

$$= \det\left(\frac{\delta\alpha}{\delta F}\right)_{F\left(A_\mu^{(\alpha)}\right)=0} . \qquad (12.94)$$

We note that since $\Delta_{\text{FP}}^{-1}[A_\mu]$ is gauge invariant we can make an inverse gauge transformation to make $F[A_\mu] = 0$ in the above derivation. On the other hand, for gauge fields which satisfy the condition

$$F(A_\mu) = 0, \qquad (12.95)$$

we have

$$\alpha(x) = 0. \qquad (12.96)$$

Thus, we determine

$$\Delta_{\text{FP}}[A_\mu] = \det\left(\frac{\delta F\left(A_\mu^{(\alpha)}\right)}{\delta\alpha}\right)_{\alpha=0} . \qquad (12.97)$$

The Faddeev-Popov determinant can, therefore, be thought of as the Jacobian that goes with a particular gauge choice. We see that the Faddeev-Popov determinant can be calculated simply by restricting to infinitesimal gauge transformations. (Here we are completely ignoring the problem of Gribov ambiguity associated with large gauge transformations.)

We can further generalize our derivation by noting that a general equation of the hypersurface has the form (Physical results are not sensitive to the choice of the hypersurface.)

$$F(A_\mu(x)) = f(x). \qquad (12.98)$$

Here $f(x)$ is independent of A_μ. Then we can insert the identity

$$\Delta_{\text{FP}}[A_\mu] \int \prod_x d\alpha(x) \, \delta\left(F\left(A_\mu^{(\alpha)}(x)\right) - f(x)\right) = 1, \qquad (12.99)$$

into the functional integral. The Faddeev-Popov determinant is unchanged by this modification because $f(x)$ does not depend on $A_\mu(x)$. Thus the generating functional in this case is given by

$$Z[J] = N \int \mathcal{D}A_\mu \Delta_{\text{FP}}[A_\mu] \, \delta\left(F\left(A_\mu(x)\right) - f(x)\right) e^{iS^{(J)}[A_\mu]}. \quad (12.100)$$

Following 't Hooft, we can now do the following. We note that physical quantities are independent of $f(x)$. Hence we can multiply the generating functional by a weight factor and integrate over all $f(x)$. Thus, the generating functional becomes

$$Z[J] = N \int \mathcal{D}A_\mu \Delta_{\text{FP}}[A_\mu] \int \mathcal{D}f \delta\left(F\left(A_\mu(x)\right) - f(x)\right)$$

$$\times e^{-\frac{i}{2\xi}\int d^4x (f(x))^2} e^{iS^{(J)}}$$

$$= N \int \mathcal{D}A_\mu \Delta_{\text{FP}}[A_\mu] e^{i\left[S^{(J)} - \frac{1}{2\xi}\int d^4x (F(A_\mu(x)))^2\right]}$$

$$= N \int \mathcal{D}A_\mu \Delta_{\text{FP}}[A_\mu] e^{i\left(S^{(J)} + S_{\text{GF}}\right)},$$

where

$$S_{\text{GF}} = \int d^4x \, \mathcal{L}_{\text{GF}} = -\frac{1}{2\xi} \int d^4x \left(F\left(A_\mu(x)\right)\right)^2, \quad (12.101)$$

and ξ is known as the gauge fixing parameter.

We furthermore, note that since

$$\Delta_{\text{FP}}[A_\mu] = \det\left(\frac{\delta F\left(A_\mu^{(\alpha)}\right)}{\delta\alpha}\right)_{\alpha=0}, \quad (12.102)$$

we can write this as

$$\Delta_{\text{FP}}[A_\mu] = \det\left(\frac{\delta F\left(A_\mu^\alpha(x)\right)}{\delta\alpha(y)}\right)_{\alpha=0}$$

$$= \int \mathcal{D}\bar{c}\mathcal{D}c \, e^{-i\left(\bar{c}, \left(\frac{\delta F}{\delta\alpha}\right)_{\alpha=0} c\right)}$$

$$= \int \mathcal{D}\bar{c}\mathcal{D}c \ e^{-i \int \mathrm{d}^4x \mathrm{d}^4y \ \bar{c}(x) \left(\frac{\delta F\left(A_\mu^\alpha(x)\right)}{\delta\alpha(y)} \right)_{\alpha=0} c(y)}$$

$$= \int \mathcal{D}\bar{c}\mathcal{D}c \ e^{iS_{\text{ghost}}},$$

where

$$S_{\text{ghost}} = - \int \mathrm{d}^4x \mathrm{d}^4y \ \bar{c}(x) \left(\frac{\delta F\left(A_\mu^\alpha(x)\right)}{\delta\alpha(y)} \right)_{\alpha=0} c(y). \tag{12.103}$$

Here we have introduced two independent fictitious fields $c(x)$ and $\bar{c}(x)$ to write the determinant in the form of an action. We note here that this is possible only if the ghost fields $c(x)$ and $\bar{c}(x)$ anticommute (Ghosts have the same Lorentz structure as the parameters of transformation, but opposite statistics.), i.e.,

$$[c(x), c(y)]_+ = 0 \,,$$

$$[\bar{c}(x), \bar{c}(y)]_+ = 0 \,,$$

$$[c(x), \bar{c}(y)]_+ = 0 \,. \tag{12.104}$$

Thus although these behave as scalar objects under Lorentz transformations, they obey anti-commutation rules. These fields are known as Faddeev-Popov ghosts and as is obvious from their anticommutation relations, graphs involving these fictitious particles in closed loops must have an additional (-1) factor just like the fermions. Thus the generating functional now takes the form

$$Z\left[J\right] = e^{iW[J]} = N \int \mathcal{D}A_\mu \mathcal{D}\bar{c}\mathcal{D}c \ e^{iS_{\text{eff}}^{(J)}[A_\mu, c, \bar{c}]}, \tag{12.105}$$

where

$$S_{\text{eff}}^{(J)}\left[A_\mu, c, \bar{c}\right] = S^{(J)}\left[A_\mu\right] + S_{\text{GF}} + S_{\text{ghost}}$$

$$= \int \mathrm{d}^4x \ \mathcal{L}_{\text{eff}}^{(J)}\left[A_\mu, c, \bar{c}\right]. \tag{12.106}$$

Let us now look at a simple gauge fixing condition, for example, the covariant condition

$$F(A_\mu) = \partial_\mu A^\mu(x) = f(x). \tag{12.107}$$

Clearly the gauge fixing Lagrangian density has the form

$$\mathcal{L}_{GF} = -\frac{1}{2\xi} (F(A_\mu(x)))^2$$

$$= -\frac{1}{2\xi} (\partial_\mu A^\mu(x))^2. \tag{12.108}$$

It is clear that this provides the longitudinal components of the quadratic terms in fields and hence breaks gauge invariance. To obtain the corresponding ghost Lagrangian density for this gauge choice, we note that

$$F\left(A_\mu^{(\alpha)}\right) = \partial_\mu A^{\mu(\alpha)}(x) = \partial_\mu (A^\mu + \partial^\mu \alpha(x)), \tag{12.109}$$

so that

$$\left.\frac{\delta F\left(A_\mu^{(\alpha)}(x)\right)}{\delta \alpha(y)}\right|_{\alpha=0} = \Box \delta^4(x-y). \tag{12.110}$$

In this case, the ghost action is obtained to be

$$S_{ghost} = -\int d^4x d^4y \, \bar{c}(x) \left.\frac{\delta F\left(A_\mu^\alpha(x)\right)}{\delta \alpha(y)}\right|_{\alpha=0} c(y)$$

$$= -\int d^4x d^4y \, \bar{c}(x) \left(\Box \delta^4(x-y)\right) c(y)$$

$$= \int d^4x \, \partial_\mu \bar{c}(x) \partial^\mu c(x) = \int d^4x \, \mathcal{L}_{ghost}, \tag{12.111}$$

where we have neglected total divergence terms.

Thus our effective Lagrangian density for this choice of gauge condition becomes

$$\mathcal{L}_{eff} = -\frac{1}{4} F_{\mu\nu} F^{\mu\nu} - \frac{1}{2\xi} (\partial_\mu A^\mu)^2 + \partial_\mu \bar{c}(x) \partial^\mu c(x) + J^\mu A_\mu. \tag{12.112}$$

We note here that the ghost fields are noninteracting in the case of the Abelian theory and, therefore, we may omit them and then for $\xi = 1$ we recognize that our effective Lagrangian density is nothing other than Maxwell's theory in the Feynman-Fermi gauge. In non-Abelian gauge theories, however, the ghost fields are interacting and have to be present. Furthermore, since these fields are really necessary for the unitarity of the S-matrix, we cannot neglect them even if they are noninteracting if we are doing calculations at finite temperature. It is also true that when Maxwell's theory is coupled to gravitational fields, the ghost fields automatically couple to the geometry also. Hence omitting the ghost Lagrangian density in such a case would lead to nonsensical results.

One way of looking at the ghost fields is as if they are there to subtract out the unphysical field degrees of freedom. For example, the A_μ field has four helicity components. On the other hand each of the ghost fields, being a scalar, has only one helicity component. Hence one can think of the effective Lagrangian density as having two helicity states ($4 - 2 \times 1 = 2$) effectively. This naive counting works pretty well as we will see later. (The ghost degrees of freedom subtract because they have the unphysical statistics.)

Let us now apply these ideas of path integral quantization to the case of non-Abelian gauge theories. Let us recall from our study in the Abelian theory that gauge invariance puts a very strong constraint on the structure of the Lagrangian density for the gauge field. In particular, the coefficient matrix of the quadratic terms in the Lagrangian density is singular and, therefore, non-invertible. As a result, if we take $\mathcal{L}_{\text{inv}}$ as describing the dynamics of the theory, then, we cannot define propagators and the entire philosophy of doing perturbative calculations with Feynman diagrams breaks down. In order to circumvent this difficulty, we add to the gauge invariant Lagrangian density a term which breaks gauge invariance and thereby allows us to define the propagator for the gauge field. Such a term is called a gauge fixing term and any term which makes the coefficient matrix of the quadratic terms nonsingular and maintains various global symmetries of the theory is allowed for this purpose. On the other hand, adding a gauge fixing Lagrangian density changes the theory, in general, and to compensate for that we have to add a Lagrangian density for the ghost fields following the prescription of Faddeev and Popov.

In the case of the non-Abelian gauge theory, as we have seen, the Lagrangian density for the gauge fields

$$\mathcal{L}_{\text{inv}} = -\frac{1}{4} F^a_{\mu\nu} F^{\mu\nu a} , \qquad (12.113)$$

is invariant under the infinitesimal gauge transformation

$$A^a_\mu(x) \rightarrow A^{(\alpha)a}_\mu(x) = A^a_\mu(x) + D_\mu \theta^a(x) , \qquad (12.114)$$

where $\theta^a(x)$ is the infinitesimal parameter of transformation and

$$D_\mu \alpha^a(x) = \partial_\mu \theta^a(x) + f^{abc} A^b_\mu(x) \theta^c(x),$$

$$F^a_{\mu\nu}(x) = \partial_\mu A^a_\nu(x) - \partial_\nu A^a_\mu(x) + f^{abc} A^b_\mu A^c_\nu . \qquad (12.115)$$

The standard covariant gauge fixing, in a non-Abelian gauge theory, consists of adding to the invariant Lagrangian density a term of the form

$$\mathcal{L}_{\text{GF}} = -\frac{1}{2\xi} (\partial_\mu A^{\mu a})^2 , \qquad (12.116)$$

which would correspond to a gauge fixing condition of the form

$$F^a[A] = \partial_\mu A^{\mu a}(x) = f^a(x) . \qquad (12.117)$$

Here ξ represents an arbitrary constant parameter known as the gauge fixing parameter. Following the prescription of Faddeev and Popov, we can write the ghost action corresponding to this gauge choice as

$$S_{\text{ghost}} = \int dx \, \mathcal{L}_{\text{ghost}} = - \int dx dy \, \overline{c}^a(x) \frac{\delta F^a[A^{(\alpha)}(x)]}{\delta \alpha^b(y)} c^b(y) . \qquad (12.118)$$

Here we have left the dimensionality of space-time arbitrary since our discussions apply in any dimension. We note that for the covariant gauge choice that we are assuming, we can write

$$F^a[A^{(\alpha)}(x)] = \partial^\mu A^{(\alpha)a}_\mu = \partial^\mu \left(A^a_\mu(x) + D_\mu \alpha^a(x) \right) . \qquad (12.119)$$

Consequently, the ghost Lagrangian density, for this choice of gauge fixing, follows to be

$$\mathcal{L}_{\text{ghost}} = -\int \mathrm{d}y \; \bar{c}^a(x) \left(\partial_x^\mu D_{\mu,x}^{ab} \delta(x-y) \right) c(y)^b$$

$$= -\bar{c}^a(x) \partial^\mu \left(D_\mu c(x) \right)^a$$

$$= \partial^\mu \bar{c}^a(x) \left(D_\mu c(x) \right)^a , \tag{12.120}$$

where we have dropped a total derivative term in the last line of the above equation and have identified

$$D_{\mu,x}^{ab} = \delta^{ab} \partial_{\mu,x} + f^{acb} A_\mu^c(x) . \tag{12.121}$$

With all these modifications, the total Lagrangian density for a non-Abelian gauge theory has the form

$$\mathcal{L}_{\text{TOT}} = \mathcal{L}_{\text{inv}} + \mathcal{L}_{\text{GF}} + \mathcal{L}_{\text{ghost}} , \tag{12.122}$$

which for the covariant gauge that we are considering is given by

$$\mathcal{L}_{\text{TOT}} = -\frac{1}{4} F_{\mu\nu}^a F^{\mu\nu a} - \frac{1}{2\xi} (\partial_\mu A^{\mu a})^2 + \partial^\mu \bar{c}^a \left(D_\mu c \right)^a . \tag{12.123}$$

The path integral can now be defined with this $\mathcal{L}_{\text{TOT}}$ and will be well behaved.

12.4 BRST Invariance

As we have mentioned earlier, the gauge fixing and the ghost Lagrangian densities modify the original theory in a compensating manner which allows us to define the Feynman rules of the theory and carry out any perturbative calculation. In a deeper sense, the gauge fixing and the ghost Lagrangian densities, in the path integral formulation, merely correspond to a multiplicative factor of unity which does not change the physical content of the theory. However, without going into too much technical details, we can see that these additional terms in the Lagrangian density have no physical content in the following way.

The total Lagrangian density has been gauge fixed and, therefore, does not have the gauge invariance of the original theory. However, the total Lagrangian density, with gauge fixing and ghost terms, develops a global fermionic symmetry which, in some sense, remembers the gauge invariance of the original theory. It is easy to check that the total Lagrangian density is invariant under the global transformations

$$\delta A_\mu^a = \omega \left(D_\mu c \right)^a ,$$

$$\delta c^a = -\frac{\omega}{2} f^{abc} c^b c^c ,$$

$$\delta \bar{c}^a = -\frac{\omega}{\xi} \left(\partial_\mu A^{\mu a} \right) , \qquad (12.124)$$

where ω is the constant anti-commuting parameter of the global transformations. The invariance of the Lagrangian density can be seen by first noting that

$$\delta \left(D_\mu c^a \right) = D_\mu \delta c^a + f^{abc} \delta A_\mu^b c^c$$

$$= -\frac{\omega}{2} D_\mu \left(f^{abc} c^b c^c \right) + \omega f^{abc} \left(D_\mu c^b \right) c^c = 0 ,$$

$$\delta \left(\frac{1}{2} f^{abc} c^b c^c \right) = f^{abc} \delta c^b c^c = -\frac{\omega}{2} f^{abc} f^{bpq} c^p c^q c^c$$

$$= -\frac{\omega}{6} \left(f^{abc} f^{bpq} + f^{abp} f^{bqc} + f^{abq} f^{bcp} \right) c^p c^q c^c$$

$$= 0 . \qquad (12.125)$$

Here we have used the Jacobi identity for the symmetry algebra. Similarly, we obtain

$$\delta \left(\partial_\mu A^{\mu a} \right) = \omega \, \partial^\mu D_\mu c^a = 0 , \qquad (12.126)$$

when the ghost equations of motion are used. This shows that

$$\delta_2 \delta_1 \phi^a = 0 , \qquad (12.127)$$

for $\phi^a = A_\mu^a, c^a, \bar{c}^a$ independent of the parameters of the transformations where $\delta_{1,2}$ correspond to transformations with parameters

$\omega_{1,2}$ respectively. We note that the nilpotency of the transformations holds off-shell only for the fields A_μ^a, c^a, while for $\bar{c}^a$ it is true only on-shell (when the ghost equations of motion are used).

The invariance of the Lagrangian density can now be easily checked. First, we note that the transformation for A_μ^a is really a gauge transformation with the parameter $\alpha^a = \omega c^a$ and, therefore, the invariant Lagrangian density is trivially invariant under these transformations, namely,

$$\delta\mathcal{L}_{\text{inv}} = 0. \tag{12.128}$$

Consequently, we need to worry only about the gauge fixing and the ghost Lagrangian densities which lead to

$$\delta\left(\mathcal{L}_{\text{GF}} + \mathcal{L}_{\text{ghost}}\right) = -\frac{1}{\xi}\left(\partial_\nu A^{\nu a}\right)\left(\partial^\mu \delta A_\mu^a\right) + \left(\partial^\mu \delta\bar{c}^a\right)D_\mu c^a$$

$$= -\frac{\omega}{\xi}\left(\partial_\nu A^{\nu a}\right)\partial^\mu D_\mu c^a - \frac{\omega}{\xi}\,\partial^\mu\left(\partial_\nu A^{\nu a}\right)D_\mu c^a$$

$$= -\partial^\mu\left(\frac{\omega}{\xi}\left(\partial_\nu A^{\nu a}\right)D_\mu c^a\right), \tag{12.129}$$

so that the action is invariant. In this derivation, we have used the fact that $\delta\left(D_\mu c^a\right) = 0$ which we have seen earlier. This shows that the total Lagrangian density is invariant under the global transformations with an anti-commuting constant parameter. This is known as the BRST transformation for a gauge theory and arises in the presence the gauge fixing and the ghost Lagrangian densities. The present formulation of the BRST symmetry, however, is slightly unpleasant in the sense that the nilpotency of the anti-ghost field transformation holds only on-shell. This is a reflection of the fact that the theory is lacking some auxiliary fields and once the correct auxiliary fields are incorporated, the symmetry algebra will close off-shell.

We can write the gauge fixing Lagrangian density by introducing an auxiliary field which will also be quite useful for our later discussions. Let us rewrite

$$\mathcal{L}_{\text{GF}} = \partial^\mu F^a A_\mu^a + \frac{\xi}{2}F^a F^a, \tag{12.130}$$

where F^a is an auxiliary field. It is clear from the form of $\mathcal{L}_{GF}$ that the equation of motion for the auxiliary field takes the form

$$\xi F^a = \partial^\mu A^a_\mu, \tag{12.131}$$

and when we eliminate F^a using this equation, we recover the original gauge fixing Lagrangian density (if we ignore a total divergence term). Among other things $\mathcal{L}_{GF}$ as written above, allows us to take such gauge choices as the Landau gauge which corresponds to simply taking the limit $\xi = 0$. The total Lagrangian density can now be written as

$$\mathcal{L}_{TOT} = \mathcal{L}_{inv} + \mathcal{L}_{GF} + \mathcal{L}_{ghost}$$

$$= -\frac{1}{4} F^a_{\mu\nu} F^{\mu\nu a} + \partial^\mu F^a A^a_\mu + \frac{\xi}{2} F^a F^a + \partial^\mu \bar{c}^a D_\mu c^a . \tag{12.132}$$

In this case, the BRST transformations take the form

$$\delta A^a_\mu = \omega \left(D_\mu c \right)^a ,$$

$$\delta c^a = -\frac{\omega}{2} f^{abc} c^b c^c ,$$

$$\delta \bar{c}^a = -\omega \, F^a ,$$

$$\delta F^a = 0 , \tag{12.133}$$

and it is straightforward to check that these transformations are nilpotent off-shell, namely,

$$\delta_2 \delta_1 \phi^a = 0 , \tag{12.134}$$

for $\phi^a = A^a_\mu, F^a, c^a, \bar{c}^a$. Therefore, F^a represents the missing auxiliary field that we had alluded to earlier.

We note that $\mathcal{L}_{inv}$ is invariant under the BRST transformation as we had argued earlier and the auxiliary field, F^a, does not transform at all which leads to

$$\delta \mathcal{L}_{TOT} = \delta \left(\mathcal{L}_{GF} + \mathcal{L}_{ghost} \right)$$

$$= \partial^\mu F^a \delta A^a_\mu + \partial^\mu \delta \bar{c}^a \left(D_\mu c \right)^a$$

$$= \omega \partial^\mu F^a \left(D_\mu c \right)^a - \omega \partial^\mu F^a \left(D_\mu c \right)^a$$

$$= 0 . \tag{12.135}$$

Unlike the formulation without auxiliary fields, here we see that the total Lagrangian density is invariant under the BRST transformations (as opposed to changing by a total divergence).

These transformations, known as BRST transformations, which define a residual global symmetry of the full theory, in some sense, replace the original gauge invariance of the theory, and play a fundamental role in the study of non-Abelian gauge theories. There is also a second set of transformations involving the anti-ghost fields of the form

$$\bar{\delta} A_\mu^a = \overline{\omega} \left(D_\mu \bar{c} \right)^a ,$$

$$\bar{\delta} c^a = \overline{\omega} \left(F^a - f^{abc} c^b \bar{c}^c \right) ,$$

$$\bar{\delta} \bar{c}^a = -\frac{\overline{\omega}}{2} f^{abc} \bar{c}^b \bar{c}^c ,$$

$$\bar{\delta} F^a = \overline{\omega} f^{abc} F^b \bar{c}^c , \tag{12.136}$$

which leave the total Lagrangian density invariant. These are known as the anti-BRST transformations. However, since these do not lead to any new information beyond what the BRST invariance provides, we will ignore this symmetry for the rest of our discussions. We note here that the BRST and the anti-BRST transformations are not quite symmetric in the ghost and the anti-ghost fields which is a reflection of the asymmetric nature of these fields in the ghost Lagrangian density. It is also worth noting here that these symmetries arise naturally in a superspace formulation of gauge theories.

In addition to these two anti-commuting symmetries, the total Lagrangian density is also invariant under the infinitesimal bosonic global symmetry transformations

$$\delta c^a = \theta c^a ,$$

$$\delta \bar{c}^a = -\theta \bar{c}^a , \tag{12.137}$$

with all other fields remaining inert. Here θ represents a constant, bosonic infinitesimal parameter and the generator of the symmetry merely counts the ghost number of the fields. (The fact that these transformations are like scale transformations and not like phase transformations has to do with the particular hermiticity properties that the ghost and the anti-ghost fields must satisfy for consistent quantization of the theory.)

The vector space of the full theory, as is clear by now, contains many more states than the physical states alone. The physical Hilbert space, therefore, needs to be properly selected for a discussion of physical questions. Furthermore, the physical space must be selected in such a way that it remains invariant under the time evolution of the system. In the covariant gauge in QED, for example, we have seen that the states in the physical space are selected as satisfying the Gupta-Bleuler condition

$$\partial^\mu A_\mu^{(+)}(x)|\text{phys}\rangle = 0, \qquad (12.138)$$

where the superscript, $(+)$, stands for the positive frequency part of the field. We recognize that even though this looks like one condition, in reality it is an infinite number of conditions since it has to hold for every value of the coordinates. In QED, the Gupta-Bleuler condition works because $\partial^\mu A_\mu$ satisfies the free Klein-Gordon equation in the covariant gauge and hence the physical space so selected remains invariant under time evolution. The corresponding operator in a non-Abelian theory, on the other hand, does not satisfy a free equation as we have seen and hence is not suitable for identifying the physical subspace. On the other hand, the generators of the BRST symmetry, Q_{BRST}, and the ghost scaling symmetry, Q_c, are conserved and hence can be used to define a physical Hilbert space which would remain invariant under the time evolution of the system. (Q_{BRST} and Q_c are the charges constructed from the Nöether current for the respective transformations whose explicit forms can be obtained from the Nöether procedure.) Thus, we can identify the physical space of the gauge theory as satisfying

$$Q_{\text{BRST}}|\text{phys}\rangle = 0,$$

$$Q_c|\text{phys}\rangle = 0. \qquad (12.139)$$

Note that even in the case of QED, these would appear to correspond to only two conditions and not an infinite number of conditions as we have seen is the case with the Gupta-Bleuler condition. It is, therefore, not clear *a priori* if these conditions are sufficient to reproduce the Gupta-Bleuler conditions in the case of QED.

To see that these conditions indeed lead to the Gupta-Bleuler conditions in QED, let us note that the Nöether current associated with the BRST transformations has the form

$$
\begin{aligned}
J_{\mathrm{BRST}}^{(\omega)\mu}(x) &= -F^{\mu\nu a}\delta A_\nu^a + \delta\bar{c}^a D^\mu c^a + (\partial^\mu \bar{c}^a)\,\delta c^a \\
&= -\omega\left(F^{\mu\nu a}D_\nu c^a + F^a D^\mu c^a - \frac{1}{2}\,f^{abc}\,(\partial^\mu \bar{c}^a)\,c^b c^c\right),
\end{aligned}
$$
$$(12.140)$$

where we have used the fact that the auxiliary field does not transform under the BRST transformations. From this, we can obtain the current without the parameter of transformations

$$
J_{\mathrm{BRST}}^\mu = F^{\mu\nu a}D_\nu c^a + F^a D^\mu c^a - \frac{1}{2}\,f^{abc}\,(\partial^\mu \bar{c}^a)\,c^b c^c, \qquad (12.141)
$$

and, in particular, for the Abelian theory where $f^{abc} = 0$, the BRST charge can be obtained to be

$$
\begin{aligned}
Q_{\mathrm{BRST}} &= \int d\mathbf{x}\,\left(F^{0\nu}\partial_\nu c + F\partial^0 c\right) \\
&= \int d\mathbf{x}\,\left(F^{0i}\partial_i c + F\dot{c}\right) = \int d\mathbf{x}\,\left(\partial_i\left(F^{0i}c\right) + F\dot{c}\right) \\
&= \int d\mathbf{x}\,F\dot{c}, \qquad (12.142)
\end{aligned}
$$

where we have used Maxwell's equations in the intermediate steps. If we use the field decomposition for the fields and normal order the BRST charge (so that the annihilation operators are to the right), this has the explicit form

$$
Q_{\mathrm{BRST}} = i \int d\mathbf{k}\,\left(c^{(-)}(-k)F^{(+)}(k) - F^{(-)}(-k)c^{(+)}(k)\right). \quad (12.143)
$$

We recognize that the condition

$$
Q_c|\mathrm{phys}\rangle = 0, \qquad (12.144)
$$

implies that the physical states must have zero ghost number. This allows, in principle, states containing equal numbers of ghost and anti-ghost particles. Denoting the physical states of the theory as

$$|\text{phys}\rangle = |A_\mu\rangle \otimes |c, \bar{c}\rangle, \tag{12.145}$$

we note that if the physical states have to further satisfy the condition

$$Q_{\text{BRST}}|\text{phys}\rangle = i \int d\mathbf{k} \left(c^{(-)}(-k)F^{(+)}(k) - F^{(-)}(-k)c^{(+)}(k) \right)$$

$$\times |A_\mu\rangle \otimes |c, \bar{c}\rangle$$

$$= 0, \tag{12.146}$$

then this implies that

$$c^{(+)}(k)|c, \bar{c}\rangle = 0 = F^{(+)}(k)|A_\mu\rangle. \tag{12.147}$$

Namely, the physical states should have no ghost particles,

$$|\text{phys}\rangle = |A_\mu\rangle \otimes |0, 0\rangle = |A_\mu\rangle, \tag{12.148}$$

and must satisfy

$$F^{(+)}(k)|\text{phys}\rangle = \left(\partial_\mu A^{\mu(+)}(k) \right) |\text{phys}\rangle = 0, \tag{12.149}$$

where we have used the equation of motion for the auxiliary field. This is precisely the Gupta-Bleuler condition in momentum space and this derivation shows how a single condition can give rise to an infinite number of conditions (in this case, for every momentum mode k). Thus, we feel confident that the physical state conditions are the right ones even for a non-Abelian theory.

We are now ready to show that the gauge fixing and the ghost Lagrangian densities lead to no physical consequences. We note that these extra terms in the Lagrangian density can, in fact, be written as a BRST variation (without the parameter of transformation), namely,

$$\mathcal{L}_{\text{GF}} + \mathcal{L}_{\text{ghost}} = \delta \left(-\partial^\mu \bar{c}^a A_\mu^a - \frac{\xi}{2} \bar{c}^a F^a \right)$$

$$= \left[Q_{\text{BRST}}, \left(-\partial^\mu \bar{c}^a A_\mu^a - \frac{\xi}{2} \bar{c}^a F^a \right) \right]_+. \tag{12.150}$$

Here we have used the fact that the BRST charge is the generator of the BRST transformations so that the transformations for fermionic operators are generated through the anti-commutator of the operators with the generator. It now follows from the physical condition that

$$\langle \text{phys}| \left(\mathcal{L}_{\text{GF}} + \mathcal{L}_{\text{ghost}} \right) |\text{phys}'\rangle$$

$$= -\langle \text{phys}| \left[Q_{\text{BRST}}, \left(\partial^\mu \bar{c}^a A_\mu^a + \frac{\xi}{2} \bar{c}^a F^a \right) \right]_+ |\text{phys}'\rangle$$

$$= 0 . \tag{12.151}$$

This shows that the terms added to the original Lagrangian density have no contribution to the physical matrix elements of the theory. We can also show that all the physical matrix elements are independent of the choice of the gauge fixing parameter ξ in the following manner (the BRST variation denoted is without the parameter of transformation)

$$\frac{\partial}{\partial \xi} \langle 0|0 \rangle^J = \frac{\partial Z[J]}{\partial \xi}$$

$$= \frac{i}{2} \langle 0| \int dx \, F^a F^a |0\rangle^J = -\frac{i}{2} \int dx \, \langle |\delta \left(\bar{c}^a F^a \right) |0\rangle^J$$

$$= -\frac{i}{2} \int dx \, \langle 0| [Q_{\text{BRST}}, \bar{c}^a F^a]_+ |0\rangle^J = 0 , \tag{12.152}$$

where we have used the fact that the vacuum belongs to the physical Hilbert space of the theory and as such is annihilated by the BRST charge.

12.5 Ward Identities

The BRST invariance of the full theory leads to many relations between various scattering amplitudes of the theory. These are known as the Ward identities or the Slavnov-Taylor identities of the theory and are quite essential in establishing the renormalizability of gauge theories. These identities are best described within the context of path integrals which we will do next.

Let us consider an effective Lagrangian density which consists of $\mathcal{L}_{\text{TOT}}$ as well as source terms as follows

$$\mathcal{L}_{\text{eff}} = \mathcal{L}_{\text{TOT}} + J^{\mu a} A_\mu^a + J^a F^a + i \left(\overline{\eta}^a c^a - \overline{c}^a \eta^a \right)$$
$$+ K^{\mu a} \left(D_\mu c \right)^a + K^a \left(-\frac{1}{2} f^{abc} c^b c^c \right). \tag{12.153}$$

Here, we have not only introduced sources for all the field variables in the theory, but, in addition, we have also added sources for the composite variations under the BRST transformation. The need for this will become clear shortly. Denoting all the fields and the sources generically by A and J respectively, we can write the generating functional for the theory as

$$Z[J] = e^{iW[J]} = N \int \mathcal{D}A \, e^{i \int \mathrm{d}x \, \mathcal{L}_{\text{eff}}}. \tag{12.154}$$

The vacuum expectation values of operators, in the presence of sources, can now be written as

$$\langle A_\mu^a \rangle = A_\mu^{(c)a} = \frac{\delta W}{\delta J^{\mu a}},$$

$$\langle F^a \rangle = F^{(c)a} = \frac{\delta W}{\delta J^a},$$

$$\langle c^a \rangle = c^{(c)a} = -i \frac{\delta W}{\delta \overline{\eta}^a},$$

$$\langle \overline{c}^a \rangle = \overline{c}^{(c)a} = -i \frac{\partial W}{\delta \eta^a},$$

$$\langle (D_\mu c)^a \rangle = \frac{\delta W}{\delta K^{\mu a}},$$

$$\left\langle \left(-\frac{1}{2} f^{abc} c^b c^c \right) \right\rangle = \frac{\delta W}{\delta K^a}. \tag{12.155}$$

Here, we have assumed the convention of a left derivative for the anti-commuting fields. The fields $A^{(c)}$ are known as the classical fields and in what follows, we will ignore the superscript (c) for notational simplicity.

When the external sources are held fixed, the effective Lagrangian density is no longer invariant under the BRST transformations. In fact, recalling that $\mathcal{L}_{\text{TOT}}$ is BRST invariant and that the BRST transformations are nilpotent, we obtain the change in $\mathcal{L}_{\text{eff}}$ (remember that the parameter is anti-commuting) to be

$$\delta\mathcal{L}_{\text{eff}} = J^{\mu a}\delta A_\mu^a + J^a \delta F^a + i\left(\overline{\eta}^a \delta c^a - \delta\overline{c}^a \eta^a\right)$$

$$= \omega\left[J^{\mu a}\left(D_\mu c\right)^a + i\left(\frac{1}{2}f^{abc}\overline{\eta}^a c^b c^c + F^a \eta^a\right)\right]. \quad (12.156)$$

On the other hand, the generating functional is defined by integrating over all possible field configurations. Therefore, with a redefinition of the fields under a BRST transformation inside the path integral, the generating functional should be invariant. This immediately leads to

$$\delta Z[J] = 0 = N \int \mathcal{D}A \left(i\int \mathrm{d}x\, \delta\mathcal{L}_{\text{eff}}\right) e^{i\int \mathrm{d}x\, \mathcal{L}_{\text{eff}}}$$

$$= i\omega \int \mathrm{d}x \left(J^{\mu a}\langle(D_\mu c)^a\rangle + i\overline{\eta}^a\langle\frac{g}{2}f^{abc}c^b c^c\rangle + i\eta^a\langle F^a\rangle\right). \quad (12.157)$$

The measure can be easily checked to be invariant under such a fermionic transformation and we note that the above relation can also be written as

$$\int \mathrm{d}x \left(J^{\mu a}(x)\frac{\delta W}{\delta K^{\mu a}(x)} - i\overline{\eta}^a(x)\frac{\delta W}{\delta K^a(x)} + i\eta^a(x)\frac{\delta W}{\delta J^a(x)}\right) = 0. \quad (12.158)$$

This is the Master equation from which one can derive all the identities relating the connected Green's functions of the theory. It is here that the usefulness of the sources for the composite operators becomes evident.

Most often, however, we are interested in the proper (1PI) vertices of the theory. These can be obtained by passing from the generating functional for the connected Green's functions, $W[J]$, to the generating functional for the proper vertices, $\Gamma[A]$, through a Legendre transformation. Defining a Legendre transformation involving (only) the field variables of the theory (The field variables are really the classical fields and we are dropping the superscript (c) for simplicity.), we have

$$\Gamma[A, K] = W[J, K] - \int \mathrm{d}x \left(J^{\mu a} A_\mu^a + J^a F^a + i \left(\overline{\eta}^a c^a - \overline{c}^a \eta^a \right) \right),$$

$$(12.159)$$

where K stands generically for the sources for the composite variations. From the definition of the generating functional for the proper vertices, it is clear that

$$\frac{\delta \Gamma}{\delta A_\mu^a} = -J^{\mu a},$$

$$\frac{\delta \Gamma}{\delta F^a} = -J^a,$$

$$\frac{\delta \Gamma}{\delta c^a} = i \overline{\eta}^a,$$

$$\frac{\delta \Gamma}{\delta \overline{c}^a} = i \eta^a,$$

$$\frac{\delta \Gamma}{\delta K_\mu^a} = \frac{\delta W}{\delta K_\mu^a},$$

$$\frac{\delta \Gamma}{\delta K^a} = \frac{\delta W}{\delta K^a}. \qquad (12.160)$$

Using these definitions, we see that we can rewrite the Master equation in terms of the generating functional of the proper vertices as

$$\int \mathrm{d}x \left(\frac{\delta \Gamma}{\delta A_\mu^a(x)} \frac{\delta \Gamma}{\delta K^{\mu a}(x)} + \frac{\delta \Gamma}{\delta c^a(x)} \frac{\delta \Gamma}{\delta K^a(x)} - F^a(x) \frac{\delta \Gamma}{\delta \overline{c}^a(x)} \right) = 0.$$

$$(12.161)$$

This is the Master equation from which we can derive all the relations, between various (1PI) proper vertices, resulting from the BRST invariance of the theory. This, in turn, is essential in proving the renormalizability of gauge theories. Thus, for example, let us note that we can write the Master identity in the momentum space as

$$\int dk \left(\frac{\delta\Gamma}{\delta A_\mu^a(-k)} \frac{\delta\Gamma}{\delta K^{\mu a}(k)} + \frac{\delta\Gamma}{\delta c^a(-k)} \frac{\delta\Gamma}{\delta K^a(k)} \right.$$

$$\left. - F^a(k) \frac{\delta\Gamma}{\delta \bar{c}^a(k)} \right) = 0. \tag{12.162}$$

Taking derivative of this with respect to $\frac{\delta^2}{\delta F^b(p)\delta c^c(-p)}$ and setting all field variables to zero gives

$$\frac{\delta^2\Gamma}{\delta F^b(p)\delta A_\mu^a(-p)} \frac{\delta^2\Gamma}{\delta c^c(-p)\delta K^{\mu a}(p)} - \frac{\delta^2\Gamma}{\delta c^c(-p)\delta \bar{c}^b(p)} = 0. \tag{12.163}$$

A simple analysis, then, shows that this relates the mixed two point function involving $F\text{-}A_\mu$ with the two point function for the ghost fields and, consequently, the counter terms (quantum corrections) should satisfy such a relation. The BRST invariance, in this way, is very fundamental in the study of gauge theories as far as renormalizability and gauge independence are concerned.

12.6   References

Becchi, C., A. Rouet and R. Stora, Communications in Mathematical Physics, **42**, 127 (1975).

Bleuler, K., Helv. Phys. Acta **23**, 567 (1950).

Das, A., "Finite Temperature Field Theory", World Scientific Publishing (1997).

Faddeev, L. D. and V. N. Popov, Physics Letters **25B**, 29 (1967).

Gribov, V. N., Nuclear Physics **B139**. 1 (1978).

Gupta, S. N., Proc. Phys. Soc. (London) **A63**, 681 (1950).

Mohapatra, R. N., Physical Review **D4**, 378 (1971).

Chapter 13

Anomalies

We have seen earlier that a continuous global symmetry in a quantum field theory leads to a current which is conserved and the Ward identities of the theory reflect this conservation law. Sometimes, however, the classical (tree level) conservation of the current is violated by quantum corrections (loop effects). This can happen, for example, if the regularization used to define divergent amplitudes does not respect the symmetry that leads to the conservation law. In this case, the divergence of the current density no longer vanishes and one says that there is an anomaly in the conservation law. In turn, this modifies the tree level Ward identities of the theory to new ones that are known as anomalous Ward identities of the theory (which reflect the modified conservation law). Anomalies are quite important from a physical point of view. Global anomalies such as the chiral anomaly have direct physical consequences. On the other hand, an anomaly in a gauge theory (namely, in the conservation of the gauge current) can render the theory inconsistent and, therefore, needs to be taken care of before one can make sense of such a theory. In this chapter, we will study some issues associated with such phenomena.

13.1 Anomalous Ward Identity

Let us consider a non Abelian gauge theory with massless fermions. If we choose the gauge group to be $SU(3)$, then the fermions can be thought of as quarks and the theory would describe quantum chromodynamics. For a general gauge group of $SU(n)$, the theory is described by the Lagrangian density

$$\mathcal{L} = -\frac{1}{4}F^a_{\mu\nu}F^{\mu\nu,a} + i\bar{\psi}^i\gamma^\mu D_\mu\psi^i, \qquad (13.1)$$

where $a = 1, 2, \cdots, n^2 - 1$, $i = 1, 2, \cdots, N$, and the covariant derivative for the fermions is defined as

$$D_\mu\psi^i = \partial_\mu\psi^i - iA^a_\mu\left(T^a\right)^{ij}\psi^j. \qquad (13.2)$$

Note that this theory in addition to having the non-Abelian local gauge invariance

$$\delta A^a_\mu = D_\mu\theta^a(x),$$
$$\delta\psi^i = i\theta^a(x)\left(T^a\psi\right)^i,$$
$$\delta\bar{\psi}^i = -i\theta^a(x)\left(\bar{\psi}T^a\right)^i, \qquad (13.3)$$

where θ^a is an infinitesimal parameter of the gauge transformation, is also invariant under the global transformation

$$\delta\psi^i = i\lambda\gamma_5\psi^i,$$
$$\delta\bar{\psi}^i = i\lambda\bar{\psi}^i\gamma_5. \qquad (13.4)$$

Here λ is an infinitesimal space-time independent parameter and γ_5 denotes the (pseudoscalar) Dirac matrix (defined to be Hermitian). This global symmetry, which corresponds to an Abelian group of transformations $(U(1))$, is commonly known as the chiral symmetry of the theory. The invariances of the theory, of course, lead to conserved currents. Thus for example for the chiral symmetry, we have

$$\partial_\mu j^\mu_5 = 0, \qquad (13.5)$$

where

$$j^\mu_5 = \bar{\psi}^i\gamma_5\gamma^\mu\psi^i. \qquad (13.6)$$

The Ward identities are nothing but generalizations of the fact that the currents are conserved.

We note here that, if the quarks were massive, i.e., if there is a mass term in the Lagrangian density of the form

$$\mathcal{L}_m = -m\bar{\psi}\psi\,, \tag{13.7}$$

then the global chiral transformations in Eq. (13.4) will not be a symmetry of the theory and as a result, the chiral current in (13.6) would not be conserved. Rather, it will satisfy the relation

$$\partial_\mu j_5^\mu = -2im\bar{\psi}\gamma_5\psi\,. \tag{13.8}$$

For massive fermions, the relation (13.8) would have related various matrix elements and would have led to the axial (chiral) Ward Identities. However, when we calculate various matrix elements regularized in a gauge invariant way, we discover that the axial Ward Identities are violated at the loop level. More specifically, the modification to the tree level identities arise only at the one loop level.

There are various ways to understand this violation. Let us first begin with an analysis in the path integral formalism where we can write the generating functional as

$$Z_{\text{TOT}} = \int \mathcal{D}\bar{\psi}\mathcal{D}\psi\mathcal{D}A_\mu\, e^{iS_{\text{TOT}}}\,, \tag{13.9}$$

and here we have kept the mass term just for generality so that

$$S_{\text{TOT}} = \int \mathrm{d}^4x \left(-\frac{1}{4}F_{\mu\nu}^a F^{\mu\nu,a} + i\bar{\psi}^i\gamma^\mu D_\mu\psi^i - m\bar{\psi}^i\psi^i\right)\,. \tag{13.10}$$

Let us now make a local chiral transformation of the form (see (13.4))

$$\delta\psi^i = i\lambda(x)\gamma_5\psi^i\,,$$
$$\delta\bar{\psi}^i = i\lambda(x)\bar{\psi}^i\gamma_5\,. \tag{13.11}$$

The generating functional should be invariant under such a field redefinition since we are integrating over all field configurations. Furthermore, for this analysis, we can treat the A_μ field as an external field (since it does not transform under the chiral transformations) and look at the only fermionic part of the action. Namely, we take

$$Z = \int \mathcal{D}\bar{\psi}\mathcal{D}\psi \; e^{i\int d^4x\left(i\bar{\psi}^i\gamma^\mu D_\mu\psi^i - m\bar{\psi}^i\psi^i\right)}, \qquad (13.12)$$

so that under the redefinition (13.11)

$$\delta Z = 0 = \int \mathcal{D}\bar{\psi}\mathcal{D}\psi \left[\int d^4x\{i\partial_\mu\lambda(x)\bar{\psi}^i\gamma_5\gamma^\mu\psi + 2\lambda(x)m\bar{\psi}^i\gamma_5\psi^i\}\right]e^{iS}$$

$$= \int \mathcal{D}\bar{\psi}\mathcal{D}\psi \left[i\int d^4x\lambda(x)\left(-\partial_\mu\bar{\psi}^i\gamma_5\gamma^\mu\psi^i - 2im\bar{\psi}^i\gamma_5\psi^i\right)\right]e^{iS}.$$

$$(13.13)$$

Therefore, this implies that

$$\left\langle \int d^4x \left(-\partial_\mu\bar{\psi}^i\gamma_5\gamma^\mu\psi^i - 2im\bar{\psi}^i\gamma_5\psi^i\right)\lambda(x)\right\rangle = 0$$

$$\text{or,} \quad \langle\partial_\mu j_5^\mu + 2im\bar{\psi}^i\gamma_5\psi^i\rangle = 0. \qquad (13.14)$$

This is the naive (tree level) Ward Identity and the question is why does it fail at the loop level. To understand this, let us note that in this derivation we have assumed that the integration measure is insensitive to changes in the field redefinitions. Therefore, we should analyze the behavior of the functional measure under such a field redefinition.

To understand the question of the measure, let us assume that we can solve for the eigenstates of the gauge invariant Dirac operator,

$$i\gamma^\mu D_\mu\phi_n = \lambda_n\phi_n. \qquad (13.15)$$

Here ϕ_n's represent the eigenstates of the Dirac operator and λ_n's are the corresponding eigenvalues. We have assume that these are discrete states just for simplicity. Furthermore, if these sates define an orthonormal complete basis (the Dirac operator is a Hermitian operator), then we have

$$\int d^4x \; \phi_n^\dagger(x)\phi_m(x) = \delta_{nm},$$

$$\sum_n \phi_n(x)\phi_n^\dagger(y) = \delta^4(x-y). \qquad (13.16)$$

We can expand the fermion fields in the basis of these eigenfunctions as

$$\psi(x) = \sum_n a_n \phi_n(x) \,,$$

$$\bar{\psi}(x) = \sum_n \phi_n^\dagger(x) b_n \,, \tag{13.17}$$

where we are suppressing the index i for simplicity. This can, for example, be thought of as a transformation from a wave function basis to the occupation number basis or any other discrete quantum number basis. In terms of bra and ket states we can write

$$\psi(x) = \langle x|\psi\rangle = \sum_n \langle x|n\rangle\langle n|\psi\rangle$$

$$= \sum_n \phi_n(x)a_n = \sum_n a_n\phi_n(x) \,, \tag{13.18}$$

and so on. The integration measure is now clearly understood to be

$$\mathcal{D}\bar{\psi}(x)\mathcal{D}\psi(x) = \prod_n \mathrm{d}b_n \mathrm{d}a_n. \tag{13.19}$$

Let us now make an infinitesimal chiral transformation (see (13.11))

$$\psi(x) \to \psi'(x) = \psi(x) + i\lambda(x)\gamma_5\psi(x) \,,$$

$$\bar{\psi}(x) \to \bar{\psi}'(x) = \bar{\psi}(x) + i\lambda\bar{\psi}(x)\gamma_5 \,. \tag{13.20}$$

We can also expand the new variables $\psi'(x), \bar{\psi}'(x)$ in the same eigen-basis. For example, we can write

$$\psi'(x) \sum_n a'_n\phi_n(x) = \sum_{n,m} c_{nm}a_m\phi_n(x) \,, \tag{13.21}$$

where the matrix c_{nm} is given by

$$c_{nm} = \delta_{nm} + i\int \mathrm{d}^4x \; \phi_n^\dagger(x)\lambda(x)\gamma_5\phi_m(x) \,. \tag{13.22}$$

This can be seen easily in the following way

$$\psi'(x) = \langle x|\psi'\rangle = \sum_n \langle x|n\rangle \langle n|\psi'\rangle$$

$$= \sum_n a'_n \phi_n(x) = \sum_n \langle x|n\rangle \langle n|c|\psi\rangle, \qquad (13.23)$$

where c denotes the infinitesimal chiral transformation. Thus

$$\psi'(x) = \sum_n a'_n \phi_n(x) = \sum_n \langle x|n\rangle \langle n|c|\psi\rangle$$

$$= \sum_{n,m} \langle x|n\rangle \langle n|c|m\rangle \langle m|\psi\rangle = \sum c_{nm} a_m \phi_n(x), \qquad (13.24)$$

where

$$c_{nm} = \langle n|c|m\rangle = \int \mathrm{d}^4 x \mathrm{d}^4 x' \langle n|x\rangle \langle x|c|x'\rangle \langle x'|m\rangle$$

$$= \int \mathrm{d}^4 x \mathrm{d}^4 x' \ \phi_n^\dagger(x) c(x, x') \phi_m(x'). \qquad (13.25)$$

We note that the chiral transformation (13.11) is a local transformation so that $c(x, x')$ is a local function, i.e.,

$$\psi'(x) = c(x)\psi(x) = \left(\mathbf{1} + i\lambda(x)\gamma_5\right)\psi(x). \qquad (13.26)$$

If we use this fact then we obtain

$$c_{nm} = \int \mathrm{d}^4 x \mathrm{d}^4 x' \phi_n^\dagger(x) c(x) \delta(x - x') \phi_m(x')$$

$$= \int \mathrm{d}^4 x \ \phi_n^\dagger(x) c(x) \phi_m(x)$$

$$= \int \mathrm{d}^4 x \ \phi_n^\dagger(x) \left(\mathbf{1} + i\lambda(x)\gamma_5\right) \phi_m(x)$$

$$= \delta_{nm} + i \int \mathrm{d}^4 x \ \lambda(x) \phi_n^\dagger(x) \gamma_5 \phi_m(x). \qquad (13.27)$$

The above analysis also shows that under the chiral transformation

$$\mathrm{d}a_n \to \mathrm{d}a'_n = (\det c_{nm})^{-1} \mathrm{d}a_n, \qquad (13.28)$$

and we recognize that the inverse of the determinant, rather than the determinant appears in this case simply because a_n's are fermionic in character (see chapter 5). Similarly, under the chiral transformation (13.11)

$$\mathrm{d}b_n \rightarrow \mathrm{d}b'_n = (\det c_{nm})^{-1}\,\mathrm{d}b_n. \qquad (13.29)$$

Therefore, it is clear that under the chiral transformation the fermionic integration measure changes and contributes a nontrivial amount given by

$$\mathcal{D}\bar{\psi}\mathcal{D}\psi \rightarrow (\det c_{nm})^{-2}\,\mathcal{D}\bar{\psi}\mathcal{D}\psi\,, \qquad (13.30)$$

where

$$
\begin{aligned}
\det c_{nm} &= \det\left(\delta_{nm} + i\int \mathrm{d}^4x\,\lambda(x)\phi_n^\dagger(x)\gamma_5\phi_m(x)\right)\\
&= \exp\left(\mathrm{Tr}\,\ln\left(\delta_{nm} + i\int \mathrm{d}^4x\,\lambda(x)\phi_n^\dagger(x)\gamma_5\phi_m(x)\right)\right)\\
&= \exp\left(\mathrm{Tr}\left(i\int \mathrm{d}^4x\,\lambda(x)\phi_n^\dagger(x)\gamma_5\phi_m(x)\right)\right)\\
&= \exp\left(i\sum_n \int \mathrm{d}^4x\,\lambda(x)\phi_n^\dagger(x)\gamma_5\phi_n(x)\right). \qquad (13.31)
\end{aligned}
$$

Since n runs over infinitely many values, this sum is ill defined. Therefore, we regularize the expression in the exponent as

$$
\begin{aligned}
&\sum_n \int \mathrm{d}^4x\,\lambda(x)\phi_n^\dagger(x)\gamma_5\phi_n(x)\\
&= \lim_{M^2\to\infty}\sum_n \int \mathrm{d}^4x\,\lambda(x)\phi_n^\dagger(x)\gamma_5\phi_n(x)e^{-\left(\frac{\lambda_n}{M}\right)^2}\\
&= \lim_{M^2\to\infty}\sum_n \int \mathrm{d}^4x\,\lambda(x)\phi_n^\dagger(x)\gamma_5 e^{-\left(\frac{i\slashed{D}}{M}\right)^2}\phi_n(x)\\
&= \lim_{M^2\to\infty}\sum_n \int \mathrm{d}^4x\frac{\mathrm{d}^4k}{(2\pi)^4}\frac{\mathrm{d}^4k'}{(2\pi)^4}\lambda(x)\tilde{\phi}_n^\dagger(k')e^{-ik'\cdot x}\gamma_5 e^{-\left(\frac{i\slashed{D}}{M}\right)^2}e^{ik\cdot x}\tilde{\phi}_n(k).
\end{aligned}
$$

$$(13.32)$$

If we use the completeness relation for the basis functions in momentum space, namely,

$$\sum_n \tilde{\phi}_n(k)\tilde{\phi}_n^\dagger(k') = (2\pi)^4 \delta^4(k - k'),\qquad (13.33)$$

then, Eq. (13.32) takes the form

$$\sum_n \int \mathrm{d}^4x \; \lambda(x)\phi_n^\dagger \gamma_5 \phi_n(x)$$

$$= \lim_{M^2 \to \infty} \int \mathrm{d}^4x \; \frac{\mathrm{d}^4k}{(2\pi)^4}\frac{\mathrm{d}^4k'}{(2\pi)^4}(2\pi)^4\delta^4(k - k')$$

$$\times \; \mathrm{Tr}\,\lambda(x)\, e^{-ik'\cdot x}\gamma_5 e^{-\left(\frac{i\slashed{D}}{M}\right)^2} e^{ik\cdot x}$$

$$= \lim_{M^2 \to \infty} \int \mathrm{d}^4x \; \frac{\mathrm{d}^4k}{(2\pi)^4}\lambda(x)\, \mathrm{Tr}\, e^{-ik\cdot x}\gamma_5 e^{-\left(\frac{i\slashed{D}}{M}\right)^2} e^{ik\cdot x}.\quad (13.34)$$

To simplify this expression, we note that using the properties of the Dirac matrices, we can write

$$(i\slashed{D})^2 = -\gamma^\mu\gamma^\nu D_\mu D_\nu$$

$$= -\left(\frac{1}{2}\{\gamma^\mu,\gamma^\nu\} + \frac{1}{2}[\gamma^\mu,\gamma^\nu]\right)D_\mu D_\nu$$

$$= \left(-\eta^{\mu\nu} - \frac{1}{2}[\gamma^\mu,\gamma^\nu]\right)D_\mu D_\nu$$

$$= -D_\mu D^\mu - \gamma^\mu\gamma^\nu\frac{1}{2}[D_\mu, D_\nu]$$

$$= -D_\mu D^\mu + \frac{i}{2}\gamma^\mu\gamma^\nu F_{\mu\nu},\qquad (13.35)$$

where we are using the matrix representation for the gauge fields, namely, $A_\mu = A_\mu^a T^a$, $F_{\mu\nu} = F_{\mu\nu}^a T^a$ with the generators T^a in the representation of the fermions. It is clear, therefore, that the regularized exponent in (13.34) can be written as

$$\sum_n \int \mathrm{d}^4x \; \lambda(x)\phi_n^\dagger(x)\gamma_5\phi_n(x)$$

$$= \lim_{M^2\to\infty} \int \mathrm{d}^4x \; \frac{\mathrm{d}^4k}{(2\pi)^4}\lambda(x) \; \mathrm{Tr} \; e^{-ik\cdot x}\gamma_5 e^{-\frac{1}{M^2}\left(-D_\mu D^\mu+\frac{i}{2}\gamma^\mu\gamma^\nu F_{\mu\nu}\right)}e^{ik\cdot x}$$

$$= \lim_{M^2\to\infty} \int \mathrm{d}^4x \; \frac{\mathrm{d}^4k}{(2\pi)^4}\lambda(x) \; \mathrm{Tr} \; \gamma_5\frac{1}{2!}\left(\frac{i}{2M^2}\gamma^\mu\gamma^\nu F_{\mu\nu}\right)$$

$$\times \left(\frac{i}{2M^2}\gamma^\lambda\gamma^\rho F_{\lambda\rho}\right)e^{-\frac{k_\mu k^\mu}{M^2}}$$

$$= \lim_{M^2\to\infty} \int \mathrm{d}^4x \; \frac{\mathrm{d}^4k}{(2\pi)^4}\lambda(x)\left(-\frac{1}{8M^4}\right) \; \mathrm{Tr} \left(\gamma_5\gamma^\mu\gamma^\nu\gamma^\lambda\gamma^\rho\right)$$

$$\times \mathrm{Tr} F_{\mu\nu}F_{\lambda\rho}e^{-\frac{k^2}{M^2}}$$

$$= \lim_{M^2\to\infty} \int \mathrm{d}^4x\left(-\frac{1}{8M^4}\right)\lambda(x) \; \mathrm{Tr}F_{\mu\nu}F_{\lambda\rho}4i\epsilon^{\mu\nu\lambda\rho}\int \frac{\mathrm{d}^4k}{(2\pi)^4}e^{-\frac{k^2}{M^2}}$$

$$= \lim_{M^2\to\infty} \int \mathrm{d}^4x \left(-\frac{1}{2M^4}\right)\lambda(x)\epsilon^{\mu\nu\lambda\rho}F_{\mu\nu}(x)F_{\lambda\rho}(x)\frac{\left(\sqrt{\pi M^2}\right)^4}{(2\pi)^4}$$

$$= \lim_{M^2\to\infty} \int \mathrm{d}^4x \left(-\frac{1}{2M^4}\right)\lambda(x)\epsilon^{\mu\nu\lambda\rho}F_{\mu\nu}(x)F_{\lambda\rho}(x)\frac{M^4}{16\pi^2}$$

$$= -\frac{i}{32\pi^2}\int \mathrm{d}^4x \; \lambda(x)F_{\mu\nu}(x)F_{\lambda\rho}(x)\epsilon^{\mu\nu\lambda\rho} . \tag{13.36}$$

We note here that in the intermediate steps we have used the fact that the "Trace" is defined over the Dirac indices as well as the gauge indices and that in our discussions we have set the coupling constant for the gauge theory to unity. Therefore, we have determined

$$\det c_{nm} = \exp\left(i\sum_n \int \mathrm{d}^4x \; \lambda(x)\phi_n^\dagger(x)\gamma_5\phi_n(x)\right)$$

$$= \exp\left(\frac{1}{32\pi^2}\int \mathrm{d}^4x \; \lambda(x)\epsilon^{\mu\nu\lambda\rho}F_{\mu\nu}F_{\lambda\rho}\right) . \tag{13.37}$$

In other words, the functional integration measure changes under the chiral redefinition of the fields and the change in the fermionic measure under an infinitesimal chiral transformation is given by

$$
\begin{aligned}
\delta\left(\mathcal{D}\bar{\psi}\mathcal{D}\psi\right) &= \mathcal{D}\bar{\psi}'\mathcal{D}\psi' - \mathcal{D}\bar{\psi}\mathcal{D}\psi \\
&= \left((\det c_{nm})^{-2} - 1\right)\mathcal{D}\bar{\psi}\mathcal{D}\psi \\
&= \mathcal{D}\bar{\psi}\mathcal{D}\psi\left[-\frac{1}{16\pi^2}\int \mathrm{d}^4x\,\lambda(x)\epsilon^{\mu\nu\lambda\rho}F_{\mu\nu}(x)F_{\lambda\rho}(x)\right].
\end{aligned}
\tag{13.38}
$$

As a result, we note that the change of the generating functional under an infinitesimal chiral transformation is correctly given by

$$
\begin{aligned}
\delta Z = 0 = \int \mathcal{D}\bar{\psi}\mathcal{D}\psi\Big[i\int \mathrm{d}^4x\,\lambda(x)\Big\{\frac{i}{16\pi^2}\epsilon^{\mu\nu\lambda\rho}F_{\mu\nu}F_{\lambda\rho} \\
-\partial_\mu j_5^\mu - 2im\bar{\psi}\gamma_5\psi\Big\}\Big]e^{iS},
\end{aligned}
\tag{13.39}
$$

which leads to the relation

$$
\left\langle\int \mathrm{d}^4x\,\lambda(x)\left(\partial_\mu j_5^\mu + 2im\bar{\psi}\gamma_5\psi - \frac{i}{16\pi^2}\epsilon^{\mu\nu\lambda\rho}F_{\mu\nu}F_{\lambda\rho}\right)\right\rangle = 0.
\tag{13.40}
$$

This corresponds to the operator identity

$$
\begin{aligned}
\partial_\mu j_5^\mu &= -2im\bar{\psi}\gamma_5\psi + \frac{i}{16\pi^2}\epsilon^{\mu\nu\lambda\rho}F_{\mu\nu}F_{\lambda\rho} \\
&= -2im\bar{\psi}\gamma_5\psi + \frac{i}{16\pi^2}F_{\mu\nu}\tilde{F}^{\mu\nu},
\end{aligned}
\tag{13.41}
$$

where the dual field strength tensor is defined as

$$
\tilde{F}_{\mu\nu} = \epsilon_{\mu\nu\lambda\rho}F^{\lambda\rho}.
\tag{13.42}
$$

The anomalies have observable effect. For example, the correct pion life time can be calculated only if the chiral anomaly is taken into account. Furthermore, if we have a theory where there is both vectorial and axial vectorial coupling of the fermions (corresponding to local symmetries) , the model cannot be renormalized unless the anomaly is cancelled. In grand unified theories where we treat both quarks and leptons as massless, they have precisely this kind of coupling. Therefore, one cannot make a consistent grand unified theory unless one can cancel out the chiral anomaly (associated with the local invariances) and this leads to the study of groups and representations which are anomaly free. In string theories that are currently fashionable similar anomaly free considerations fix the gauge group to be uniquely SO(32) or E(8) × E(8).

13.2 Schwinger Model

The Schwinger model describes quantum electrodynamics (QED) in 1+1 dimension. This is a model that can be explicitly solved and the structure of the resulting effective gauge theory is directly deduced from the anomaly in the theory. Let us see how this arises. The Lagrangian density describing the interactions of massless charged fermions with an Abelian gauge field is given by

$$\mathcal{L} = -\frac{1}{4}F_{\mu\nu}F^{\mu\nu} + i\bar{\psi}\gamma^{\mu}\left(\partial_{\mu} - ieA_{\mu}\right)\psi\,, \qquad (13.43)$$

where e denotes the electric charge of the fermions. This theory is invariant under local $U(1)$ transformations as well as chiral $U(1)$ transformations (since fermions are massless). We would examine the model in two different approaches. But first let us note some of the special features of two dimensional space time.

First, note that in two dimensions the photon does not have any true degrees of freedom since there cannot be any transverse polarization. Secondly, the electromagnetic coupling in two dimensional QED is dimensional unlike in four dimensional QED where it is a dimensionless constant. This can be seen as follows. In natural units, the action is dimensionless

$$[S] = \left[\int \mathrm{d}^2x\, \mathcal{L}\right] = 0\,,$$

which implies that
$$[\mathcal{L}] = 2 \,.$$

This determines the canonical dimension of fields to be
$$[A_\mu] = 0 \,,$$

and
$$[\psi] = [\bar\psi] = \frac{1}{2} \,.$$

In turn, this determines
$$[e] = 1 \,. \tag{13.44}$$

Furthermore, by power counting alone it is easily seen that this theory has no infinities. So this is a completely finite theory.

Note also that in two dimensions if we choose our Dirac matrices to satisfy
$$\gamma^{0\dagger} = \gamma^0 \,, \quad \left(\gamma^0\right)^2 = \mathbf{1} \,,$$
$$\left(\gamma^1\right)^\dagger = -\gamma^1 \,, \quad \left(\gamma^1\right)^2 = -\mathbf{1} \,,$$

and
$$\gamma_5 = \gamma^0 \gamma^1$$

so that
$$\gamma_5^\dagger = \gamma^{1\dagger} \gamma^{0\dagger} = -\gamma^1 \gamma^0 = \gamma^0 \gamma^1 = \gamma_5 \,,$$
$$\gamma_5^2 = \gamma^0 \gamma^1 \gamma^0 \gamma^1 = -\left(\gamma^0\right)^2 \left(\gamma^1\right)^2 = \mathbf{1} \,, \tag{13.45}$$

then a possible representation for these matrices can be given by
$$\gamma^0 = \begin{pmatrix} 0 & 1 \\ 1 & 0 \end{pmatrix} \,,$$
$$\gamma^1 = \begin{pmatrix} 0 & -1 \\ 1 & 0 \end{pmatrix} \,,$$

$$\gamma_5 = \begin{pmatrix} 1 & 0 \\ 0 & -1 \end{pmatrix}.$$

Furthermore, note that in addition to the second rank metric tensor which is diagonal and has the form

$$\eta^{\mu\nu} = (+, -), \tag{13.46}$$

in two dimensions we have another second rank tensor given by

$$\epsilon^{\mu\nu} = -\epsilon^{\nu\mu}. \tag{13.47}$$

This is known as the Levi-Civita tensor of two dimensions. It is completely anti-symmetric and is chosen as

$$\epsilon^{01} = 1 = -\epsilon^{10}. \tag{13.48}$$

Similarly, for the covariant Levi-Civita tensor, we have

$$\epsilon_{01} = -1 = -\epsilon_{10}. \tag{13.49}$$

Using this tensor, we can write

$$\gamma_5 = -\frac{1}{2}\epsilon_{\mu\nu}\gamma^{\mu}\gamma^{\nu} = -\frac{1}{2}\epsilon^{\mu\nu}\gamma_{\mu}\gamma_{\nu}$$

$$= -\frac{1}{4}\epsilon^{\mu\nu}\left[\gamma_{\mu}, \gamma_{\nu}\right] = -\frac{1}{2}\epsilon^{\mu\nu}\sigma_{\mu\nu}, \tag{13.50}$$

where we have defined (in analogy with four dimensions)

$$\sigma_{\mu\nu} = \frac{1}{2}\left[\gamma_{\mu}, \gamma_{\nu}\right]. \tag{13.51}$$

This leads to

$$\epsilon^{\lambda\rho}\gamma_5 = -\frac{1}{2}\epsilon^{\lambda\rho}\epsilon^{\mu\nu}\sigma_{\mu\nu}$$

$$= -\frac{1}{2}\left(-\eta^{\lambda\mu}\eta^{\rho\nu} + \eta^{\lambda\nu}\eta^{\rho\mu}\right)\sigma_{\mu\nu}$$

$$\text{or,} \quad \epsilon^{\lambda\rho}\gamma_5 = \sigma^{\lambda\rho}$$

$$\text{or,} \quad \sigma^{\mu\nu} = \epsilon^{\mu\nu}\gamma_5. \tag{13.52}$$

It is clear now that we can write

$$\gamma^\mu \gamma^\nu = \frac{1}{2}\{\gamma^\mu, \gamma^\nu\} + \frac{1}{2}[\gamma^\mu, \gamma^\nu] = \eta^{\mu\nu} + \epsilon^{\mu\nu}\gamma_5\,.$$

Furthermore,

$$\sigma^{\mu\nu}\gamma_\nu = \epsilon^{\mu\nu}\gamma_5\gamma_\nu,$$

$$\gamma^\mu = \epsilon^{\mu\nu}\gamma_5\gamma_\nu\,.$$

Conversely,

$$\gamma_5\gamma^\mu = \epsilon^{\mu\nu}\gamma_\nu\,. \tag{13.53}$$

As a result of these identities, we note that in two dimensions, the vector and the axial vector currents defined as

$$j^\mu = e\bar\psi\gamma^\mu\psi\,,$$

and

$$j_5^\mu = e\bar\psi\gamma_5\gamma^\mu\psi\,,$$

are related as

$$j^\mu = \epsilon^{\mu\nu}j_{5\nu},$$

$$\text{or,} \quad j_5^\mu = \epsilon^{\mu\nu}j_\nu\,. \tag{13.54}$$

Let us now calculate the fermion two point function in the presence of an external electromagnetic field. Denoting by

$$iS_F^A(x - x') = \langle T\left(\psi(x)\bar\psi(x')\right)\rangle^A\,, \tag{13.55}$$

we note that this satisfies the differential equation

$$i\gamma^\mu\left(\partial_\mu - ieA_\mu\right)S_F^A(x - x') = \delta^2(x - x')\,. \tag{13.56}$$

The free fermion two point function can be related to the two point function for the massless scalar field as

$$iS_F^{(0)}(x - x') = i\slashed\partial\Delta_F^{(0)}(x - x')\,,$$

where

$$i\Delta_F^{(0)}(x - x') = \langle T\left(\phi(x)\phi(x')\right)\rangle, \qquad (13.57)$$

and satisfies the equation

$$\Box \Delta_F^{(0)}(x - x') = -\delta^2(x - x'). \qquad (13.58)$$

To solve for the fermion Green's function we note that the most general solution can be written as

$$S_F^A(x - x') = e^{i(\phi(x) - \phi(x'))} S_F^{(0)}(x - x'), \qquad (13.59)$$

where $S_F^{(0)}(x - x')$ is the free Green's function satisfying

$$i\partial\!\!\!/ S_F^{(0)}(x - x') = \delta^2(x - x'), \qquad (13.60)$$

and $\phi(x)$ is a 2×2 matrix function satisfying

$$i\gamma^\mu\left(\partial_\mu(i\phi(x)) - ieA_\mu\right) = 0$$

$$\text{or,} \quad -\gamma^\mu \partial_\mu \phi(x) + e\gamma^\mu A_\mu = 0$$

$$\text{or,} \quad \partial\!\!\!/\phi(x) = e\gamma^\mu A_\mu$$

$$\text{or,} \quad \Box\phi(x) = e\gamma^\mu\gamma^\nu \partial_\mu A_\nu = e\left(\eta^{\mu\nu} + \epsilon^{\mu\nu}\gamma_5\right)\partial_\mu A_\nu$$

$$= e\partial_\mu A^\mu + \frac{e}{2}\gamma_5 \epsilon^{\mu\nu} F_{\mu\nu},$$

so that

$$\Box \,\text{Tr}\, \phi(x) = 2e\partial_\mu A^\mu, \qquad (13.61)$$

and

$$\Box \,\text{Tr}\, \gamma_5\phi(x) = e\epsilon^{\mu\nu} F_{\mu\nu}. \qquad (13.62)$$

Let us note here that the whole point of this exercise is to show that in two dimensions one can always find a matrix function $\phi(x)$ such that

$$\gamma^\mu A_\mu = \gamma^\mu \partial_\mu \phi(x). \qquad (13.63)$$

That this is true can also be seen from the fact that in two dimension we can always decompose a vector as

$$A_\mu = \partial_\mu \sigma + \epsilon_{\mu\nu}\partial^\nu\eta, \qquad (13.64)$$

so that

$$\gamma^\mu A_\mu = \gamma^\mu \partial_\mu \sigma + \gamma^\mu \epsilon_{\mu\nu} \partial^\nu \eta$$

$$= \gamma^\mu \partial_\mu \sigma + \gamma^\mu \gamma_5 \partial_\mu \eta$$

$$= \gamma^\mu \partial_\mu (\sigma + \gamma_5 \eta) = \gamma^\mu \partial_\mu \phi,$$

with

$$\phi(x) = \sigma(x) + \gamma_5 \eta(x) . \qquad (13.65)$$

To obtain the complete Green's functions let us note that the time ordered scalar Green's function can be written in terms of positive and negative frequency functions as

$$\Delta_F^{(0)}(x) = \theta\left(x^0\right) i\Delta^{(+)}(x^0, x) - \theta\left(-x^0\right) i\Delta^{(-)}(x^0, x),$$

where

$$i\Delta^\pm(x^0, x) = \pm \frac{1}{2(2\pi)} \int_{-\infty}^\infty \frac{dk}{|\mathbf{k}|} e^{-i|\mathbf{k}|x^0 + ikx}$$

$$= \pm \frac{1}{2(2\pi)} \int_0^\infty \frac{dk}{k} e^{-ikx^0} \left(e^{ikx} + e^{-ikx} \right)$$

$$= \pm \frac{1}{2\pi} \int_0^\infty \frac{dk}{k} e^{-ikx^0} \cos kx .$$

Therefore, we have

$$\Delta_F^{(0)}(x^0, x) = \frac{1}{2\pi} \int_0^\infty \frac{dk}{k} e^{-ikx^0} \cos kx, \quad x^0 \geq 0,$$

and for $x^0 \geq 0$, we have

$$iS_F^{(0)}(x^0, x) = i\gamma^\mu \partial_\mu \left(\frac{1}{2\pi} \int_0^\infty \frac{dk}{k} e^{-ikx^0} \cos kx \right)$$

$$= \frac{i}{2\pi} \int_0^\infty \frac{dk}{k} \left(-i\gamma^0 k \cos kx - \gamma^1 k \sin kx \right) e^{-ikx^0}$$

$$= -\frac{i\gamma^1}{2\pi} \int_0^\infty dk \left(-i\gamma^1 \gamma^0 \cos kx + \sin kx \right) e^{-ikx^0}$$

$$= -\frac{i\gamma^1}{2\pi} \int_0^\infty dk \, (i\gamma_5 \cos kx + \sin kx) \, e^{-ikx^0}$$

$$= \frac{\gamma^1 \gamma_5}{2\pi} \int_0^\infty dk \, (\cos kx - i\gamma_5 \sin kx) \, e^{-ikx^0}$$

$$= \frac{\gamma^1 \gamma_5}{2\pi} \int_0^\infty dk \, e^{-ikx^0 - i\gamma_5 kx} . \qquad (13.66)$$

For $x^0 = 0$, this leads to

$$iS_F^{(0)}(x) = \frac{\gamma^1 \gamma_5}{2\pi} \int_0^\infty dk \, e^{-i\gamma_5 kx} = \frac{\gamma^1 \gamma_5}{2\pi} \frac{1}{i\gamma_5 x} = \frac{\gamma^1}{2\pi i x} . \qquad (13.67)$$

Thus for $x^0 = x_0'$, we determine that

$$iS_F^A(x - x') = e^{i(\phi(x) - \phi(x'))} \frac{\gamma^1}{2\pi i (x - x')}. \qquad (13.68)$$

The expectation value of the current is defined in a gauge invariant way as

$$j^\mu(x) = \lim_{x \to x', \, x^0 = x_0'} e\bar{\psi}(x')\gamma^\mu \psi(x) e^{ie \int_x^{x'} dx^\lambda A_\lambda}. \qquad (13.69)$$

This leads explicitly to

$$-\langle j^\mu(x) \rangle = \lim_{x \to x', \, x^0 = x_0'} e\mathrm{Tr} \, \gamma^\mu \langle \psi(x)\bar{\psi}(x') e^{ie \int_x^{x'} dx^\lambda A_\lambda}$$

$$= \lim_{x \to x', \, x^0 = x_0'} e\mathrm{Tr} \left(\gamma^\mu \left[1 + i(x - x')\frac{\partial}{\partial x}\phi(x) \right] \right.$$

$$\times \left. \frac{\gamma^1}{2\pi i(x - x')} \left[1 + ie(x' - x)A_1(x) \right] \right)$$

$$= \lim_{x \to x'} \mathrm{Tr} \left[\frac{e\gamma^\mu \gamma^1}{2\pi i(x - x')} + \frac{e}{2\pi}\gamma^\mu \partial_1 \phi \gamma^1 - \frac{e^2}{2\pi}\gamma^\mu \gamma^1 A_1 \right] . \qquad (13.70)$$

Neglecting the uninteresting infinite part we obtain

$$\langle j^\mu(x) \rangle = -\frac{e}{2\pi} \mathrm{Tr} \, \gamma^\mu \left(\partial_1 \phi \gamma^1 - e\gamma^1 A_1 \right) . \qquad (13.71)$$

Furthermore, note that

$$\phi(x) = \frac{1}{2} \text{Tr} \ (\phi(x)) + \frac{1}{2}\gamma_5 \ \text{Tr} \ (\gamma_5\phi(x)) \ , \qquad (13.72)$$

so that we can write

$$\langle j^\mu(x) \rangle = -\frac{e}{2\pi} \ \text{Tr} \ \left(\frac{1}{2}\gamma^\mu\gamma^1\partial_1 \ \text{Tr} \ \phi(x) \right.$$

$$\left. -\frac{1}{2}\gamma^\mu\gamma^1\gamma_5\partial_1 \ \text{Tr} \ \gamma_5\phi(x) - e\gamma^\mu\gamma^1 A_1 \right)$$

$$= -\frac{e}{2\pi} \left[\eta^{\mu 1}\partial_1 \ (\text{Tr} \ \phi(x)) - \epsilon^{\mu 1}\partial_1 \ \text{Tr} \ (\gamma_5\phi(x)) - 2e\eta^{\mu 1} A_1 \right] \ .$$

$$(13.73)$$

Furthermore, since

$$\not\partial\phi(x) = e\not A$$

$$\text{or,} \ \not\partial \left[\frac{1}{2} \ \text{Tr} \ \phi(x) + \frac{1}{2}\gamma_5 \ \text{Tr} \ \gamma_5\phi(x) \right] = e\not A,$$

we obtain

$$\text{Tr} \ \gamma^0\not\partial \left[\frac{1}{2} \ \text{Tr} \ \phi(x) + \frac{1}{2}\gamma_5 \ \text{Tr} \ \gamma_5\phi(x) \right] = \text{Tr} \ e\gamma^0\not A$$

$$\text{or,} \ \partial_0 \ \text{Tr} \ \phi(x) + \partial_1 \ \text{Tr} \ (\gamma_5\phi(x)) = 2eA^0,$$

and

$$\text{Tr} \ \gamma^1\not\partial \left[\frac{1}{2} \ \text{Tr} \ \phi(x) + \frac{1}{2}\gamma_5 \ \text{Tr} \ \gamma_5\phi(x) \right] = \text{Tr} \ e\gamma^1\not A$$

$$\text{or,} \ \partial_1 \ \text{Tr} \ \phi(x) - \partial_0 \ \text{Tr} \ (\gamma_5\phi(x)) = 2eA^1 \ . \qquad (13.74)$$

Therefore, using these relations as well as the definition in (13.73) we can write

$$\langle j^0(x) \rangle = \frac{e}{2\pi}\partial_1 \ \text{Tr} \ \gamma_5\phi(x)$$

$$= -\frac{e}{2\pi} \left(\partial_0 \ \text{Tr} \ \phi(x) - 2eA^0 \right)$$

$$= -\frac{e}{2\pi} \left(\partial^0 \ \text{Tr} \ \phi(x) - 2eA^0 \right) \ ,$$

and

$$\langle j^1(x) \rangle = \frac{e}{2\pi} \left(\partial_1 \, \text{Tr} \, \phi(x) - 2eA_1 \right)$$

$$= -\frac{e}{2\pi} \left(\partial^1 \, \text{Tr} \, \phi(x) - 2eA^1 \right) . \tag{13.75}$$

Covariantly, therefore, we can write

$$\langle j^\mu(x) \rangle = -\frac{e}{2\pi} \left(\partial^\mu \, \text{Tr} \, \phi(x) - 2eA^\mu \right) . \tag{13.76}$$

Putting in the solution from (13.61)

$$\text{Tr} \, \phi(x) = 2e \Box^{-1} \partial_\mu A^\mu , \tag{13.77}$$

we obtain

$$\langle j^\mu(x) \rangle = -\frac{e}{2\pi} \left(2e \partial^\mu \Box^{-1} \partial^\nu A_\nu - 2eA^\mu \right)$$

$$= \frac{e^2}{\pi} \left(\eta^{\mu\nu} - \partial^\mu \Box^{-1} \partial^\nu \right) A_\nu . \tag{13.78}$$

This is a gauge invariant current which is conserved as is readily seen from its form. However, it is also clear from this relation that

$$\partial_\mu j_5^\mu = \partial_\mu \epsilon^{\mu\nu} j_\nu = \frac{e^2}{2\pi} \epsilon^{\mu\nu} F_{\mu\nu} \neq 0. \tag{13.79}$$

Namely, the chiral current is no longer conserved.

Let us next study the generating functional for the Schwinger model. Going over to Euclidean space, we can write

$$Z_{\text{TOT}} = \int \mathcal{D}A_\mu \mathcal{D}\psi \, e^{-S_E^{(\text{TOT})}},$$

where

$$S_E^{(\text{TOT})} = \int \mathrm{d}^2 x_E \left(\frac{1}{4} F_{\mu\nu} F_{\mu\nu} + i\bar{\psi}\gamma_\mu D_\mu \psi \right) . \tag{13.80}$$

The integration over the fermion fields can be easily done (the action is quadratic in the variables) and the result is

$$Z = \int \mathcal{D}A_\mu \, \det\left(i\partial\!\!\!/ + eA\!\!\!/\right) e^{-\int d^2 x_E \left(\frac{1}{4}F_{\mu\nu}F_{\mu\nu}\right)}$$

$$= \int \mathcal{D}A_\mu \, e^{\text{Tr} \ln(i\partial\!\!\!/ + eA\!\!\!/)} e^{-\int d^2 x_E \left(\frac{1}{4}F_{\mu\nu}F_{\mu\nu}\right)}$$

$$= \int \mathcal{D}A_\mu \, e^{-S_{\text{eff}}}, \tag{13.81}$$

where

$$S_{\text{eff}} = \int d^2 x_E \left(\frac{1}{4}F_{\mu\nu}F_{\mu\nu}\right) - \ln\left(i\partial\!\!\!/ + eA\!\!\!/\right). \tag{13.82}$$

To understand the meaning of this better, let us look at the Euler Lagrange equation satisfied by the photon field.

$$\partial_\mu \frac{\partial S_{\text{eff}}}{\partial \partial_\mu A_\nu} - \frac{\partial S_{\text{eff}}}{\partial A_\nu} = 0$$

$$\text{or,} \quad \partial_\mu F_{\mu\nu} - \left(-\text{Tr} \, \left(i\partial\!\!\!/ + eA\!\!\!/\right)^{-1} e\gamma_\nu\right) = 0$$

$$\text{or,} \quad \partial_\mu F_{\mu\nu} + e \, \text{Tr} \, S_F^A(x,x)\gamma_\nu = 0$$

$$\text{or,} \quad \partial_\mu F_{\mu\nu} - \langle j_\nu(x) \rangle = 0$$

$$\text{or,} \quad \partial_\mu F_{\mu\nu} - \frac{e^2}{\pi}\left(\delta_{\nu\lambda} - \partial_\nu \Box_E^{-1}\partial^\lambda\right) A_\lambda = 0. \tag{13.83}$$

This now describes a massive photon with the mass of the photon given by

$$m_{\text{ph}}^2 = \frac{e^2}{\pi}. \tag{13.84}$$

This can be readily seen by choosing a covariant gauge such as the Landau gauge

$$\partial_\mu A_\mu = 0, \tag{13.85}$$

in which case the Euler-Lagrange equation becomes

$$\partial_\mu F_{\mu\nu} - m_{\text{ph}}^2 A_\nu = 0$$

$$\text{or,} \quad \Box_E A_\nu - m_{\text{ph}}^2 A_\nu = 0$$

$$\text{or,} \quad \left(\Box_E - m_{\text{ph}}^2\right) A_\nu = 0, \tag{13.86}$$

which is the equation for a massive gauge field.

The effective action for the photon field in this case becomes

$$S_{\text{eff}} = \frac{1}{4} F_{\mu\nu} F_{\mu\nu} + \frac{m_{\text{ph}}^2}{2} A_\mu \left(\delta_{\mu\nu} - \partial_\mu \Box_E^{-1} \partial_\nu \right) A_\nu, \qquad (13.87)$$

which again shows that the photon in the effective theory has become massive.

Another way of seeing the same result is to look at the generating functional for the fermions (in Mankowski space)

$$Z = \int \mathcal{D}\bar{\psi}\mathcal{D}\psi \, e^{iS},$$

where

$$S = \int \mathrm{d}^2 x \, i\bar{\psi} \left(\slashed{\partial} - ie\slashed{A} \right) \psi.$$

It follows now that

$$\frac{\delta Z}{\delta A_\mu(x)} = \int \mathcal{D}\bar{\psi}\mathcal{D}\psi \, ie\bar{\psi}\gamma^\mu \psi(x) e^{iS}$$

$$= \int \mathcal{D}\bar{\psi}\mathcal{D}\psi \, ij^\mu(x) e^{iS}$$

$$= i\langle j^\mu(x) \rangle Z$$

$$= \frac{ie^2}{\pi} \left(\eta^{\mu\nu} - \partial^\mu \Box^{-1} \partial^\nu \right) A_\nu Z.$$

Integrating this, we obtain

$$Z = e^{\frac{ie^2}{2\pi} \int \mathrm{d}^2 x A_\mu \left(\eta^{\mu\nu} - \partial^\mu \Box^{-1} \partial^\nu \right) A_\nu}$$

$$= e^{\frac{im_{\text{ph}}^2}{2} \int \mathrm{d}^2 x A_\mu \left(\eta^{\mu\nu} - \partial^\mu \Box^{-1} \partial^\nu \right) A_\nu}.$$

Therefore, we can write

$$Z_{\text{TOT}} = \int \mathcal{D}A_\mu \mathcal{D}\bar{\psi}\mathcal{D}\psi \, e^{i \int \mathrm{d}^2 x \left(-\frac{1}{4} F_{\mu\nu} F^{\mu\nu} \right) + iS}$$

$$= \int \mathcal{D}A_\mu \, e^{i \int \mathrm{d}^2 x \left(-\frac{1}{4} F_{\mu\nu} F^{\mu\nu} + \frac{m_{\text{ph}}^2}{A_\mu} \left(\eta^{\mu\nu} - \partial^\mu \Box^{-1} \partial^\nu \right) A_\nu \right)}$$

$$= \int \mathcal{D}A_\mu \, e^{iS_{\text{eff}}}. \qquad (13.88)$$

The phenomenon of the gauge field becoming massive through spontaneous breaking of a local symmetry is quite well understood. However, the novel feature in this case is that there is no scalar field in the theory. In this sense it is called a dynamical symmetry breaking and the symmetry that is broken in this case is the chiral symmetry.

Note that the nonlocal effective Lagrangian density

$$\mathcal{L}_{\text{eff}} = -\frac{1}{4}F_{\mu\nu}F^{\mu\nu} + \frac{m_{\text{ph}}^2}{2}A_\mu \left(\eta^{\mu\nu} - \partial^\mu\Box^{-1}\partial^\nu\right)A_\nu, \qquad (13.89)$$

can be written in a local manner by introducing a scalar field $B(x)$ as

$$\mathcal{L}_{\text{eff}} = -\frac{1}{4}F_{\mu\nu}F^{\mu\nu} + \frac{1}{2}\partial_\mu B(x)\partial^\mu B(x) - m_{\text{ph}}\epsilon^{\mu\nu}A_\mu\partial_\nu B(x). \quad (13.90)$$

Integrating out the scalar field $B(x)$ (on which the action depends quadratically) we obtain the earlier form of the effective action. On the other hand once we accept this form of the effective action given, we can integrate out the A_μ field completely since the Lagrangian density depends on the gauge fields at most quadratically. The resulting Lagrangian density in this case is given by

$$\mathcal{L}_{\text{eff}} = \frac{1}{2}\partial_\mu B(x)\partial^\mu B(x) - \frac{m_{\text{ph}}^2}{2}B(x)B(x). \qquad (13.91)$$

That is, the complete Schwinger model including all quantum corrections is equivalent to a free, massive scalar field. This is often referred to as the bosonization of the model. The fermions have completely disappeared from the theory.

Let us now analyze the Schwinger model from a different point of view. The Euclidean generating functional, as we have seen, is given by

$$Z = \int \mathcal{D}A_\mu \mathcal{D}\bar{\psi}\mathcal{D}\psi\, e^{-S_E^{(\text{TOT})}}, \qquad (13.92)$$

where the Euclidean action has the form

$$S_E^{(\text{TOT})} = \int d^2x_E \left[\frac{1}{4}F_{\mu\nu}F_{\mu\nu} + i\bar{\psi}\gamma_\mu\left(\partial_\mu - ieA_\mu\right)\psi\right]. \qquad (13.93)$$

Since the theory has a local gauge invariance let us choose a covariant gauge condition such as the Landau gauge condition

$$\partial_\mu A_\mu = 0 \,. \tag{13.94}$$

Let us recall that in two dimensions we can write

$$A_\mu = \partial_\mu \sigma + \epsilon_{\mu\nu} \partial_\nu \eta \,. \tag{13.95}$$

Therefore, the Landau gauge condition

$$\partial_\mu A_\mu = 0 \quad \Longrightarrow \quad \sigma = 0 \,. \tag{13.96}$$

In other words, in this gauge we can write $A_\mu = \epsilon_{\mu\nu} \partial_\nu \eta$ and the fermion Lagrangian density takes the form

$$\begin{aligned}
\mathcal{L} &= i\bar{\psi}\gamma_\mu \left(\partial_\mu - ieA_\mu\right)\psi \\
&= i\bar{\psi}\gamma_\mu \left(\partial_\mu - ie\epsilon_{\mu\nu}\partial^\nu\eta\right)\psi \\
&= i\bar{\psi}\gamma_\mu \left(\partial_\mu - ie\gamma_5\partial_\mu\eta\right)\psi \,.
\end{aligned} \tag{13.97}$$

Let us note next that if we make a local chiral transformation of the form

$$\begin{aligned}
\psi(x) &= e^{ie\gamma_5\alpha(x)}\psi'(x)\,, \\
\bar{\psi}(x) &= \bar{\psi}'(x)e^{ie\gamma_5\alpha(x)}\,,
\end{aligned} \tag{13.98}$$

then the fermion Lagrangian density becomes

$$\begin{aligned}
\mathcal{L} &= i\bar{\psi}'e^{ie\gamma_5\alpha(x)}\gamma^\mu \left(\partial_\mu - ie\gamma_5\partial_\mu\eta\right)e^{ie\gamma_5\alpha(x)}\psi' \\
&= i\bar{\psi}'\gamma^\mu \left(\partial_\mu + ie\gamma_5\partial_\mu\alpha(x) - ie\gamma_5\partial_\mu\eta\right)\psi' \,.
\end{aligned} \tag{13.99}$$

Therefore, if we choose $\alpha(x) = \eta(x)$, the fermion would decouple and the theory would appear to consist of free photons and free massless fermions. This is quite different from what we have seen in the earlier analysis. The real story lies in the change of the functional measure.

In this two dimensional case the change in the measure has to be calculated very carefully. To do this, let us assume that the finite transformation with $\alpha(x)$ is obtained by making N successive infinitesimal transformations with parameter $\epsilon(x)$ such that

$$\lim_{\epsilon \to 0,\ N \to \infty} N\epsilon(x) = \alpha(x). \tag{13.100}$$

The infinitesimal chiral transformations are given by

$$\psi(x) = \psi'(x) + ie\gamma_5\epsilon(x)\psi'(x),$$
$$\bar{\psi}(x) = \bar{\psi}'(x) + ie\epsilon(x)\bar{\psi}'(x)\gamma_5, \tag{13.101}$$

which can be inverted to give

$$\psi'(x) = \psi(x) - ie\gamma_5\epsilon\psi,$$
$$\bar{\psi}'(x) = \bar{\psi}(x) - ie\epsilon\bar{\psi}\gamma_5. \tag{13.102}$$

Furthermore, let us assume that we have already made n infinitesimal transformations. The fermion Lagrangian density in this case is given by

$$\mathcal{L} = i\bar{\psi}\gamma^\mu \left(\partial_\mu + ie\gamma_5\partial_\mu(n\epsilon(x)) - ie\gamma_5\partial_\mu\eta \right) \psi.$$

Let us define the eigenvalue equation

$$i\slashed{D}\phi_k(x) = i\gamma^\mu \left(\partial_\mu + ie\gamma_5\partial_\mu(n\epsilon) - ie\gamma_5\eta \right) \phi_k(x)$$
$$= \lambda_k\phi_k(x). \tag{13.103}$$

If the ϕ_k's form a complete set, then as discussed earlier, they satisfy

$$\int \mathrm{d}^2x_E \; \phi_k^\dagger(x)\phi_m(x) = \delta_{km},$$
$$\sum_k \phi_k(x)\phi_k^\dagger(y) = \delta^2(x-y). \tag{13.104}$$

We can also expand the functions ψ and $\bar{\psi}$ as

$$\psi(x) = \sum_m a_m\phi_m(x),$$
$$\bar{\psi}(x) = \sum_m b_m\phi_m^\dagger(x), \tag{13.105}$$

so that

$$\mathcal{D}\bar{\psi}\mathcal{D}\psi = \prod_m \mathrm{d}b_m \mathrm{d}a_m. \qquad (13.106)$$

If we make an infinitesimal transformation, as before, we can determine that

$$\psi(x) \to \psi'(x) = \sum_m a'_m \phi_m(x) = \sum_{m,\ell} c_{m\ell} a_\ell \phi_m(x), \qquad (13.107)$$

where

$$c_{m\ell} = \delta_{m\ell} - ie \int \mathrm{d}^2 x_E \epsilon(x)\phi_m^\dagger(x)\gamma_5\phi_\ell(x). \qquad (13.108)$$

Therefore,

$$\prod_m \mathrm{d}a_m = \det c_{m\ell} \prod_m \mathrm{d}a'_m. \qquad (13.109)$$

Similarly,

$$\prod_m \mathrm{d}b_m = \det c_{m\ell} \prod_m \mathrm{d}b'_m, \qquad (13.110)$$

such that

$$\mathcal{D}\bar{\psi}\mathcal{D}\psi = (\det c_{m\ell})^2 \mathcal{D}\bar{\psi}'\mathcal{D}\psi'. \qquad (13.111)$$

Let us, therefore, calculate the determinant,

$$\det c_{m\ell} = \det\left(\delta_{m\ell} - ie \int \mathrm{d}^2 x_E \, \epsilon(x)\phi_m^\dagger\gamma_5\phi_\ell\right)$$

$$= \exp\left(\mathrm{Tr}\,\ln\left(\delta_{m\ell} - ie \int \mathrm{d}^2 x_E \, \epsilon(x)\phi_m^\dagger\gamma_5\phi_\ell\right)\right)$$

$$= \exp\left(-ie\sum_m \int \mathrm{d}^2 x_E \, \epsilon(x)\phi_m^\dagger(x)\gamma_5\phi_m\right). \qquad (13.112)$$

The exponent can be calculated in a regularized manner as before

$$-ie \sum_m \int d^2 x_E \, \epsilon(x) \phi_m^\dagger(x) \gamma_5 \phi_m(x)$$

$$= \lim_{M^2 \to \infty} -ie \sum_m \int d^2 x_E \, \epsilon(x) \phi_m^\dagger(x) \gamma_5 \phi_m(x) \, e^{-\frac{\lambda_m^2}{M^2}}$$

$$= \lim_{M^2 \to \infty} -ie \sum_m \int d^2 x_E \, \epsilon(x) \phi_m^\dagger(x) \gamma_5 \, e^{\frac{\slashed{D}^2}{M^2}} \phi_m(x)$$

$$= \lim_{M^2 \to \infty} -ie \sum_m \int d^2 x_E \frac{d^2 k_E}{(2\pi)^2} \frac{d^2 k'_E}{(2\pi)^2} \epsilon(x) \tilde\phi_m^\dagger(k')$$

$$\times \, e^{-ik' \cdot x} \gamma_5 e^{\frac{\slashed{D}^2}{M^2}} e^{ik \cdot x} \tilde\phi_m(k) \,. \tag{13.113}$$

Using the completeness relation

$$\sum_m \tilde\phi_m(k) \tilde\phi_m^\dagger(k') = (2\pi)^2 \delta^2(k - k') \,, \tag{13.114}$$

the regularized exponent in the above relation takes the form

$$= \lim_{M^2 \to \infty} -ie \int d^2 x_E \frac{d^2 k_E}{(2\pi)^2} \epsilon(x) \, \mathrm{Tr} \, e^{-ik \cdot x} \gamma_5 \, e^{\frac{\slashed{D}^2}{M^2}} \, e^{ik \cdot x} \,. \tag{13.115}$$

Let us note the two dimensional identity

$$\slashed{D}^2 = \gamma_\mu D_\mu \gamma_\nu D_\nu$$

$$= \gamma_\mu \gamma_\nu \left(\partial_\mu - ie\gamma_5 n \partial_\mu \epsilon + ie\gamma_5 \partial_\mu \eta \right)$$

$$\times \left(\partial_\nu + ie\gamma_5 n \partial_\nu \epsilon - ie\gamma_5 \partial_\nu \eta \right)$$

$$= \partial_\mu \partial_\mu + e^2 \partial_\mu (n\epsilon - \eta) \partial_\mu (n\epsilon - \eta)$$

$$- 2ie\epsilon_{\mu\nu} \partial_\mu (n\epsilon - \eta) \partial_\nu + ie\gamma_5 \partial_\mu \partial_\mu (n\epsilon - \eta) \,. \tag{13.116}$$

Using this we obtain

$$
-ie \sum_m \int d^2x_E \; \epsilon(x) \phi_m^\dagger(x) \gamma_5 \phi_m'(x)
$$

$$
= \lim_{M^2 \to \infty} -ie \int d^2x_E \frac{d^2k}{(2\pi)^2} \epsilon(x) \frac{2ie}{M^2} \partial_\mu \partial^\mu (n\epsilon - \eta) \; e^{-\frac{k^2}{M^2}}
$$

$$
= \lim_{M^2 \to \infty} \frac{2e^2}{M^2} \int d^2x_E \; \epsilon(x) \partial_\mu \partial^\mu (n\epsilon - \eta) \times \frac{\pi M^2}{4\pi^2}
$$

$$
= \frac{e^2}{2\pi} \int d^2x_E \; \epsilon(x) \partial_\mu \partial^\mu (n\epsilon - \eta). \tag{13.117}
$$

Thus, we see that under this single infinitesimal transformation, the measure changes as

$$
\mathcal{D}\bar{\psi}\mathcal{D}\psi = (\det c_{m\ell})^2 \, \mathcal{D}\bar{\psi}'\mathcal{D}\psi'
$$

$$
= \exp\left(\frac{e^2}{\pi} \int d^2x_E \; \epsilon(x) \partial_\mu \partial^\mu (n\epsilon(x) - \eta)\right) \mathcal{D}\bar{\psi}'\mathcal{D}\psi'
$$

$$
= \left(1 + \frac{e^2}{\pi} \int d^2x_E \; \epsilon(x) \partial_\mu \partial^\mu (n\epsilon(x) - \eta)\right) \mathcal{D}\bar{\psi}'\mathcal{D}\psi'. \tag{13.118}
$$

As we make N infinitesimal transformations, the measure changes as

$$
\mathcal{D}\bar{\psi}\mathcal{D}\psi = \prod_{n=0}^{N} \left(1 + \frac{e^2}{\pi} \int d^2x_E \; \epsilon(x) \partial_\mu \partial^\mu (n\epsilon - \eta)\right) \mathcal{D}\bar{\psi}'\mathcal{D}\psi'. \tag{13.119}
$$

Using the result that if

$$
x = \prod_{n=0}^{N} \left(1 + a\epsilon + nb\epsilon^2\right), \tag{13.120}
$$

then,

$$\ln x = \sum_{n=0}^{N} \ln \left(1 + a\epsilon + nb\epsilon^2\right)$$

$$= \sum_{n=0}^{N} \left(a\epsilon + nb\epsilon^2\right)$$

$$= aN\epsilon + \frac{b}{2}N(N+1)\epsilon^2 . \qquad (13.121)$$

Therefore, in the limit $\epsilon \to 0, N \to \infty, N\epsilon = \alpha$, we obtain

$$\ln x = a\alpha + \frac{b}{2}\alpha^2, \quad \Longrightarrow \quad x = e^{a\alpha + \frac{b}{2}\alpha^2}. \qquad (13.122)$$

This gives the change in the measure for a finite chiral transformation as

$$\mathcal{D}\bar{\psi}\mathcal{D}\psi = \exp\left(\frac{e^2}{\pi}\int d^2x_E\ \alpha(x)\partial_\mu\partial_\mu\left(\frac{1}{2}\alpha(x) - \eta\right)\right)\mathcal{D}\bar{\psi}'\mathcal{D}\psi'. \tag{13.123}$$

Furthermore, if we choose $\alpha(x) = \eta(x)$, then we obtain

$$\mathcal{D}\bar{\psi}\mathcal{D}\psi = \exp\left(\frac{e^2}{\pi}\int d^2x_E\ \eta(x)\partial_\mu\partial^\mu\left(\frac{1}{2}\eta(x) - \eta(x)\right)\right)\mathcal{D}\bar{\psi}'\mathcal{D}\psi'$$

$$= \exp\left(-\frac{e^2}{2\pi}\int d^2x_E\ \eta(x)\partial_\mu\partial_\mu\eta(x)\right)\mathcal{D}\bar{\psi}'\mathcal{D}\psi'$$

$$= \exp\left(\frac{e^2}{2\pi}\int d^2x_E\ \partial_\mu\eta(x)\partial_\mu\eta(x)\right)\mathcal{D}\bar{\psi}'\mathcal{D}\psi'$$

$$= \exp\left(-\frac{e^2}{2\pi}\int d^2x_E\ \epsilon_{\mu\nu}\partial_\nu\eta(x)\epsilon_{\mu\lambda}\partial_\lambda\eta(x)\right)\mathcal{D}\bar{\psi}'\mathcal{D}\psi'$$

$$= \exp\left(-\frac{e^2}{2\pi}\int d^2x_E\ A_\mu(x)A_\mu(x)\right)\mathcal{D}\bar{\psi}'\mathcal{D}\psi'. \qquad (13.124)$$

Thus if we make a finite chiral transformation to decouple the fermions, there is a change in the functional measure leading to the generating functional

$$Z = \int \mathcal{D}A_\mu \mathcal{D}\bar{\psi}' \mathcal{D}\psi' \; e^{-S_{\text{eff}}} , \qquad (13.125)$$

where

$$S_{\text{eff}} = \int \mathrm{d}^2 x_E \left[\frac{1}{4} F_{\mu\nu} F_{\mu\nu} + \frac{e^2}{2\pi} A_\mu A_\mu + i\bar{\psi}' \gamma_\mu \partial_\mu \psi' \right] . \qquad (13.126)$$

This shows again that the photon becomes massive in this theory and the fermions decouple completely from the spectrum.

In this method, we see that the solubility of the model is closely related to the Jacobian of the chiral transformation that decouples the fermions. This, in turn, is connected with the anomaly of the system. This method can be extended to a very general class of two dimensional Abelian models which reduce to different known models in different limits.

13.3   References

Adler, S., Phys. Rev. **177**, 2426 (1969).

Bell, J. and R. Jackiw, Nuovo Cimento **60 A**, 47 (1969).

Das, A. and M. Hott, Zeit. Phys. **C 67**, 707 (1995).

Das, A. and V. S. Mathur, Phys. Rev. **D 33**, 489 (1986).

Fujikawa, K., Phys. Rev. Lett. **42**, 1195 (1979); Phys. Rev. **D 21**, 2848 (1980).

Roskies, R. and F. Schaposnik, Phys. Rev. **D23**, 558 (1981).

Schwinger, J., Phys. Rev. **128**, 2425 (1962).

Chapter 14

Systems at Finite Temperature

14.1 Statistical Mechanics

Let us review very briefly various concepts from statistical mechanics. Let us consider not one quantum mechanical system, but a whole collection of identical quantum systems-an ensemble. Thus, for example, it can be an ensemble of oscillators or any other physical system. Let us further assume for simplicity that the physical system under consideration has discrete eigenvalues of energy. Each system in this ensemble can, of course, be in any eigenstate of energy. Thus, we can define p_n to represent the probability of finding a system in the ensemble to be in an energy eigenstate $|n\rangle$. This is, of course, completely statistical in the sense that p_n can be identified with the number of physical systems in the state $|n\rangle$ divided by the total number of systems in the ensemble. Such a situation is quite physical as we know from our studies in statistical mechanics. Namely, we may have an ensemble of physical systems in thermal equilibrium with a heat bath. For a given ensemble, the value of any observable quantity averaged over the entire ensemble will take the form

$$\langle A \rangle = \bar{A} = \sum_n p_n \langle n|A|n \rangle = \sum_n p_n A_n \,, \qquad (14.1)$$

where we are assuming that the energy eigenstates are normalized and that

$$A_n = \langle n|A|n \rangle \,, \qquad (14.2)$$

denotes the expectation value of the operator in the quantum mechanical state $|n\rangle$. Thus, there are two kinds of averaging involved here. First, we have the average in a quantum state (expectation value) and second, we have the averaging with respect to the probability distribution of systems in the ensemble.

Being a probability, p_n has to satisfy certain conditions. Namely,

$$1 \geq p_n \geq 0, \quad \text{for all } n,$$

$$\sum_n p_n = 1. \tag{14.3}$$

It is in general very difficult to determine the probability distribution for an ensemble. However, if we are dealing with a thermodynamic ensemble, namely, an ensemble interacting with a large heat bath, and if we allow sufficient time to achieve thermal equilibrium, then we know that the probability distribution, in this case, is given by the Maxwell-Boltzmann distribution. Namely, in this case, we can write

$$p_n = \frac{1}{Z} e^{-\frac{E_n}{kT}}. \tag{14.4}$$

Here E_n is the energy of the nth quantum state, k is the Boltzmann constant and T the temperature of the system. The normalization factor Z can be determined from the relations for the probabilities in Eq. (14.3) as

$$\sum_n p_n = 1$$

$$\text{or,} \quad \frac{1}{Z} \sum_n e^{-\frac{E_n}{kT}} = 1$$

$$\text{or,} \quad Z = \sum_n e^{-\frac{E_n}{kT}} = \sum_n \langle n | e^{-\beta H} | n \rangle$$

$$\text{or,} \quad Z(\beta) = \text{Tr}\, e^{-\beta H}, \tag{14.5}$$

where we have defined

$$\beta = \frac{1}{kT}. \tag{14.6}$$

$Z(\beta)$ is known as the partition function of the system and plays the most fundamental role in deriving the thermodynamic properties of the system.

For a statistical ensemble, it is easy to see that the thermodynamic average of any quantity defined in Eq. (14.1) will be given by

$$\langle A \rangle_\beta = \sum_n p_n \langle n|A|n \rangle$$

$$= \frac{1}{Z(\beta)} \sum_n e^{-\beta E_n} \langle n|A|n \rangle$$

$$= \frac{1}{Z(\beta)} \sum_n \langle n|e^{-\beta H} A|n \rangle$$

$$= \frac{1}{Z(\beta)} \operatorname{Tr}\left(e^{-\beta H} A\right)$$

$$= \frac{\operatorname{Tr}\left(e^{-\beta H} A\right)}{\operatorname{Tr} e^{-\beta H}}. \tag{14.7}$$

In particular, the average energy associated with the system follows from Eq. (14.7) to be

$$\langle H \rangle_\beta = U = \frac{\operatorname{Tr}\left(e^{-\beta H} H\right)}{\operatorname{Tr} e^{-\beta H}}$$

$$= \frac{-\frac{\partial}{\partial \beta} \operatorname{Tr} e^{-\beta H}}{\operatorname{Tr} e^{-\beta H}}$$

$$= \frac{1}{Z(\beta)} \left(-\frac{\partial Z(\beta)}{\partial \beta}\right)$$

$$= -\frac{\partial}{\partial \beta} \ln Z(\beta). \tag{14.8}$$

The amount of order or the lack of it, for an ensemble, is defined through the entropy as

$$S = -\sum_n p_n \ln p_n = -\langle \ln p \rangle. \tag{14.9}$$

By definition, it is clear that the entropy is always positive semi-definite since $0 \leq p_n \leq 1$. Furthermore, its value is zero for a pure ensemble for which

$$p_n = \delta_{nm}, \quad \text{for a fixed } m. \tag{14.10}$$

For such an ensemble, all the individual systems are in the same energy state and, therefore, it is an ordered ensemble. On the other hand, the larger the number of states the physical system can be in, the more disordered the ensemble becomes and the entropy increases. For a thermodynamic ensemble, as we have seen in Eq. (14.4),

$$p_n = \frac{1}{Z(\beta)} e^{-\beta E_n}.$$

Therefore, we can calculate the entropy of the ensemble to be

$$S = -\sum_n p_n \ln p_n$$

$$= -\sum_n p_n (-\beta E_n - \ln Z(\beta))$$

$$= \beta \sum_n p_n E_n + \ln Z(\beta) \sum_n p_n$$

$$= \beta U + \ln Z(\beta)$$

$$= -\beta \frac{\partial}{\partial \beta} \ln Z(\beta) + \ln Z(\beta)$$

$$= -\beta^2 \frac{\partial}{\partial \beta} \left(\frac{1}{\beta} \ln Z(\beta) \right). \tag{14.11}$$

Here we have used Eq. (14.8) and (14.3) in the intermediate steps.
Given the internal energy, U, and the entropy, S, the free energy

for an ensemble can be written as

$$F(\beta) = U - \frac{S}{\beta}$$

$$= -\frac{\partial}{\partial \beta} \ln Z(\beta) + \frac{\partial}{\partial \beta} \ln Z(\beta) - \frac{1}{\beta} \ln Z(\beta)$$

$$= -\frac{1}{\beta} \ln Z(\beta). \tag{14.12}$$

In terms of the free energy, we can define the other thermodynamical quantities as

$$U = -\frac{\partial}{\partial \beta} \ln Z(\beta) = \frac{\partial}{\partial \beta}(\beta F) = F + \beta \frac{\partial F}{\partial \beta},$$

$$S = -\beta^2 \frac{\partial}{\partial \beta}\left(\frac{1}{\beta} \ln Z(\beta)\right) = \beta^2 \frac{\partial F}{\partial \beta}. \tag{14.13}$$

It is also interesting to note from Eq. (14.12) that the partition function takes a particularly simple form when expressed in terms of the free energy. Namely, we can write

$$Z(\beta) = e^{-\beta F(\beta)}. \tag{14.14}$$

We have gone over some of these concepts in some detail in order to bring out the essential similarities with the concepts of path integral that we have been discussing so far.

One of the major interests in the study of statistical mechanics is the question of phase transitions in such systems. Phase transitions are all too familiar to us from our studies of the different phases of water. Even in solids, such as the magnets, the hysteresis effect or the effect of spontaneous magnetization provides an example of a phase transition. Namely, we know that below the Curie temperature, T_c, if a magnetic material is subjected to an external magnetic field, then the material develops a residual magnetization even when the external field is switched off. The amount of residual magnetization decreases as the temperature of the system approaches the Curie temperature and vanishes at T_c. For $T > T_c$, the system exhibits no spontaneous magnetization. The temperature $T = T_c$ is, therefore, a critical temperature separating the different phases of a magnetic material.

The behavior of physical systems near the critical point is of great significance. This can be studied from the point of view of statistical mechanics quite well. They can also be studied with equal ease using the concepts of path integrals. However, before we discuss this, let us recapitulate how one uses statistical mechanics to study critical phenomena.

14.2   Critical Exponents

To fix ideas clearly, let us discuss the critical exponents in the context of a specific model which explains the properties of magnetization quite well. This model goes under the name of planar Ising model or the Ising model in two dimensions. The crucial feature of this model is that it ascribes the magnetic properties of a material to its spin content. This should be quite familiar from our studies of atomic systems where we know that elementary particles with a nontrivial spin possess magnetic dipole moments

Let us consider a square lattice with equal spacing in both x and y directions. Let us also assume that at each lattice site labeled by $n = (n_1, n_2)$, there is a spin $S(n)$ which can either point up or down. Accordingly, we assume

$$S(n) = \begin{cases} 1 \text{ for spin up,} \\ -1 \text{ for spin down.} \end{cases} \qquad (14.15)$$

Furthermore, let us assume that the spins interact as locally as is possible. In fact, the Hamiltonian for the Ising model is taken to be

$$H = -J \sum_{n,\hat{\mu}} S(n)S(n + \hat{\mu}), \qquad (14.16)$$

where we have assumed a simplified coupling for the problem. Here $\hat{\mu}$ stands for either of the two unit vectors on the lattice. In simple language, then, the Ising model assumes nearest neighbor interaction for the spins which are supposed to be pointing only along the

z-axis. The constant, J, measures the strength of the spin-spin interaction. It is clear that if its value is positive, then a minimum of the energy will be obtained when all the spins are pointing along the same direction-either up or down. Accordingly, such a coupling is known as a ferromagnetic coupling. Conversely, if J is negative, then the coupling is known as anti-ferromagnetic. It is worth pointing out here that the Hamiltonian for the Ising model has a discrete symmetry in the sense that if we flip all the spins of the system, then the Hamiltonian does not change.

Let us next subject this spin system to a constant external magnetic field B. In this case, the Hamiltonian becomes

$$H = -J \sum_{n,\hat{\mu}} S(n)S(n+\hat{\mu}) + B \sum_n S(n) \,. \tag{14.17}$$

The partition function defined in Eqs. (14.5) and (14.14), for the present case, takes the form

$$Z(\beta, B) = e^{-\beta F(\beta, B)} = \mathrm{Tr}\, e^{-\beta H} = \sum_{\text{config}} e^{-\beta H} \,. \tag{14.18}$$

The summation, here, is over all possible spin configurations of the system. Let us note that at every lattice site, the spin can take two possible values. Consequently, if N denotes the total number of lattice points, then there are 2^N possible spin configurations over which the summation in Eq. (14.18) has to be carried out. The true partition function, of course, has to be calculated in the thermodynamic limit when $N \to \infty$.

Let us note now from Eqs. (14.17) and (14.18) that, in this case, we have

$$\frac{\partial Z}{\partial B} = -\beta \frac{\partial F}{\partial B} Z = \mathrm{Tr}\, \left(-\beta \sum_n S(n)\, e^{-\beta H} \right)$$

$$\text{or,} \quad \frac{\partial F}{\partial B} = \frac{1}{Z} \mathrm{Tr}\, \left(\sum_n S(n)\, e^{-\beta H} \right) = \left\langle \sum_n S(n) \right\rangle_\beta \,. \tag{14.19}$$

Using the translation invariance of the theory, we can write

$$\langle S(n) \rangle_\beta = \langle S(0) \rangle_\beta \,, \tag{14.20}$$

so that we obtain

$$\frac{\partial F}{\partial B} = N \langle S(0) \rangle_\beta. \tag{14.21}$$

Thus, from Eq. (14.21), the mean magnetization per site can be obtained to be

$$M(\beta, B) = \frac{1}{N} \langle \sum_n S(n) \rangle_\beta = \langle S(0) \rangle_\beta = \frac{1}{N} \frac{\partial F}{\partial B}. \tag{14.22}$$

This is, of course, a function of both the temperature and the applied magnetic field and its value can be calculated once we know the free energy or the partition function.

The magnetic susceptibility is proportional to the rate of change of magnetization with the applied field and is defined to be

$$\chi = -\left.\frac{\partial M}{\partial B}\right|_{B=0} = -\frac{1}{N} \left.\frac{\partial^2 F}{\partial B^2}\right|_{B=0}$$

$$= \frac{\beta}{N} \left(\sum_{n,m} \langle S(n)S(m) \rangle_\beta - \langle \sum_n S(n) \rangle_\beta^2 \right)_{B=0}$$

$$= \frac{\beta}{N} \left(N \sum_n \langle S(n)S(0) \rangle_\beta - N^2 \langle S(0) \rangle_\beta^2 \right)_{B=0}, \tag{14.23}$$

where we have used Eqs. (14.19) and (14.20). Thus, we see that the magnetic susceptibility is related to the fluctuations in the spin. It is large at those temperatures where the correlation between the spins is large. Note that if the system has no net magnetization, i.e., no spontaneous magnetization, namely, if

$$\langle S(0) \rangle_\beta|_{B=0} = 0, \tag{14.24}$$

then, the magnetic susceptibility is completely determined by the spin-spin correlation function. Namely, in this case, we have

$$\chi = \beta \sum_n \langle S(n)S(0) \rangle_\beta \Big|_{B=0}. \tag{14.25}$$

If, for some temperature $\beta > \beta_c$, we find in our spin system that

$$\langle S(0) \rangle_\beta |_{B=0} \neq 0 , \qquad (14.26)$$

then, the system shows spontaneous magnetization or residual magnetization. In this case, we note that the discrete symmetry of the system is spontaneously broken. The spontaneous magnetization vanishes as we approach the critical temperature and for $\beta < \beta_c$, the system will show no spontaneous magnetization simply because the thermal motion will dominate. The critical temperature and the behavior of spontaneous magnetization near the critical temperature, namely, how the magnetization vanishes as the temperature approaches the critical temperature

$$M|_{B=0} \sim (T - T_c)^\beta , \qquad (14.27)$$

can be calculated once we know the partition function. Let us emphasize here that the parameter β in the exponent is not $\frac{1}{kT}$ which was defined earlier but represents a critical exponent. (The notation is unfortunate, but this is the convention.) Furthermore, the spontaneous magnetization defines an order parameter in the sense that its value separates the two different phases.

We can similarly calculate the correlation length between the spins, $\xi(T)$, at any temperature by analyzing the magnetic susceptibility. For very high temperatures, it is clear that the thermal motion will not allow any appreciable correlation between the spins. However, as the temperature of the system is lowered to the critical temperature, the system may develop long range correlations and the behavior of the correlation length near the critical temperature is parameterized by another critical exponent of the form

$$\xi(T) \sim (T - T_c)^{-\nu} . \qquad (14.28)$$

The magnetic susceptibility may similarly become large at this point and its behavior near the critical temperature is parameterized by yet another critical exponent as

$$\chi(T) \sim (T - T_c)^{-\gamma} . \qquad (14.29)$$

Similarly, other thermodynamic quantities in the system such as the specific heat defined as

$$C = -T \frac{\partial^2 F}{\partial T^2}, \tag{14.30}$$

may also display a singular behavior at the critical point and all these can be calculated once we know the partition function.

14.3 Harmonic Oscillator

The calculation of the partition function for the one dimensional quantum harmonic oscillator is quite straightforward. We know that for an oscillator with a natural frequency ω, the energy levels are given by

$$E_n = \left(n + \frac{1}{2}\right)\omega, \quad n = 0, 1, 2, \cdots, \tag{14.31}$$

where we have set $\hbar = 1$. For this system then, the partition function can be derived using Eqs. (14.5) and (14.31) to be

$$\begin{aligned}
Z(\beta) &= \operatorname{Tr} e^{-\beta H} \\
&= \sum_n e^{-\beta E_n} \\
&= \sum_n e^{-\beta\left(n+\frac{1}{2}\right)\omega} \\
&= e^{-\frac{\beta\omega}{2}} \left(\sum_{n=0}^{\infty} e^{-n\beta\omega}\right) \\
&= e^{-\frac{\beta\omega}{2}} \left(\frac{1}{1 - e^{-\beta\omega}}\right) \\
&= \frac{1}{e^{\frac{\beta\omega}{2}} - e^{-\frac{\beta\omega}{2}}} = \frac{1}{2\sinh\frac{\beta\omega}{2}}.
\end{aligned} \tag{14.32}$$

Since we know the partition function, we can calculate the thermodynamic properties of the system.

Let us next see how we can calculate the partition function for the harmonic oscillator through the path integral method. Let us recall that we have already calculated the transition amplitude for the harmonic oscillator which has the form (see Eq. (3.66))

$$\langle x_f, T | x_i, 0 \rangle = \langle x_f | e^{-iHT} | x_i \rangle = N \int \mathcal{D}x \, e^{iS[x]}$$

$$\text{or,} \quad \langle x_f | e^{-iHT} | x_i \rangle = \left(\frac{m\omega}{2\pi i \sin \omega T} \right)^{\frac{1}{2}} e^{iS[x_{cl}]}, \tag{14.33}$$

whereas we have noted earlier (see Eq. (3.84) with $J = 0$)

$$S[x_{cl}] = \frac{m\omega}{2 \sin \omega T} \left[(x_i^2 + x_f^2) \cos \omega T - 2 x_i x_f \right]. \tag{14.34}$$

We note here that we have set $\hbar = 1$, $J = 0$ in the above equations and T here denotes the time interval between the initial and the final points of the trajectory.

From the definition of the partition function,

$$Z(\beta) = \text{Tr} \, e^{-\beta H},$$

we note that the trace can be taken in any basis. In particular, if we choose the coordinate basis in the Schrödinger picture, then we can write

$$Z(\beta) = \int dx \, \langle x | e^{-\beta H} | x \rangle. \tag{14.35}$$

We now recognize the integrand in Eq. (14.35) merely as the transition amplitude (see Eq. (1.43)) for the harmonic oscillator with the identification

$$T = -i\beta,$$
$$x_f = x_i = x. \tag{14.36}$$

In other words, the integrand really is the transition amplitude between the same coordinate state in the Euclidean time with $\beta = \frac{1}{kT}$ (T is the temperature here) playing the role of the Euclidean time interval. Using this, then, we obtain from Eq. (14.35)

$$Z(\beta) = \int dx \left(\frac{m\omega}{2\pi i(-i\sinh\beta\omega)} \right)^{\frac{1}{2}}$$

$$\times \, e^{\left(\frac{im\omega}{2(-i\sinh\beta\omega)} (2x^2\cosh\beta\omega - 2x^2) \right)}$$

$$= \left(\frac{m\omega}{2\pi\sinh\beta\omega} \right)^{\frac{1}{2}} \int dx \, e^{-\left(\frac{m\omega}{\sinh\beta\omega}(\cosh\beta\omega - 1)x^2 \right)}$$

$$= \left(\frac{m\omega}{2\pi\sinh\beta\omega} \right)^{\frac{1}{2}} \int dx \, e^{-\left(m\omega\tanh\frac{\beta\omega}{2} x^2 \right)}$$

$$= \left(\frac{m\omega}{2\pi\sinh\beta\omega} \right)^{\frac{1}{2}} \left(\frac{\pi}{m\omega\tanh\frac{\beta\omega}{2}} \right)^{\frac{1}{2}}$$

$$= \left(\frac{1}{2\sinh\beta\omega\tanh\frac{\beta\omega}{2}} \right)^{\frac{1}{2}}$$

$$= \left(\frac{1}{4\sinh^2\frac{\beta\omega}{2}} \right)^{\frac{1}{2}} = \frac{1}{2\sinh\frac{\beta\omega}{2}} \, . \tag{14.37}$$

This is, of course, the partition function which we had found by a direct calculation in Eq. (14.32). Let us note next that since

$$Z(\beta) = e^{-\beta F} = \frac{1}{2\sinh\frac{\beta\omega}{2}} \, , \tag{14.38}$$

we obtain

$$F = -\frac{1}{\beta}\ln Z = \frac{1}{\beta}\left(\ln 2 + \ln\sinh\frac{\beta\omega}{2} \right) . \tag{14.39}$$

Consequently, we note from Eq. (14.13) that

$$\langle H \rangle_\beta = U = \frac{\partial}{\partial \beta}(\beta F) = -\frac{\partial \ln Z}{\partial \beta}$$

$$= \frac{\omega}{2} \frac{\cosh \frac{\beta\omega}{2}}{\sinh \frac{\beta\omega}{2}}$$

$$= \frac{\omega}{2} \frac{e^{\frac{\beta\omega}{2}} + e^{-\frac{\beta\omega}{2}}}{e^{\frac{\beta\omega}{2}} - e^{-\frac{\beta\omega}{2}}}$$

$$= \frac{\omega}{2} \frac{1 + e^{-\beta\omega}}{1 - e^{-\beta\omega}}$$

$$= \frac{\omega}{2} + \frac{\omega e^{-\beta\omega}}{1 - e^{-\beta\omega}}$$

$$\text{or,} \quad \langle H \rangle_\beta = U = \frac{\omega}{2} + \frac{\omega}{e^{\beta\omega} - 1}. \tag{14.40}$$

This is exactly what we would have obtained from Planck's law (remember that $\hbar = 1$). Among other things, it tells us that for low temperatures or large β, we have

$$\langle H \rangle_\beta = U \simeq \frac{\omega}{2}. \tag{14.41}$$

Namely, in such a case, the oscillators remain in the ground state. On the other hand, for very high temperatures or small β, we get

$$\langle H \rangle_\beta = U \simeq \frac{\omega}{2} + \frac{\omega}{\beta\omega} \simeq \frac{1}{\beta} = kT. \tag{14.42}$$

This is, of course, the expression for the equipartition of energy. (We expect the system to behave in a classical manner at very high temperature.)

This analysis of the derivation of the partition function for the harmonic oscillator from the path integral is quite instructive in the sense that it shows that a $(D + 1)$-dimensional Euclidean quantum field theory can be related to a D-dimensional quantum statistical system since the Euclidean time interval can be consistently identified with $\beta = \frac{1}{kT}$ as in Eq. (14.36). In fact, the relation between the two for a bosonic field theory can be simply obtained as

$$Z(\beta) = N \int_{\phi(0)=\phi(\beta)} \mathcal{D}\phi \, e^{-S_E}$$

$$= N \int_{\phi(0)=\phi(\beta)} \mathcal{D}\phi \, e^{-\int_0^\beta dt \int d^3x \mathcal{L}_E} . \qquad (14.43)$$

Here we are assuming integration over the end points which is equivalent to taking the trace. Furthermore, we note that the field variables, in this case, are assumed to satisfy a periodic boundary condition

$$\phi(t + \beta) = \phi(t) . \qquad (14.44)$$

This, as is clear, arises from the trace in the definition of the partition function. A careful analysis for the fermions shows that the partition function, in such a case, can be written exactly in the same manner but with anti-periodic boundary conditions. Namely, for fermions, we have

$$Z(\beta) = N \int \mathcal{D}\bar{\psi}\mathcal{D}\psi \, e^{-S_E[\psi,\bar{\psi}]}$$

$$= N \int \mathcal{D}\bar{\psi}\mathcal{D}\psi \, e^{-\int_0^\beta dt \int d^3x \, \mathcal{L}_E[\psi,\bar{\psi}]} , \qquad (14.45)$$

with the boundary conditions

$$\psi(0) = -\psi(\beta),$$

$$\bar{\psi}(0) = -\bar{\psi}(\beta) . \qquad (14.46)$$

These boundary conditions can be shown to be related to the question of quantum statistics associated with the different systems. This way of describing a quantum statistical system in equilibrium through a Euclidean path integral is known as the Matsubara formalism or the imaginary time formalism (since we rotate to imaginary time). Let us note here without going into details that there exist other formalisms which allow for the presence of both time and temperature in the theory simultaneously. These go under the name of real time formalisms.

The Matsubara formalism also suggests that a $(D+1)$-dimensional bosonic Euclidean quantum field theory can be related to a $(D+1)$-dimensional classical statistical system in the following way. Let us consider a quantum mechanical system described by the Lagrangian

$$L = \frac{1}{2} m\dot{x}^2 - V(x) \,. \tag{14.47}$$

Then, the generating functional for such a system will have the form

$$Z = N \int \mathcal{D}x \, e^{\frac{i}{\hbar} S[x]} = N \int \mathcal{D}x \, e^{\frac{i}{\hbar} \int_0^t dt \, L} \,. \tag{14.48}$$

Here we have put back Planck's constant for reasons which will be clear shortly. If we rotate to Euclidean time, the generating functional takes the form

$$Z = N \int \mathcal{D}x \, e^{-\frac{1}{\hbar} \int_0^t dt \left(\frac{1}{2} m\dot{x}^2 + V(x) \right)} \,. \tag{14.49}$$

This has precisely the form of a classical partition function if we identify

$$H = \int_0^t dt \left(\frac{1}{2} m\dot{x}^2 + V(x) \right) \tag{14.50}$$

as governing the dynamics of the system and

$$\hbar = kT = \frac{1}{\beta} \,. \tag{14.51}$$

Here we should note that the variable, t, in this case should be treated as a space variable and not as a time. Let us also recall that the Planck's constant measures quantum fluctuations in a quantum mechanical system whereas temperature measures thermal fluctuations in a statistical system. The identification of the two above, therefore, relates the quantum fluctuations in a quantum mechanical system with the thermal fluctuations in a corresponding classical statistical system. This connection can be simply extended to a field theory where the Euclidean action would act as the Hamiltonian for the corresponding classical statistical system.

14.4 Fermionic Oscillator

Let us next calculate the partition function for a fermionic oscillator with a natural frequency ω, both using the path integrals and the standard methods. We note from our discussion in section 5.1 that the Hilbert space for the fermionic oscillator is quite simple. In fact, from Eqs. (5.12) and (5.13), we note that it is like a two level system with energy eigenvalues

$$E_0 = -\frac{\omega}{2},$$

$$E_1 = \frac{\omega}{2}. \tag{14.52}$$

Once again, we have set $\hbar = 1$ for simplicity here. It, then, follows from the definition of the partition function in Eq. (14.5) that for the fermionic oscillator we have using Eq. (14.52)

$$
\begin{aligned}
Z(\beta) &= \operatorname{Tr} e^{-\beta H} \\
&= e^{-\beta E_0} + e^{-\beta E_1} \\
&= e^{\frac{\beta\omega}{2}} + e^{-\frac{\beta\omega}{2}} \\
&= e^{\frac{\beta\omega}{2}}\left(1 + e^{-\beta\omega}\right) = 2\cosh\frac{\beta\omega}{2}.
\end{aligned} \tag{14.53}
$$

The evaluation of the partition function for the fermionic oscillator follows from the form of the transition amplitude derived in Eq. (5.91). Following our discussion in the last section, we note that in the case of fermions, we have to impose anti-periodic boundary conditions (see Eq. (14.46)), namely, for the calculation of the partition function, we require

$$\psi_f = -\psi_i,$$

$$\bar{\psi}_f = -\bar{\psi}_i. \tag{14.54}$$

We can now calculate the partition function for the system with the identifications in Eqs. (14.36) and (14.54) as well as the result in

Eq. (5.91) as

$$Z(\beta) = \int d\bar{\psi}_i d\psi_i \left(e^{\frac{\beta\omega}{2}} e^{(-e^{-\beta\omega}\bar{\psi}_i\psi_i - \bar{\psi}_i\psi_i)} \right)$$

$$= e^{\frac{\beta\omega}{2}} \int d\bar{\psi}_i d\psi_i \left(1 - (1 + e^{-\beta\omega})\bar{\psi}_i\psi_i \right)$$

$$= e^{\frac{\beta\omega}{2}} \left(1 + e^{-\beta\omega} \right)$$

$$= 2\cosh\frac{\beta\omega}{2} . \tag{14.55}$$

Here we have used the nilpotency properties of Grassmann variables (see Eq. (5.17)) as well as the integration rules given in Eqs. (5.26) and (5.27).

This is exactly the same result which we had obtained earlier in Eq. (14.53) for the partition function of the system. We note now from the definition in Eq. (14.12) that the free energy for the fermionic oscillator is given by

$$F(\beta) = -\frac{1}{\beta} \ln Z(\beta) = -\frac{1}{\beta} \left(\ln 2 + \ln \cosh\frac{\beta\omega}{2} \right) . \tag{14.56}$$

The average energy for the ensemble can now be calculated from Eq. (14.13) to be

$$\langle H \rangle_\beta = U = \frac{\partial}{\partial\beta}(\beta F(\beta))$$

$$= -\frac{\omega}{2} \frac{\sinh\frac{\beta\omega}{2}}{\cosh\frac{\beta\omega}{2}}$$

$$= -\frac{\omega}{2} \frac{e^{\frac{\beta\omega}{2}} - e^{-\frac{\beta\omega}{2}}}{e^{\frac{\beta\omega}{2}} + e^{-\frac{\beta\omega}{2}}}$$

$$\text{or,} \quad \langle H \rangle_\beta = U = -\frac{\omega}{2} + \frac{\omega}{e^{\beta\omega} + 1} . \tag{14.57}$$

It now follows that for low temperatures or large β ($\beta = \frac{1}{kT}$), we have

$$\langle H \rangle_\beta = U \simeq -\frac{\omega}{2} . \tag{14.58}$$

Namely, the system likes to remain in the ground state for low temperatures whereas for high temperatures or small β, we obtain

$$\langle H \rangle_\beta = U \simeq -\frac{\beta\omega^2}{4} = -\frac{\omega^2}{4kT}. \tag{14.59}$$

In this case, we see that the average energy of the system goes to zero inversely with the temperature which amounts to saying that the system tries to populate equally the two available energy states.

14.5 References

Das, A., "Finite Temperature Field Theory", World Scientific Publishing.

Huang, K., "Statistical Mechanics", John Wiley Publishing.

Kogut, J., Rev. Mod. Phys., **51**, 659 (1979).

Matsubara, T., Prog. Theor. Phys., **14**, 351 (1955).

Umezawa, H. et al, "Thermofield Dynamics and Condensed States", North Holland Publishing.

Chapter 15

Ising Model

15.1 One Dimensional Ising Model

Let us pursue the ideas of statistical mechanics, which we have developed in the last chapter, with the example of the one dimensional Ising model. The Hamiltonian for spins interacting through nearest neighbors on a one dimensional lattice (chain) is given by

$$H = -J \sum_{i=1}^{N} s_i s_{i+1} + B \sum_{i=1}^{N} s_i. \tag{15.1}$$

Here we have assumed that the total number of lattice sites is N and that the spin system is being subjected to an external magnetic field, B, which is a constant. The classical partition function for this system is, by definition,

$$Z(\beta) = \sum_{s_i = \pm 1} e^{-\beta H} = \sum_{s_i = \pm 1} e^{\left(\beta J \sum_{i=1}^{N} s_i s_{i+1} - \beta B \sum_{i=1}^{N} s_i\right)}. \tag{15.2}$$

To study this system, let us assume periodic boundary condition on the lattice (cyclicity condition), namely,

$$s_{i+N} = s_i, \tag{15.3}$$

and ask whether there exists a quantum mechanical system whose Euclidean generating functional will give rise to the partition function for the one dimensional Ising model.

Let us consider the quantum mechanical system described by the Hamiltonian

$$H_q = -\alpha\,\sigma_1 + \gamma\,\sigma_3\,, \qquad (15.4)$$

where σ_1 and σ_3 are the two Pauli matrices and α and γ are two arbitrary constant parameters at this point. Let $|s\rangle$ denote the two component eigenstates of σ_3 such that

$$\sigma_3|s\rangle = s|s\rangle, \quad s = \pm 1\,. \qquad (15.5)$$

We can now calculate the Euclidean transition amplitude for the quantum system described by Eq. (15.4) between two eigenstates of σ_3, which is defined to be

$$\langle s_{\text{fin}}|e^{-TH_q}|s_{\text{in}}\rangle\,. \qquad (15.6)$$

Dividing the time interval into N steps of infinitesimal length ϵ such that (for large N)

$$N\epsilon = T\,, \qquad (15.7)$$

and introducing a complete set of eigenstates of σ_3 at every intermediate point, we obtain

$$\langle s_{\text{fin}}|e^{-TH_q}|s_{\text{in}}\rangle$$
$$= \sum_{s_i=\pm 1} \langle s_M|e^{-\epsilon H_q}|s_N\rangle \langle s_N|e^{-\epsilon H_q}|s_{N-1}\rangle \cdots \langle s_2|e^{-\epsilon H_q}|s_1\rangle\,,$$

$$(15.8)$$

where the intermediate sums are for the values $i = 2, 3, \ldots, N$. Furthermore, we have also identified

$$s_{\text{in}} = s_1, \quad s_{\text{fin}} = s_M\,. \qquad (15.9)$$

Note that if ϵ is small, then we can expand the individual exponents in Eq. (15.8) and write

$$\langle s_{i+1}|e^{-\epsilon H_q}|s_i\rangle \simeq \langle s_{i+1}|(1 + \epsilon\alpha\sigma_1 - \epsilon\gamma\sigma_3)|s_i\rangle\,. \qquad (15.10)$$

Using the relations,

$$\langle s_{i+1}|s_i\rangle = \left(\frac{1}{2}(s_i + s_{i+1})\right)^2,$$

$$\langle s_{i+1}|\sigma_1|s_i\rangle = \left(\frac{1}{2}(s_i - s_{i+1})\right)^2,$$

$$\langle s_{i+1}|\sigma_3|s_i\rangle = \frac{1}{2}(s_i + s_{i+1}), \qquad (15.11)$$

which can be explicitly checked, we obtain

$$\langle s_{i+1}|e^{-\epsilon H_q}|s_i\rangle$$

$$\simeq \left(\frac{1}{2}(s_i + s_{i+1})\right)^2 + \epsilon\alpha\left(\frac{1}{2}(s_i - s_{i+1})\right)^2 - \epsilon\gamma\frac{1}{2}(s_i + s_{i+1}).$$

$$(15.12)$$

From the fact that for any i, $s_i = \pm 1$, we also have the following identities.

$$\left(\frac{1}{2}(s_i + s_{i+1})\right)^{2n} = \left(\frac{1}{2}(s_i + s_{i+1})\right)^2, \quad \text{for } n \geq 1,$$

$$\left(\frac{1}{2}(s_i + s_{i+1})\right)^{2n+1} = \frac{1}{2}(s_i + s_{i+1}), \quad \text{for } n \geq 0,$$

$$\left(\frac{1}{2}(s_i - s_{i+1})\right)^{2n} = \left(\frac{1}{2}(s_i - s_{i+1})\right)^2, \quad \text{for } n \geq 1,$$

$$\left(\frac{1}{2}(s_i + s_{i+1})\right)^n \left(\frac{1}{2}(s_i - s_{i+1})\right)^m = 0, \quad \text{for } n, m \geq 1. \qquad (15.13)$$

Using the relations in Eq. (15.13), then, we can obtain (for constant parameters δ and Δ)

$$e^{\left(\Delta\left(\frac{1}{2}(s_i-s_{i+1})\right)^2+\delta\frac{1}{2}(s_i+s_{i+1})\right)}$$

$$= 1 + \sum_{n=1}^{\infty} \frac{1}{n!}\left[\Delta\left(\frac{1}{2}(s_i-s_{i+1})\right)^2 + \delta\frac{1}{2}(s_i+s_{i+1})\right]^n$$

$$= 1 + \sum_{n=1}^{\infty}\left[\frac{\Delta^n}{n!}\left(\frac{1}{2}(s_i-s_{i+1})\right)^{2n} + \frac{\delta^n}{n!}\left(\frac{1}{2}(s_i+s_{i+1})\right)^n\right]$$

$$= 1 + \sum_{n=1}^{\infty}\frac{\Delta^n}{n!}\left(\frac{1}{2}(s_i-s_{i+1})\right)^2$$

$$+ \sum_{n=0}^{\infty}\frac{\delta^{2n+1}}{(2n+1)!}\left(\frac{1}{2}(s_i+s_{i+1})\right)^{2n+1}$$

$$+ \sum_{n=1}^{\infty}\frac{\delta^{2n}}{(2n)!}\left(\frac{1}{2}(s_i+s_{i+1})\right)^{2n}$$

$$= 1 + (e^{\Delta}-1)\left(\frac{1}{2}(s_i-s_{i+1})\right)^2 + \sum_{n=0}^{\infty}\frac{\delta^{2n+1}}{(2n+1)!}\frac{1}{2}(s_i+s_{i+1})$$

$$+ \sum_{n=1}^{\infty}\frac{\delta^{2n}}{(2n)!}\left(\frac{1}{2}(s_i+s_{i+1})\right)^2$$

$$= 1 + (e^{\Delta}-1)\left(\frac{1}{2}(s_i-s_{i+1})\right)^2 + \sinh\delta\left(\frac{1}{2}(s_i+s_{i+1})\right)$$

$$+ (\cosh\delta-1)\left(\frac{1}{2}(s_i+s_{i+1})\right)^2 . \tag{15.14}$$

Let us now use the algebraic relation

$$\left(\frac{1}{2}(s_i+s_{i+1})\right)^2 + \left(\frac{1}{2}(s_i-s_{i+1})\right)^2 = 1 . \tag{15.15}$$

Then, we can write Eq. (15.14) also as

$$e^{\left(\Delta\left(\frac{1}{2}(s_i-s_{i+1})\right)^2+\delta\frac{1}{2}(s_i+s_{i+1})\right)}$$

$$= \cosh\delta\,\left(\frac{1}{2}(s_i+s_{i+1})\right)^2 + e^{\Delta}\left(\frac{1}{2}(s_i-s_{i+1})\right)^2$$

$$+ \sinh\delta\,\left(\frac{1}{2}(s_i+s_{i+1})\right). \tag{15.16}$$

We note that this has precisely the same form as the transition amplitude between two neighboring sites in Eq. (15.12) provided we make the identification

$$\cosh\delta = \frac{e^{\delta}+e^{-\delta}}{2} = 1\,,$$

$$e^{\Delta} = \epsilon\alpha\,,$$

$$\sinh\delta = \frac{e^{\delta}-e^{-\delta}}{2} = -\epsilon\gamma. \tag{15.17}$$

Equivalently, with the identification

$$e^{\Delta} = \epsilon\alpha,$$

$$e^{\delta} = 1 - \epsilon\gamma\,, \tag{15.18}$$

we can write

$$\langle s_{i+1}|e^{-\epsilon H_q}|s_i\rangle = e^{\left(\Delta\left(\frac{1}{2}(s_i-s_{i+1})\right)^2+\delta\frac{1}{2}(s_i+s_{i+1})\right)}. \tag{15.19}$$

Consequently, using this identification, we obtain

$$\mathrm{Tr}\,e^{-TH_q} = \sum_{s_i=\pm 1} \langle s_1|e^{-\epsilon H_q}|s_N\rangle\,\langle s_N|e^{-\epsilon H_q}|s_{N-1}\rangle \cdots \langle s_2|e^{-\epsilon H_q}|s_1\rangle$$

$$= \sum_{s_i=\pm 1} e^{\left(\Delta\sum_{i=1}^{N}\left(\frac{1}{2}(s_i-s_{i+1})\right)^2+\delta\sum_{i=1}^{N}\frac{1}{2}(s_i+s_{i+1})\right)}$$

$$= \sum_{s_i=\pm 1} e^{\left(\Delta\sum_{i=1}^{N}\frac{1}{2}(1-s_is_{i+1})+\delta\sum_{i=1}^{N}s_i\right)}$$

$$= e^{\frac{N\Delta}{2}} \sum_{s_i=\pm 1} e^{\left(-\frac{\Delta}{2}\sum_{i=1}^{N} s_i s_{i+1}+\delta \sum_{i=1}^{N} s_i\right)}$$

$$= e^{\frac{N\Delta}{2}} \sum_{s_i=\pm 1} e^{\left(\beta J\sum_{i=1}^{N} s_i s_{i+1}-\beta B \sum_{i=1}^{N} s_i\right)}$$

$$= e^{\frac{N\Delta}{2}} \sum_{s_i=\pm 1} e^{-\beta H} , \tag{15.20}$$

provided we identify

$$\Delta = -2\beta J, \quad \delta = -\beta B . \tag{15.21}$$

With these identifications, then we see that we can write

$$\mathrm{Tr}\, e^{-TH_q} = e^{\frac{N\Delta}{2}} Z(\beta) , \tag{15.22}$$

where $Z(\beta)$ represents the partition function for the one dimensional ising model. Once again, this shows that the quantum fluctuations in a quantum theory can be related to the thermal fluctuations of a classical statistical system.

15.2 The Partition Function

To evaluate the partition function for the one dimensional Ising model explicitly, let us rewrite the exponent in Eq. (15.2) in the partition function in a way that is easy to use. Note that

$$e^{\left(\beta J s_i s_{i+1}-\beta B\frac{1}{2}(s_i+s_{i+1})\right)}$$

$$= e^{\left(-2\beta J\times\frac{1}{2}(-1+1-s_i s_{i+1})-\beta B\times\frac{1}{2}(s_i+s_{i+1})\right)}$$

$$= e^{\beta J} e^{\left(-2\beta J\left(\frac{1}{2}(s_i-s_{i+1})\right)^2-\beta B\frac{1}{2}(s_i+s_{i+1})\right)}$$

$$= e^{\beta J}\left[\cosh \beta B \left(\frac{1}{2}(s_i+s_{i+1})\right)^2 + e^{-2\beta J}\left(\frac{1}{2}(s_i-s_{i+1})\right)^2\right.$$

$$\left. - \sinh \beta B \frac{1}{2}(s_i+s_{i+1})\right] , \tag{15.23}$$

where we have used the identities in Eq. (15.13). It is now clear that
if we define a matrix operator

$$K = e^{\beta J} \left[\cosh \beta B + e^{-2\beta J} \sigma_1 - \sinh \beta B \, \sigma_3 \right] , \qquad (15.24)$$

then, the matrix element of this operator between the eigenstates
$|s_i\rangle$ and $|s_{i+1}\rangle$ of the σ_3 operator will be obtained using Eq. (15.11)
to be

$$\langle s_{i+1} | K | s_i \rangle = e^{\left(\beta J s_i s_{i+1} - \beta B \frac{1}{2} (s_i + s_{i+1}) \right)} . \qquad (15.25)$$

Note that K is a 2×2 matrix and has the explicit form

$$K = \begin{pmatrix} e^{\beta J} (\cosh \beta B - \sinh \beta B) & e^{-\beta J} \\ e^{-\beta J} & e^{\beta J} (\cosh \beta B + \sinh \beta B) \end{pmatrix}$$

$$= \begin{pmatrix} e^{\beta(J-B)} & e^{-\beta J} \\ e^{-\beta J} & e^{\beta(J+B)} \end{pmatrix} . \qquad (15.26)$$

From this analysis, it is clear that we can derive the partition
function for the one dimensional Ising model explicitly as follows.

$$Z(\beta) = \sum_{s_i = \pm 1} e^{\left(\beta J \sum_{i=1}^{N} s_i s_{i+1} - \beta B \sum_{i=1}^{N} \frac{(s_i + s_{i+1})}{2} \right)}$$

$$= \sum_{s_i = \pm 1} \langle s_1 | K | s_N \rangle \langle s_N | K | s_{N-1} \rangle \cdots \langle s_2 | K | s_1 \rangle$$

$$= \operatorname{Tr} K^N = \lambda_1^N + \lambda_2^N , \qquad (15.27)$$

where λ_1 and λ_2 are the two eigenvalues of the matrix K in
Eq. (15.26).

The eigenvalues of the matrix K can be easily obtained from

$$\det(K - \lambda I) = 0$$

$$\text{or, } \det \begin{pmatrix} e^{\beta(J-B)} - \lambda & e^{-\beta J} \\ e^{-\beta J} & e^{\beta(J+B)} - \lambda \end{pmatrix} = 0$$

$$\text{or, } \lambda^2 - 2\lambda e^{\beta J} \cosh \beta B + e^{2\beta J} - e^{-2\beta J} = 0$$

$$\text{or, } \lambda^2 - 2\lambda e^{\beta J} \cosh \beta B + 2 \sinh 2\beta J = 0 . \qquad (15.28)$$

This is a quadratic equation whose solutions are easily obtained to be

$$\lambda = e^{\beta J} \cosh \beta B \pm \left(e^{2\beta J} \cosh^2 \beta B - 2 \sinh 2\beta J \right)^{\frac{1}{2}}$$

$$= e^{\beta J} \cosh \beta B \pm \left(e^{2\beta J} (1 + \sinh^2 \beta B) - e^{2\beta J} + e^{-2\beta J} \right)^{\frac{1}{2}}$$

$$= e^{\beta J} \cosh \beta B \pm \left(e^{2\beta J} \sinh^2 \beta B + e^{-2\beta J} \right)^{\frac{1}{2}}. \tag{15.29}$$

If we identify the two eigenvalues as

$$\lambda_1 = e^{\beta J} \cosh \beta B + \left(e^{2\beta J} \sinh^2 \beta B + e^{-2\beta J} \right)^{\frac{1}{2}},$$

$$\lambda_2 = e^{\beta J} \cosh \beta B - \left(e^{2\beta J} \sinh^2 \beta B + e^{-2\beta J} \right)^{\frac{1}{2}}, \tag{15.30}$$

then, we note that since $\lambda_1 > \lambda_2$, for large N, we can approximately write

$$Z(\beta) = \mathrm{Tr}\, K^N$$
$$= \lambda_1^N + \lambda_2^N \simeq \lambda_1^N$$
$$= \left[e^{\beta J} \cosh \beta B + (e^{2\beta J} \sinh^2 \beta B + e^{-2\beta J})^{\frac{1}{2}} \right]^N. \tag{15.31}$$

This method of evaluating the partition function is known as the matrix method and we recognize K as the transfer matrix for the system (see also section 3.3).

We can now derive various quantities of thermodynamic interest. Let us note from Eq. (15.31) that we can write

$$\ln Z(\beta) = N \ln \left[e^{\beta J} \cosh \beta B + (e^{2\beta J} \sinh^2 \beta B + e^{-2\beta J})^{\frac{1}{2}} \right]$$

$$= N \left[\beta J + \ln \left\{ \cosh \beta B + (\sinh^2 \beta B + e^{-4\beta J})^{\frac{1}{2}} \right\} \right]. \tag{15.32}$$

Therefore, the average magnetization per site defined in Eq. (14.22) can now be derived from Eq. (15.32) to be

$$M = \frac{1}{N}\frac{\partial F}{\partial B} = \frac{1}{N}\frac{\partial}{\partial B}\left(-\frac{1}{\beta}\ln Z(\beta)\right)$$

$$= -\frac{1}{N\beta}\frac{\partial \ln Z(\beta)}{\partial B}$$

$$= -\frac{1}{N\beta}\frac{N\left(\beta\sinh\beta B + \frac{1}{2}\frac{2\beta\sinh\beta B\cosh\beta B}{\left(\sinh^2\beta B + e^{-4\beta J}\right)^{\frac{1}{2}}}\right)}{\cosh\beta B + \left(\sinh^2\beta B + e^{-4\beta J}\right)^{\frac{1}{2}}}$$

$$= -\frac{\sinh\beta B}{\left(\sinh^2\beta B + e^{-4\beta J}\right)^{\frac{1}{2}}}. \tag{15.33}$$

It is interesting to note that when the external magnetic field is switched off, the magnetization vanishes. In this one dimensional system, therefore, there is no spontaneous magnetization and consequently, it cannot describe the properties of a magnet. The magnetic susceptibility for such a system can also be easily calculated (see Eq. (14.23)) and takes the form

$$\chi = -\frac{\partial M}{\partial B}\bigg|_{B=0} = \beta\, e^{2\beta J}. \tag{15.34}$$

This shows that for $|2\beta J| < 1$, the susceptibility obeys Curie's law. Namely, in this case,

$$\chi \simeq \beta = \frac{1}{kT}. \tag{15.35}$$

The absence of spontaneous magnetization in the present system may appear puzzling because naively, we would have expected the configurations where all the spins are "up" or "down" to correspond to minimum energy states. These, being ordered, we would have expected spontaneous magnetization for the system. The lack of magnetization can actually be understood through the instanton calculation which we discussed earlier. Let us recall that for the double-well potential (see section 7.4 as well as chapter 8)

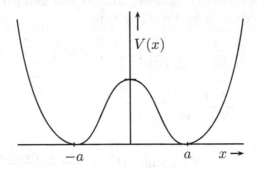

the naive ground states would give

$$\langle x \rangle = \pm a \, . \tag{15.36}$$

The true ground state, as we have seen earlier in Eq. (7.53), is a mixture of these two states (the symmetric state) such that

$$\langle x \rangle_{\text{true}} = 0 \, . \tag{15.37}$$

In this case, we showed explicitly that the tunneling or the presence of instanton states contributes significantly leading to the mixing of the states and restoring the symmetry.

In the one dimensional Ising spin system, we can correspondingly think of the following two configurations

$$\uparrow\uparrow\uparrow\uparrow\uparrow\uparrow\uparrow\uparrow\uparrow$$

$$\downarrow\downarrow\downarrow\downarrow\downarrow\downarrow\downarrow\downarrow\downarrow$$

as denoting the two ground states for which the magnetization is nonzero or

$$\langle M \rangle_\beta \neq 0 \, . \tag{15.38}$$

However, in the present case, there are other spin configurations such as

$\uparrow\uparrow\uparrow\uparrow\downarrow\downarrow\downarrow\downarrow\downarrow$ −one kink or one instanton,

$\uparrow\uparrow\uparrow\downarrow\downarrow\downarrow\uparrow\uparrow\uparrow$ −two kinks or one instanton-anti-instanton, and so on which contribute significantly. It is worth recalling that in a thermodynamic ensemble, it is the free energy which plays the dominant role. Even though these configurations have higher energy, they also are more disordered. Consequently, they will have a higher entropy and as a result can have a lower free energy. The consequence resulting from the contributions of these spin configurations is that the true ensemble average of magnetization vanishes. Namely,

$$\langle M \rangle_\beta^{\text{true}} = 0 \,. \tag{15.39}$$

This qualitative discussion can actually be made more precise through the use of the path integrals.

As we have seen, path integrals are defined by discretizing space-time variables. In fact, space-time lattices are often used to define a regularized quantum field theory. The continuum theory is, of course, obtained in the limit when the lattice spacing goes to zero. Viewed in this way, let us note that

$$H = -J \sum_i s_i s_{i+1} + B \sum_i s_i$$

$$= \frac{J}{2} \sum_i (s_{i+1} - s_i)^2 + B \sum_i s_i - NJ$$

$$\xrightarrow{\text{continuum}} \int \mathrm{d}x \left(\frac{\alpha}{2} (\partial s(x))^2 + \gamma s(x) \right) + \text{constant}. \tag{15.40}$$

where α and γ are two constants. Namely, we can think of the one dimensional Ising model as corresponding to a one dimensional free scalar field theory interacting with a constant external source in the continuum limit.

15.3 Two Dimensional Ising Model

Let us next consider a two dimensional array of spins on a square lattice interacting through nearest neighbors. Once again, let us use periodic boundary conditions along both the axes so that

$$s_i = s_{i_1,i_2} = s_{i_1+N,i_2} = s_{i_1,i_2+N} \,, \tag{15.41}$$

where we are using the notation that $i = (i_1, i_2)$ denotes a point on the two dimensional lattice and we are assuming that N denotes the total number of lattice sites along any axis.

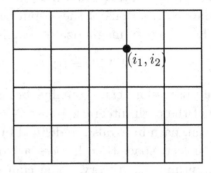

The total number of points on the lattice is then obtained to be

$$n = N^2 . \tag{15.42}$$

The spins are assumed to take only the values ± 1. That is,

$$s_i = s_{i_1, i_2} = \pm 1, \quad \text{for all } i_1, i_2 . \tag{15.43}$$

The Hamiltonian describing the interaction of the spins is given by (see also Eq. (14.17))

$$H = -J \sum_{\langle ij \rangle} s_i s_j = -J \sum_{i_1, i_2 = 1}^{N} \left(s_{i_1, i_2} s_{i_1+1, i_2} + s_{i_1, i_2} s_{i_1, i_2+1} \right). \tag{15.44}$$

The symbol $\langle ij \rangle$ is introduced as a short hand for sites which are nearest neighbors. We can also think of the sum in Eq. (15.44) as being taken over all the links of the lattice. (Remember that a link connects two nearest neighbors on a lattice.)

The partition function for the system described by the Hamiltonian in Eq. (15.44) can now be defined to be

$$Z(\beta) = \sum_{s_i = \pm 1} e^{-\beta H} = \sum_{s_i = \pm 1} e^{\left(\beta J \sum_{\langle ij \rangle} s_i s_j\right)}$$

$$= \sum_{s_i = \pm 1} e^{\left(\kappa \sum_{\langle ij \rangle} s_i s_j\right)}$$

$$= \sum_{s_i = \pm 1} \prod_{\langle ij \rangle} e^{\kappa s_i s_j}, \qquad (15.45)$$

where we have defined

$$\kappa = \beta J. \qquad (15.46)$$

Note that we are discussing the simpler case when the spin system is not interacting with an external magnetic field. This partition function, as it stands, appears to be only slightly more complicated than that for the one dimensional case in Eq. (15.2). However, as we will see, this partition function is much more difficult to evaluate in closed form. Before going into the actual evaluation of this partition function, let us discuss some of the symmetries associated with this system.

15.4 Duality

Let us note that since

$$s_i = \pm 1, \qquad (15.47)$$

we can expand the exponent in the partition function in Eq. (15.45) to obtain

$$e^{\kappa s_i s_j} = \cosh \kappa + s_i s_j \sinh \kappa$$

$$= \cosh \kappa \left(1 + s_i s_j \tanh \kappa\right). \qquad (15.48)$$

Therefore, we can also write

$$Z(\beta) = \sum_{s_i = \pm 1} \prod_{\langle ij \rangle} e^{\kappa s_i s_j}$$

$$= \sum_{s_i = \pm 1} \prod_{\langle ij \rangle} \cosh \kappa \left(1 + s_i s_j \tanh \kappa\right)$$

$$= (\cosh \kappa)^{2n} \sum_{s_i = \pm 1} \prod_{\langle ij \rangle} \left(1 + s_i s_j \tanh \kappa\right)$$

$$= (\cosh \kappa)^{2n} \sum_{s_i = \pm 1} \prod_{\langle ij \rangle} \sum_{l=0}^{1} \left(s_i s_j \tanh \kappa\right)^l$$

$$= (\cosh \kappa)^{2n} \sum_{s_i = \pm 1} \prod_{\langle ij \rangle} \sum_{l=0}^{1} (\tanh \kappa)^l (s_i s_j)^l . \qquad (15.49)$$

We see that we can simplify this expression by assigning a number $l_k = l_{ij} = l_{ji} = (0,1)$ to each link between the sites i and j and rewriting

$$Z(\beta) = (\cosh \kappa)^{2n} \sum_{l_k} (\tanh \kappa)^{l_1 + l_2 + \cdots} \sum_{s_i = \pm 1} \prod_{\langle ij \rangle} (s_i s_j)^{l_{ij}} . \qquad (15.50)$$

Let us next note that the product on the right hand side in Eq. (15.50) can simply be understood as the product of the spins at each lattice site with an exponent corresponding to the sum of the link numbers for links meeting at that site. Namely, for nearest neighbors, j,

$$\sum_{s_i = \pm 1} \prod_{\langle ij \rangle} (s_i s_j)^{l_{ij}} = \sum_{s_i = \pm 1} \prod_{i} (s_i)^{\sum_j l_{ij}} = \sum_{s_i = \pm 1} \prod_{i} (s_i)^{n_i} , \qquad (15.51)$$

where we have defined, for nearest neighbors j,

$$n_i = \sum_{j} l_{ij} . \qquad (15.52)$$

The sum over the nearest neighbors can now be done to give

$$\sum_{s_i=\pm 1} \prod_{\langle ij \rangle} (s_i s_j)^{l_{ij}} = \prod_i \sum_{s_i=\pm 1} (s_i)^{n_i} = \prod_i \left(1 + (-1)^{n_i}\right). \quad (15.53)$$

It is clear that the expression in Eq. (15.53) vanishes when n_i is odd. For even n_i, on the other hand, it has the value

$$\sum_{s_i=\pm 1} \prod_{\langle ij \rangle} (s_i s_j)^{l_{ij}} = 2^n. \quad (15.54)$$

Putting everything back in Eq. (15.50), we obtain

$$Z(\beta) = (2 \cosh^2 \kappa)^n \sum_{l_k} (\tanh \kappa)^{l_1 + l_2 + \cdots} = Z(\kappa). \quad (15.55)$$

The constraint here is that the l_k's in Eq. (15.55) must satisfy

$$\sum_j l_{ij} = 0, \quad \mod 2, \quad (15.56)$$

for any four links joining at a site. In other words, if l_1, l_2, l_3 and l_4 denote the link numbers for four links meeting at a common site, then

$$l_1 + l_2 + l_3 + l_4 = 0, \quad \mod 2. \quad (15.57)$$

Let us next consider the dual lattice associated with our original lattice. It is constructed by placing a lattice site at the center of each plaquette of the original lattice.

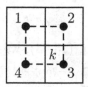

Thus, each plaquette of the dual lattice encloses a given site of the original lattice and intersects the four links originating from that site. Let us also define a dual variable σ_i at each site of the dual lattice and assume that it can take values ± 1. Denoting by $(1, 2, 3, 4)$ the sites of the dual lattice which enclose the point k of the original lattice, we note that for every link that is intersected by a dual link, we can define

$$l_1 = \frac{1}{2}(1 - \sigma_1 \sigma_2) \,,$$

$$l_2 = \frac{1}{2}(1 - \sigma_2 \sigma_3) \,,$$

$$l_3 = \frac{1}{2}(1 - \sigma_3 \sigma_4) \,,$$

$$l_4 = \frac{1}{2}(1 - \sigma_4 \sigma_1) \,. \tag{15.58}$$

We see that each of the l_k's have the value 0 or 1 as required. Furthermore, we also have

$$l_1 + l_2 + l_3 + l_4 = \frac{1}{2}(4 - \sigma_1 \sigma_2 - \sigma_2 \sigma_3 - \sigma_3 \sigma_4 - \sigma_4 \sigma_1)$$

$$= \frac{1}{2}\left(4 - (\sigma_1 + \sigma_3)(\sigma_2 + \sigma_4)\right)$$

$$= 0, \quad \text{mod } 2 \,. \tag{15.59}$$

In other words, the constraint equation in Eq. (15.57) can be naturally solved through the dual lattice variables.

Going back to the expression for the partition function in Eq. (15.55), we note using Eq. (15.58) that

$$(\tanh \kappa)^{l_1} = (\tanh \kappa)^{\frac{1}{2}(1 - \sigma_1 \sigma_2)} = e^{-\kappa^*(1 - \sigma_1 \sigma_2)} \,, \tag{15.60}$$

where we have defined

$$\tanh \kappa = e^{-2\kappa^*} \,. \tag{15.61}$$

Substituting this back into Eq. (15.55), we obtain

$$Z(\kappa) = (2 \cosh^2 \kappa)^n \sum_{\sigma_i} e^{\left(-2n\kappa^* + \kappa^* \sum_{\langle ij \rangle} \sigma_i \sigma_j\right)}$$

or, $$\frac{Z(\kappa)}{(2 \cosh^2 \kappa)^n} = e^{-2n\kappa^*} \sum_{\sigma_i} e^{\left(\kappa^* \sum_{\langle ij \rangle} \sigma_i \sigma_j\right)} = \frac{Z(\kappa^*)}{(e^{2\kappa^*})^n} \,. \tag{15.62}$$

This relation in Eq. (15.62) is quite interesting in that the relation

$$\tanh \kappa = e^{-2\kappa^*},$$

which can also be written as

$$\sinh 2\kappa \sinh 2\kappa^* = 1, \qquad (15.63)$$

defines a transformation between strong and weak couplings (or high and low temperatures (see Eq. (15.46))). And we find that the corresponding partition functions are related as well. Consequently, if there exists a single phase transition in this model (which was known from general arguments due to Peirels), it must occur at a unique point where

$$\kappa = \kappa_c = \kappa_c^*$$

$$\text{or, } \sinh^2 2\kappa_c = 1$$

$$\text{or, } \sinh 2\kappa_c = 1$$

$$\text{or, } e^{2\kappa_c} - e^{-2\kappa_c} = 2$$

$$\text{or, } e^{2\kappa_c} = \sqrt{2} + 1$$

$$\text{or, } \kappa_c = J\beta_c = \frac{1}{2} \ln \left(\sqrt{2} + 1 \right)$$

$$\text{or, } \beta_c = \frac{1}{2J} \ln \left(\sqrt{2} + 1 \right). \qquad (15.64)$$

15.5 High and Low Temperature Expansions

Quite often in statistical mechanics, the partition function cannot be evaluated exactly. In such a case, we would like to study the system at very high temperatures as well as at very low temperatures to see if any meaningful conclusion regarding the system can be obtained. In the language of field theory, we have seen in Eq. (14.51) that the temperature can be related to the Planck's constant which in some sense measures the quantum coupling. Therefore, high and low temperature expansions are also known as strong coupling and weak coupling expansions (or approximations).

Let us go back to the partition function for the 2-d Ising model. We have

$$Z(\kappa) = \sum_{s_i = \pm 1} e^{\left(\kappa \sum_{\langle ij \rangle} s_i s_j\right)}, \qquad (15.65)$$

where, as in Eq. (15.46), we have defined

$$\kappa = \beta J = \frac{J}{kT}.$$

If the temperature is high enough, then κ is small. We have seen in Eq. (15.55) that we can write

$$\frac{Z(\kappa)}{\left(2 \cosh^2 \kappa\right)^n} = \sum_{l_k = 0,1} (\tanh \kappa)^{l_1 + l_2 + \cdots}, \qquad (15.66)$$

where the link numbers are assumed to satisfy

$$l_1 + l_2 + l_3 + l_4 = 0, \quad \mod 2,$$

for any four links meeting at a lattice site. Since κ is small for high temperatures, so is $\tanh \kappa$ and the right hand side can be expanded in a power series in $\tanh \kappa$. To do that, let us note that the link numbers, l_k's, can only take values 0 or 1. Accordingly, let us postulate the rule that if $l_k = 0$, then we will not draw a bond connecting the two lattice sites whereas if $l_k = 1$, then a bond will connect the sites. With this rule then, the constraint on the link numbers simply says that there must be an even number of bonds originating from a given lattice site. Consequently, we note that the first term on the right hand side of the expansion will correspond to the case where there are no bonds on the lattice.

 $\longrightarrow 1.$

The next term in the series will be of the form

 $\longrightarrow (\tanh\kappa)^4$.

In other words, the first nontrivial term in the series will correspond to the product of the weight factor $\tanh\kappa$ over a single plaquette. The plaquette can be drawn in n-different ways on the lattice (recall the periodic boundary condition) and hence this term will come with a multiplicity of n.

The next term in the series will represent the product of the weight factor $\tanh\kappa$ over a plaquette involving two lattice lengths.

 $\longrightarrow (\tanh\kappa)^6$.

It is not hard to see that such a diagram can be drawn in $2n$ different ways and hence this term will come with a multiplicity of $2n$. At the next order the diagrams that will contribute are

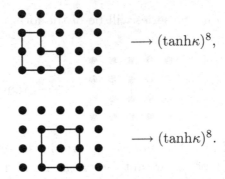

$$\longrightarrow (\tanh\kappa)^8,$$

$$\longrightarrow (\tanh\kappa)^8.$$

The combinatorics can be worked out in a straightforward manner for these graphs so that the high temperature expansion of the partition function will have the form

$$\frac{Z(\kappa)}{(2\cosh^2\kappa)^n} = 1 + n(\tanh\kappa)^4 + 2n(\tanh\kappa)^6$$

$$+ \frac{1}{2}n(n+9)(\tanh\kappa)^8 + \cdots . \qquad (15.67)$$

To obtain the low temperature expansion, let us note that when T is small,

$$\kappa = \frac{J}{kT},$$

is large. If $T = 0$, then we will expect all the spins to be frozen along one axis, say up. Therefore, the low temperature expansion would merely measure the deviation from such an ordered configuration. Namely, the low temperature expansion will be a measure of how many spins flip as T becomes nonzero but small. Thus, dividing the partition function by $e^{2n\kappa}$ (which is the value of the partition function when all the spins are pointing along one direction), we have from Eq. (15.45)

$$\frac{Z(\kappa)}{e^{2n\kappa}} = \sum_{s_i = \pm 1} e^{\left(\kappa \sum_{\langle ij \rangle}(-1+s_i s_j)\right)} . \qquad (15.68)$$

To develop the right hand side diagrammatically, let us draw a cross on the lattice to represent a flipped spin. Thus, the first term on the right hand side will correspond to a diagram of the form

The next term in the series will correspond to the case where one of the spins on the lattice has flipped and will represent a diagram of the form

In other words, a single flipped spin will interact with a nearest neighbor with weight $e^{-2\kappa}$ and since there are four nearest neighbors for any site, the term would have a weight $e^{-8\kappa}$. Furthermore, the flipped spin can occur at any lattice site and hence this term will come with a multiplicity of n.

The next term in the series will correspond to two flipped spins. Interestingly enough, this leads to two possibilities. Namely, the flipped spins can be nearest neighbors or they need not be. Diagrammatically, the two possibilities can be represented as

$$\longrightarrow e^{-12\kappa},$$

$$\longrightarrow e^{-16\kappa}.$$

In other words, in the first configuration, the interaction between the two nearest neighbor spins which are flipped does not contribute to the partition function. Furthermore, the number of ways a pair of flipped spins can occur as nearest neighbors is $2n$. The multiplicity of the second diagram, obviously, will be $\frac{1}{2}n(n-5)$. However, that is not the only kind of diagram which contributes an amount $e^{-16\kappa}$. In fact, there is another class of diagrams, namely, ones where there are three or four flipped spins which are nearest neighbors also contribute the same amount.

$$\begin{array}{ccccc} \bullet & \bullet & \bullet & \bullet & \bullet \\ \bullet & \times & \times & \times & \bullet \\ \bullet & \bullet & \bullet & \bullet & \bullet \end{array} \quad \longrightarrow e^{-16\kappa},$$

$$\begin{array}{cccc} \bullet & \bullet & \bullet & \bullet \\ \bullet & \times & \bullet & \bullet \\ \bullet & \times & \times & \bullet \\ \bullet & \bullet & \bullet & \bullet \end{array} \quad \longrightarrow e^{-16\kappa},$$

$$\begin{array}{ccccc} \bullet & \bullet & \bullet & \bullet \\ \bullet & \times & \times & \bullet & \bullet \\ \bullet & \times & \times & \bullet & \bullet \\ \bullet & \bullet & \bullet & \bullet & \bullet \end{array} \quad \longrightarrow e^{-16\kappa}.$$

Thus, we can consistently derive a low temperature expansion of the partition function which has the form

$$\frac{Z(\kappa)}{e^{2n\kappa}} = 1 + ne^{-8\kappa} + 2ne^{-12\kappa} + \frac{1}{2}n(n+9)e^{-16\kappa} + \cdots . \quad (15.69)$$

It is clear now that if we denote, for low temperatures,

$$\kappa = \kappa^*, \quad (15.70)$$

then, we can write the low temperature expansion also as

$$\frac{Z(\kappa^*)}{e^{2n\kappa^*}} = 1 + ne^{-8\kappa^*} + 2ne^{-12\kappa^*} + \frac{1}{2}n(n+9)e^{-16\kappa^*} + \cdots . \quad (15.71)$$

Thus, comparing Eqs. (15.67) and (15.71) we see that under the mapping

$$\tanh \kappa = e^{-2\kappa^*}, \tag{15.72}$$

we have

$$\frac{Z(\kappa)}{(2 \cosh^2 \kappa)^n} = \frac{Z(\kappa^*)}{e^{2n\kappa^*}}. \tag{15.73}$$

This is, of course, the duality relation that we have derived earlier in Eq. (15.62). In the present case, we see explicitly that the duality mapping (transformation) really takes us from the high temperature expansion to the low temperature expansion and vice versa.

15.6 Quantum Mechanical Model

Before finding the correspondence of the two dimensional Ising model with a quantum mechanical model, let us derive the transfer matrix for the system. Let us begin by writing the Hamiltonian for the system in a way that is better suited for our manipulations. Let us label the sites on a given row by $1 \leq i \leq N$ and the rows by $1 \leq m \leq N$.

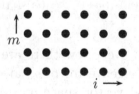

Then, we can write the interaction energy between the spins on a row, m, as

$$H(m) = -J \sum_{i=1}^{N} s_i(m)s_{i+1}(m). \tag{15.74}$$

Similarly, the interaction energy between two adjacent rows, say m and $(m+1)$, can be written as

$$H(m, m+1) = -J \sum_{i=1}^{N} s_i(m)s_i(m+1). \tag{15.75}$$

Given this, we can write (see Eq. (15.44)) the total energy of the system as

$$H = \sum_{m=1}^{N} (H(m) + H(m, m+1))$$

$$= -J \sum_{m=1}^{N} \sum_{i=1}^{N} (s_i(m)s_{i+1}(m) + s_i(m)s_i(m+1)). \quad (15.76)$$

If we desire, we can also add an external magnetic field at this point. However, let us ignore it for simplicity.

The partition function will involve the exponent

$$e^{-\beta H} = e^{\left(-\beta \sum_{m=1}^{N}(H(m)+H(m,m+1))\right)}. \quad (15.77)$$

Let us, for simplicity, concentrate only on a single factor of this exponent. Namely, let us look at

$$e^{-\beta(H(m)+H(m,m+1))}$$

$$= e^{\beta J\left(\sum_{i=1}^{N}(s_i(m)s_{i+1}(m)+s_i(m)s_i(m+1))\right)}$$

$$= \prod_{i=1}^{N} e^{\beta J s_i(m)s_i(m+1)} e^{\beta J s_i(m)s_{i+1}(m)}. \quad (15.78)$$

Let us next note that on every row there are N sites and if we introduce a two component eigenvector of σ_3 at every site, then we can define a 2^N dimensional vector space on every row through a direct product as

$$|s(m)\rangle = |s_1\rangle \otimes |s_2\rangle \otimes \cdots |s_N\rangle$$

$$= |s_1, s_2, \ldots, s_N\rangle. \quad (15.79)$$

We can define an inner product on such states as

$$\langle s(m+1)|s(m)\rangle = \langle s_1', s_2', \ldots, s_N'|s_1, s_2, \ldots, s_N\rangle$$

$$= \delta_{s_1 s_1'} \delta_{s_2 s_2'} \cdots \delta_{s_N s_N'}. \quad (15.80)$$

Similarly, the completeness relation will take the form

$$\sum_{s_i=\pm 1} |s(m)\rangle\langle s(m)| = \mathbf{1}\,, \tag{15.81}$$

where $\mathbf{1}$ denotes the $2^N \times 2^N$ identity matrix.

With these preliminaries, let us now introduce the following $2^N \times 2^N$ matrices. (There will be N of each.)

$$\sigma_1(i) = \mathbf{1} \otimes \mathbf{1} \otimes \cdots \otimes \sigma_1 \otimes \mathbf{1} \cdots \otimes \mathbf{1},$$

$$\sigma_3(i) = \mathbf{1} \otimes \mathbf{1} \otimes \cdots \otimes \sigma_3 \otimes \mathbf{1} \cdots \otimes \mathbf{1}. \tag{15.82}$$

Namely, all the entries in the above expression correspond to the trivial 2×2 identity matrix except at the i th entry. Let us also record here the product formula for matrices defined through direct products, namely,

$$(A \otimes B)(C \otimes D) = AC \otimes BD\,. \tag{15.83}$$

We also note that the σ_1 in the i th place acts on the vectors $|s_i\rangle$ and, therefore, as a 2×2 matrix, we can write

$$\langle s_i'|\alpha e^{\theta\sigma_1}|s_i\rangle = \langle s_i'|\alpha\cosh\theta + \alpha\sigma_1\sinh\theta|s_i\rangle$$

$$= \begin{pmatrix} \alpha\cosh\theta & \alpha\sinh\theta \\ \alpha\sinh\theta & \alpha\cosh\theta \end{pmatrix}, \tag{15.84}$$

where α is a constant. On the other hand, we note that a term such as

$$e^{\beta J s_i(m)s_i(m+1)},$$

can also be written as a 2×2 matrix of the form

$$e^{\beta J s_i s_i'} = \begin{pmatrix} e^{\beta J} & e^{-\beta J} \\ e^{-\beta J} & e^{\beta J} \end{pmatrix}. \tag{15.85}$$

Therefore, comparing the two relations in Eqs. (15.84) and (15.85), we note that we can identify

$$\langle s_i'|\alpha e^{\theta\sigma_1}|s_i\rangle = e^{\beta J s_i s_i'}\,, \tag{15.86}$$

provided the following relations are true, namely,

$$\alpha \cosh \theta = e^{\beta J},$$

$$\alpha \sinh \theta = e^{-\beta J}. \tag{15.87}$$

Equivalently, we can make the above identification provided

$$\tanh \theta = e^{-2\beta J},$$

$$\alpha = (2 \sinh 2\beta J)^{\frac{1}{2}}. \tag{15.88}$$

Consequently, it is clear that if we define a $2^N \times 2^N$ matrix as

$$K_1 = \alpha^N \, e^{\theta \sigma_1(1)} \times e^{\theta \sigma_1(2)} \cdots \times e^{\theta \sigma_1(N)}$$

$$= (2 \sinh 2\beta J)^{\frac{N}{2}} \, e^{\left(\theta \sum_{i=1}^{N} \sigma_1(i)\right)}, \tag{15.89}$$

with

$$\tanh \theta = e^{-2\beta J}, \tag{15.90}$$

then, we can write

$$\langle s(m+1)|K_1|s(m)\rangle = e^{\beta J \left(\sum_i s_i(m) s_i(m+1)\right)}. \tag{15.91}$$

Let us also note that

$$\langle s(m+1)|e^{\left(\tilde{\theta}\sigma_3(i+1)\sigma_3(i)\right)}|s(m)\rangle$$

$$= \langle s(m+1)| \cosh \tilde{\theta} + \sigma_3(i+1)\sigma_3(i) \sinh \tilde{\theta}|s(m)\rangle$$

$$= \langle s(m+1)|s(m)\rangle \, (\cosh \tilde{\theta} + s_{i+1} s_i \sinh \tilde{\theta})$$

$$= \langle s(m+1)|s(m)\rangle \, e^{\left(\tilde{\theta} s_{i+1} s_i\right)}$$

$$= \langle s(m+1)|s(m)\rangle \, e^{\beta J s_i s_{i+1}}, \tag{15.92}$$

provided we make the identification

$$\tilde{\theta} = \beta J . \tag{15.93}$$

Thus, defining

$$K_2 = e^{\beta J \left(\sum_{i=1}^{N} \sigma_3(i+1)\sigma_3(i) \right)}, \tag{15.94}$$

we note that we can write

$$K = K_1 K_2 = (2\sinh 2\beta J)^{\frac{N}{2}} e^{\theta \sum_{i=1}^{N} \sigma_1(i)}$$

$$\times e^{\beta J \sum_{i=1}^{N} \sigma_3(i+1)\sigma_3(i)}, \tag{15.95}$$

which defines the transfer matrix for the system. Namely,

$$\langle s(m+1)|K|s(m)\rangle = \langle s(m+1)|K_1 K_2|s(m)\rangle$$

$$= \langle s(m+1)|K_1|s(m)\rangle \, e^{\beta J \sum_{i=1}^{N} s_i s_{i+1}}$$

$$= e^{\beta J \sum_{i=1}^{N} s_i(m)s_i(m+1)} e^{\beta J \sum_{i=1}^{N} s_i(m)s_{i+1}(m)}$$

$$= e^{-\beta(H(m)+H(m,m+1))} . \tag{15.96}$$

The partition function can now be written as

$$Z(\beta) = \sum_{s_i = \pm 1} e^{-\beta H}$$

$$= \sum_{s_i = \pm 1} \langle s(1)|K|s(N)\rangle\langle s(N)|K|s(N-1)\rangle \cdots \langle s(2)|K|s(1)\rangle$$

$$= \mathrm{Tr}\, K^N . \tag{15.97}$$

This, therefore, is the starting point for the Onsager solution of the two dimensional Ising model.

In field theory language, we are looking for a quantum Hamiltonian whose Euclidean transition amplitude will yield the partition function. Furthermore, in field theory, even if we are dealing with a theory on the lattice, we would prefer the time variable to be continuous. Thus, let us identify one of the axes, say the vertical one, to correspond to time and we choose the separation between the rows to be ϵ, a very small quantity.

The continuum time limit, of course, will be obtained by choosing $\epsilon \to 0$. Let us note that we are only changing the spacing among the rows. The spacing along a row, of course, is unchanged. At first sight, this may appear bothersome. But, let us recall that if there is a critical point in the theory, then the correlation lengths become quite large in this limit and in such a limit the lattice structure becomes quite irrelevant. Let us also note here that by making the lattice asymmetric, we have actually destroyed the isotropy of the system and, consequently, the couplings along different axes can, in principle, be different. Thus, allowing for different couplings along the two axes, we can write

$$H = \sum_{m=1}^{N} \sum_{i=1}^{N} \left(-J' s_i(m) s_i(m+1) - J s_i(m) s_{i+1}(m) \right)$$

$$= \sum_{m=1}^{N} \left(H(m) + H(m, m+1) \right), \tag{15.98}$$

where

$$H(m) = -J \sum_{i=1}^{N} s_i s_{i+1},$$

$$H(m, m+1) = -J' \sum_{i=1}^{N} s_i s_i'. \tag{15.99}$$

The partition function, in this case, will have the form

$$Z(\beta) = \sum_{s_i = \pm 1} e^{-\beta H}$$

$$= \sum_{s_i = \pm 1} e^{-\beta \sum_{m=1}^{N} (H(m) + H(m,m+1))}$$

$$= \sum_{s_i = \pm 1} \prod_{m=1}^{N} e^{\left(\beta J \sum_i s_i s_{i+1} + \beta J' \sum_i s_i s_i'\right)} . \qquad (15.100)$$

In the quantum field theory language, we can write the Euclidean time interval as

$$T_{\text{Eucl.}} = N\epsilon, \qquad (15.101)$$

and assume that there exists a quantum Hamiltonian H_q such that we can write

$$Z(\beta) = \text{Tr} \, e^{-T_{Eucl.} H_q} = \text{Tr} \, e^{-N \epsilon H_q}$$

$$= \sum_{s_i = \pm 1} \langle s(1) | e^{-\epsilon H_q} | s(N) \rangle \cdots \langle s(2) | e^{-\epsilon H_q} | s(1) \rangle$$

$$= \sum_{s_i = \pm 1} \prod_{m=1}^{N} \langle s(m+1) | e^{-\epsilon H_q} | s(m) \rangle . \qquad (15.102)$$

Thus, comparing Eqs. (15.100) and (15.102), we recognize that the two can be identified if

$$\langle s(m+1) | e^{-\epsilon H_q} | s(m) \rangle = e^{\beta \sum_i (J s_i s_{i+1} + J' s_i s_i')} . \qquad (15.103)$$

From our discussion of the transfer matrix in Eqs. (15.95) and (15.96), we immediately conclude that

$$H_q = - \sum_i (\sigma_1(i) + \lambda \sigma_3(i+1)\sigma_3(i)), \qquad (15.104)$$

where λ is a constant parameter and as before, we can identify

$$\epsilon \lambda = \beta J,$$

$$\tanh \epsilon \simeq \epsilon = e^{-2\beta J'} . \qquad (15.105)$$

This relation is quite interesting in the sense that it brings out a relationship between the coupling strength as a function of the lattice spacing. In particular, we note that as we make the spacing between the rows smaller, the corresponding coupling between the rows becomes stronger. This is renormalization group behavior of the couplings in its crudest form.

15.7   Duality in the Quantum System

We have been able to relate the 2-d Ising model to a one dimensional quantum mechanical system with a Hamiltonian

$$H_q = -\sum_{i=1}^{N}(\sigma_1(i) + \lambda\sigma_3(i+1)\sigma_3(i)). \tag{15.106}$$

$$\bullet \quad \bullet \quad \bullet \cdots \quad \bullet \quad \text{original lattice}$$
$$1 \quad 2 \quad 3 \cdots \quad N$$
$$\bullet \quad \bullet \quad \bullet \cdots \quad \bullet \quad \text{dual lattice}$$

Let us next consider the dual lattice corresponding to this one dimensional lattice and define the dual operators on the dual lattice as

$$\mu_1(i) = \sigma_3(i+1)\sigma_3(i),$$

$$\mu_3(i) = \prod_{j=1}^{i} \sigma_1(j). \tag{15.107}$$

It is easy to see that

$$\mu_1^2(i) = (\sigma_3(i+1)\sigma_3(i))^2 = I,$$

$$\mu_3^2(i) = \left(\prod_{j=1}^{i} \sigma_j\right)^2 = I. \tag{15.108}$$

These results can be shown to follow from the basic commutation relations of the Pauli matrices, namely,

$$[\sigma_1(i), \sigma_1(j)] = 0 = [\sigma_3(i), \sigma_3(j)], \quad \text{if } i \neq j,$$

$$\sigma_1^2(i) = I = \sigma_3^2(i),$$

$$[\sigma_1(i), \sigma_3(i)]_+ = 0. \tag{15.109}$$

Using these, we can also derive that for $i \neq j$,

$$[\mu_1(i), \mu_1(j)] = [\sigma_3(i+1)\sigma_3(i), \sigma_3(j+1)\sigma_3(j)] = 0,$$

$$[\mu_3(i), \mu_3(j)] = \left[\prod_{k=1}^{i} \sigma_1(k), \prod_{l=1}^{j} \sigma_1(l)\right] = 0. \tag{15.110}$$

On the other hand,

$$[\mu_1(i), \mu_3(i)]_+ = \left[\sigma_3(i+1)\sigma_3(i), \prod_{j=1}^{i} \sigma_1(j)\right]_+$$

$$= \sigma_3(i+1) \prod_{j=1}^{i-1} [\sigma_3(i), \sigma_1(i)]_+ \sigma_1(j)$$

$$= 0. \tag{15.111}$$

Thus, $\mu_1(i)$ and $\mu_3(i)$ also have the same algebraic properties as the Pauli matrices on the original lattice. Furthermore, let us note that by definition,

$$\mu_3(i+1)\mu_3(i) = \sigma_1(i+1). \tag{15.112}$$

Using these, then, we note that we can write

$$H_q(\lambda) = -\sum_{i=1}^{N} (\sigma_1(i) + \lambda\sigma_3(i+1)\sigma_3(i))$$

$$= -\sum_{i=1}^{N} (\mu_3(i+1)\mu_3(i) + \lambda\mu_1(i))$$

$$= -\lambda \sum_{i=1}^{N} \left(\mu_1(i) + \lambda^{-1} \mu_3(i+1) \mu_3(i) \right)$$

$$= \lambda H_q(\lambda^{-1}) . \tag{15.113}$$

This is the self duality relation for this system. Namely, it maps the strong coupling properties of the system to its weak coupling properties. This shows, in particular, that the energy eigenvalues of this system must also satisfy the relation

$$E(\lambda) = \lambda E(\lambda^{-1}) . \tag{15.114}$$

For some finite value of λ, if there is a phase transition such that the correlation lengths become infinite or that some energy eigenvalue becomes zero (zero mode), then the above duality relation implies that this must happen at $\lambda = 1$. This is precisely how we had determined the critical temperature for the 2-d Ising model in Eq. (15.64). Let us also note here that since we have a quantum mechanical description of the 2-d Ising model, we can also develop a perturbation theory in the standard manner.

15.8   References

Huang, K., "Statistical Mechanics", John Wiley Publishing.
Kogut, J., Rev. Mod. Phys., **51**, 659 (1979).
Savit, R., Rev. Mod. Phys., **52**, 453 (1980).

Index

Abelian gauge theory, 240
Anharmonic oscillator, 71
Anomalous Ward identity, 279
Anomaly, 279
Anti-commutation relation, 75, 76
Anti-instanton, 147
Anti-periodic boundary condition, 322
Asymptotic equation, 160

Baker-Campbell-Hausdorff formula, 15
Basis states, 11
Bethe-Salpeter equation, 191
Born diagram, 193
Bosonization, 300
BRST invariance, 266

Chiral anomaly, 279
Classical field, 187, 188
Classical path, 21
Classical phase, 123, 124
Classical statistical system, 323
Classical trajectory, 28, 32, 45, 49, 130
Complex scalar field, 218
Connected diagram, 184
Connected Green's function, 184
Continuous transformation, 212

Continuum limit, 40
Coordinate basis, 8, 9
Correlation function, 59, 61, 67
Coulomb gauge, 241
Critical exponent, 314

Double well, 143
Double well potential, 335
Double-well potential, 134
Dual lattice, 341
Duality, 339
Duality in quantum systems, 356

Effective action, 67, 187, 193
Effective potential, 199
Entropy, 311
Euclidean action, 143
Euclidean equation, 144
Euclidean field theory, 321
Euclidean generating functional, 327
Euclidean rotation, 53
Euclidean space, 58, 143
Euler-Lagrange equation, 6, 32, 45

Faddeev-Popov determinant, 260
Fermi-Dirac statistics, 76
Fermi-Feynman gauge, 243
Fermionic oscillator, 75, 324

Feynman diagram, 183
Feynman Green's function, 47, 70, 91, 171
Feynman path integral, 17, 19
Feynman propagator, 86, 89, 174, 179
Feynman rules, 181, 200
Field, 1
Finite temperature, 309
Fourier series, 34
Fourier transform, 12, 46
Free energy, 312, 325
Free particle, 25, 36
Functional, 3
Functional derivative, 3
Functional integral, 25

Gauge choice, 241
Gauge theories, 239
Gaussian integral, 82
Generating functional, 53, 64, 178
Global transformation, 212
Goldstone particle, 235
Goldstone mode, 234
Goldstone theorem, 235
Grassmann variable, 78
Green's function, 45, 176
Gribov ambiguity, 254
Gupta-Bleuler quantization, 245

Half oscillator, 111
Harmonic oscillator, 31, 69, 318
Heisenberg picture, 8
Heisenberg states, 12

Imaginary time, 55, 58, 143
Instanton, 147
Instantons, 143
Interaction picture, 9
Invariance, 209
Ising model, 327
Ising model one dimensional, 327

Ising model two dimensional, 337
Isospectral, 107

Jost function, 159

Left derivative, 79, 86
legendre transformation, 210
Local transformation, 212
Loop expansion, 200

Many degrees of freedom, 167
Maxwell theory, 239
Metric, 2
Mid-point prescription, 16
Multi-instanton, 148, 163

Nicolai map, 105
Noether's theorem, 212
Non-Abelian gauge theory, 246
Normal ordering, 14

One loop effective potential, 203
One particle irreducible, 186
Operator ordering, 13, 14

Partition function, 311, 324, 332
Pauli principle, 77
Periodic boundary condition, 322
Perturbation theory, 72
Perturbative expansion, 181
Phase transformation, 219
Phase transition, 313
Proper self energy, 196

Quantum chromodynamics, 280
Quantum correction, 179
Quantum fluctuation, 33
Quantum mechanical model, 349
Quantum statistical system, 321

Regularization, 111, 113
Regularized superpotential, 115

Relativistic field theory, 170
Right derivative, 79, 86

Saddle point method, 127, 130,
    131, 148, 152
Scalar field theory, 170
Schrödinger equation, 7, 121
Schrödinger picture, 8
Schrödinger states, 12
Schwinger model, 289
Schwinger-Dyson equation, 191
Semi-classical approximation, 125
Semi-classical methods, 121, 130,
    144
Shape invariance, 105
Singular potentials, 111
Slavnov-Taylor identities, 224
Space-time translation, 215
Spontaneous magnetization, 316,
    335
Spontaneous symmetry breaking,
    226, 232
Steepest descent, 128, 130, 134
Stirling's approximation, 129
Stress tensor, 217
Superpotential, 113
Supersymmetric Oscillator, 97
Supersymmetric quantum
    mechanics, 102
Supersymmetry, 97
Supplementary condition, 245
Symmetries, 209

Temperature expansion, 343
Temperature expansion high, 346
Temperature expansion low, 346
Time evolution operator, 7
Time ordered product, 176
Time ordering, 60, 90
Time translation invariance, 148,
    154
Transfer matrix, 334
Transition amplitude, 9, 16, 32, 36,
    41, 51, 131
Tree diagram, 193

Vacuum functional, 64, 67, 85, 88,
    169
Vacuum generating functional, 170

Ward identities, 222
Ward identity, 274
Weyl ordering, 14
WKB approximation, 121, 125,
    134
WKB wave function, 125, 126,
    134, 138

Zero eigenvalue, 149
Zero mode, 150, 158, 358